ENERGY ENGINEERING

ENERGY ENGINEERING
■ ■ ■

JOHN W. MITCHELL
Department of Mechanical Engineering
The University of Wisconsin, Madison

A Wiley-Interscience Publication
JOHN WILEY & SONS
New York   Chichester   Brisbane   Toronto   Singapore

To Carol

Copyright © 1983 by John Wiley & Sons, Inc.

All rights reserved. Published simultaneously in Canada.

Reproduction or translation of any part of this work beyond that permitted by Section 107 or 108 of the 1976 United States Copyright Act without the permission of the copyright owner is unlawful. Requests for permission or further information should be addressed to the Permissions Department, John Wiley & Sons, Inc.

Library of Congress Cataloging in Publication Data:

Mitchell, John W.
   Energy engineering.

   "A Wiley-Interscience publication."
   Includes index.
   1. Power (Mechanics) 2. Energy conservation.
I. Title

TJ163.9.M57   1983   621.042   82-19977
ISBN 0-471-08772-6

PREFACE

The availability and cost of energy have become dominant factors in society today. Many schemes have been proposed for developing new energy sources and for conserving present ones. It is always possible to use less energy in any process. The first goal of the engineer is to determine the methods by which energy utilization is reduced but the output remains the same, or even increases. The second goal is to determine which methods of using less energy are cost effective.

The objective of this book is to evaluate the technical aspects of energy usage with a view toward more effective utilization. Conventional engineering techniques are used to evaluate the mechanisms of energy use. Economic considerations are of equal importance and life cycle cost and saving techniques are used to determine cost-effective measures. The evaluation focuses on those uses which are significant in the overall picture and attempts to determine those technical measures that can reduce usage and save money.

In this book, the phrases "conservation," "conserve energy," "saving energy," and the like appear. Engineers know that the First Law of Thermodynamics states that energy is always conserved in any process, and that the engineer has no choice in the matter. What is really at issue are the ideas behind the Second Law of Thermodynamics, which deal with thermodynamic availability (or available energy, or exergy, or irreversibility). Thus, the engineer is really concerned with minimizing the destruction of available energy in a given process. However, the common usage of "energy" instead of "available energy" makes for a nonrigorous phrasing at times. It is hoped that the technical content is not confused by this wording.

The book is divided into several chapters. In Chapter 1, an overview of the use and price of energy is given. Those sectors in which the energy consumption is a significant fraction of the total usage are noted. Chapter 2 reviews conventional economic techniques and introduces life cycle costing. A simple formulation suitable for energy economics is developed which accounts for fuel inflation in the future. This second chapter provides the basis for the economic evaluation in the later sections.

The remaining chapters deal with the energy use in the various sectors. Chapters 3, 4, and 5 are concerned with energy use in buildings and focus on heat loss and gain through the building envelope. The devices that produce heating and cooling (furnaces, heat pumps) are discussed in Chapters 6 and 7. The problems of electrical power production, waste heat utilization, and cogeneration are the subjects of Chapter 8. The second law availability of fuel for use in heating is the subject of Chapter 9. Chapter 10 discusses several industrial problem areas including waste heat utilization and recycling. The transportation industry is the subject of Chapter 11 and it focuses on automotive use. Chapter 12 deals briefly with the role of alternate sources such as wood, solar, and wind.

The engineering profession is slowly moving toward the International System (SI) of units, and the teaching profession should be in the foreground. However, it is the author's experience that the industrial organization and the individual home owner are much more comfortable in their own system of units, and may even resist favorable action when the analysis is presented in other terms. As a result, each chapter in this book is written in the units which appear to be in common usage in each area. The units of energy flow are, variously, Btu/hr, ton, kilowatt, or horsepower, as appropriate. The idea here is that in trying to show more effective energy utilization, it is best to deal in the terms the customer is accustomed to.

The book has evolved from a one-semester course taught at the senior-graduate level in mechanical engineering. A nominal prerequisite is a first course in heat transfer, but it is really the maturity of the student in facing and solving engineering problems which is important. Students from other disciplines (economics, business, as well as other engineering departments) have successfully taken the course.

Analyzing the energy and economic aspects of a given situation requires practice. There are problems at the end of each chapter to illustrate the features of each area. These should be taken as a guide, since both the student and the instructor may conceive of more relevant illustrations.

In addition to the problem assignments, students have undertaken term projects. The goal of these has been to find a situation in which energy might be used more effectively, to analyze the energy use of the current approach, to evaluate technically how energy reduction might be accomplished, and to determine the cost effectiveness of the proposed changes. Sometimes, these projects require information not currently available or costs that are hard to come by, and sometimes they show that the present methods are acceptable. However, this is the nature of the current energy problem, and times may change the situation. Many projects reveal that changes can be made that are both energy and cost effective. More important, all have been valuable in helping the student learn to tackle a complicated, poorly specified energy problem. This process is the real goal of this book.

It is hoped that this book provides a basic engineering and economic structure for analyzing energy related problems. It is not an encyclopedia of known solutions. Rather, it recognizes that energy supplies and costs are dynamic and that what is a satisfactory solution today may not be in the future. It is hoped that the approach illustrated here will be of value in helping resolve the energy problems of the future through more effective utilization.

Lastly, I would like to acknowledge the help and inspiration that I have received in preparing this book. The students in my classes have been a continual source of ideas, both in questioning what is presented to them and in volunteering new thoughts. Many sections in the book are the result of one of their term papers. My colleagues at the Solar Energy Laboratory, William A. Beckman, in particular, and Sanford A. Klein and John A. Duffie, have provided many insights into all aspects of the energy issue. The time they have given in discussions with me and their interest in this book is appreciated. My major professor, A. L. London of Stanford University, taught me how to critically analyze engineering problems. I appreciate his dedication and his enthusiastic and continued interest in my work.

JOHN W. MITCHELL

Madison, Wisconsin
March 1983

CONTENTS

Nomenclature — xiii

1. **INTRODUCTION** — 1
   - 1.1. Historical Energy Consumption Patterns, 2
   - 1.2. Fuel Prices, 6
   - 1.3. Projections of Future Energy Use, 7
   - 1.4. Conventional Energy Supplies for the Future, 9
   - 1.5. Summary, 10
   - Suggested Reading, 11

2. **ECONOMIC CONSIDERATIONS** — 13
   - 2.1. Basic Economic Calculations, 14
   - 2.2. Economic Examples, 16
   - 2.3. General Life Cycle Cost Formulation, 20
   - 2.4. Economic Optimum Level of Insulation, 25
   - 2.5. Summary, 28
   - Suggested Reading, 28
   - Problems, 28

3. **RESIDENTIAL AND COMMERCIAL BUILDING HEATING REQUIREMENTS** — 31
   - 3.1. Overview, 31
   - 3.2. Building Heat Losses, 34
   - 3.3. Energy Gains, 47
   - 3.4. Design Heating Requirements, 49
   - 3.5. Annual Energy Consumption, 49
   - 3.6. Example Calculation for Annual Energy Consumption, 61
   - 3.7. Regional Variations in Energy Use, 74
   - 3.8. Summary, 76
   - Suggested Reading, 76
   - Problems, 77

CONTENTS

4. RESIDENTIAL AND COMMERCIAL AIR CONDITIONING REQUIREMENTS   79

     4.1. Overview, 79
     4.2. Building Energy Gains, 80
     4.3. Design Cooling Load, 89
     4.4. Annual Cooling Energy Requirements, 91
     4.5. Example Calculation for Design and Seasonal Cooling Load, 96
     4.6. Summary, 102
    Suggested Reading, 102
    Problems, 102

5. HEATING AND COOLING OF COMMERCIAL BUILDINGS   105

     5.1. Load Profiles, 106
     5.2. Space Conditioning Systems, 108
     5.3. Heat Recovery Systems, 112
     5.4. Annual Energy Requirements, 113
     5.5. Illumination, 115
     5.6. Comfort, 119
     5.7. Summary, 123
    Suggested Reading, 124
    Problems, 124

6. COMBUSTION FURNACES   127

     6.1. Furnace Operation, 128
     6.2. Combustion Reaction, 130
     6.3. Heat Exchanger Process, 134
     6.4. Chimney Processes, 136
     6.5. Furnace Efficiency During Steady-State Operation, 144
     6.6. Seasonal Furnace Efficiency, 146
     6.7. Intermittent Combustion Furnaces, 151
     6.8. Summary of Natural Draft Furnace Performance, 152
     6.9. Summary, 155
    Suggested Reading, 155
    Problems, 155

7. HEAT PUMPS   157

     7.1. Thermodynamics of Operation, 158
     7.2. Heat Pump Performance, 161
     7.3. Calculation of System Performance, 164
     7.4. Generalized Weather Distributions for Heat Pump Calculations, 166

NOMENCLATURE

thermal conductivity; specific heat ratio
distance
life cycle costs
life cycle savings
mortgage interest rate; mass
mass flow rate
ratio of maintenance, etc. costs to investment
number of years
depreciation life time
term of a loan
minimum of n and n_L
minimum of n and n_d
number of hours
number of air changes per hour
number of transfer units
property tax rate
perimeter; pressure
present worth factor for single sum
present worth factor for annual series
present worth factor for annual costs
present worth factor for first costs
heat flow rate
heat flow per unit area, flux
total heat flow
tax credit on initial investment
compression ratio
thermal resistance; gas constant
entropy
ratio of salvage cost to investment
effective income tax rate; time
temperature
thermal conductance
volume; velocity
volume flow rate
work, weight; width
work rate, power
distance; thickness
elevation; height

CONTENTS

7.5. House Heating Requirements, 168
7.6. Example Calculation, 168
7.7. Comparison Between Systems, 171
7.8. Water Source Heat Pumps, 173
7.9. Heat Pump and Storage Systems, 175
7.10. Commercial Applications, 177
7.11. Summary, 179
Suggested Reading, 179
Problems, 179

8. ELECTRICAL POWER PRODUCTION — 181

8.1. Thermodynamic Principles, 181
8.2. Availability of Energy Sources, 184
8.3. Availability Destruction due to Combustion, 187
8.4. Cogeneration of Electricity and Heat, 191
8.5. Price of Delivered Electricity and Heat from Cogeneration Plants, 199
8.6. Load Management, 200
8.7. Industrial Cogeneration, 204
8.8. Summary, 206
Suggested Reading, 206
Problems, 206

9. AVAILABILITY OF FUEL FOR HEATING — 209

9.1. Thermodynamic Concepts, 209
9.2. System Performance, 212
9.3. Gas-Fired Heat Pumps, 215
9.4. Summary, 220
Suggested Reading, 220
Problems, 221

10. ENERGY USE IN INDUSTRY — 223

10.1. Overview, 223
10.2. Economic Optimum Insulation Levels, 225
10.3. Heat Exchangers for Waste Heat Reclamation, 228
10.4. Heat Pumps for Waste Heat Recovery, 242
10.5. Industrial Refrigeration, 248
10.6. Recycling, 253
10.7. Summary, 256
Suggested Reading, 256
Problems, 256

xii   CONTENTS

11. TRANSPORTATION ENERGY USE   259

    11.1. Overview, 259
    11.2. Automobile Transportation, 261
    11.3. Vehicle Power Requirements, 263
    11.4. Engine Characteristics, 272
    11.5. Distribution of Energy Use for Automobiles, 277
    11.6. The Effect of Speed Limits, 279
    11.7. Summary, 280
    Suggested Reading, 280
    Problems, 280

12. RENEWABLE ENERGY SOURCES   283

    12.1. Bioconversion, 283
    12.2. Wood, 285
    12.3. Wind, 287
    12.4. Solar Energy, 291
    12.5. Summary, 301
    Suggested Reading, 301

APPENDIXES   303

    Appendix A   Weather Data for Selected Cities, 304
    Appendix B   Energy Content of Fuels, 306

INDEX   307

NOMENCLA[TURE]

a	availability per unit mass; coefficient
a_{rp}	availability of reaction of fuel
A	area; availability
$\dot{A}$	rate of availability flow
b	coefficient
c	indicator for taxes for commercial build[ings]
c_p	specific heat at constant pressure
C	cost; capacitance rate
C_d	discharge or drag coefficient
C_E	equipment cost per unit installed capaci[ty]
C_F	fuel cost per unit of energy
C_I	insulation cost per unit of R value
C_P	plant cost
COP	coefficient of performance
d	market discount rate or return on invest[ment]
D	fraction of first cost
D_m	days in a month
DD_m	monthly degree days
E	equipment or investment related costs;
$\dot{E}$	energy flow rate
f	fuel inflation rate; time fraction
F	first year fuel cost; force
g	rate of energy gain into a space; gravita[tion]
G	energy gain into a space
h	convection coefficient; enthalpy
h_{fg}	latent heat of vaporization
h_{rp}	enthalpy of reaction of fuel
H	height
H_T	daily average incident solar radiation o[n]
i	annual inflation rate
I_T	instantaneous solar insolation on a tilt[ed]

k
L
LCC
LCS
m
$\dot{m}$
M
n
n_d
n_L
n_1
n_2
N
$\dot{N}$
N_{tu}
p
P
PW
PWF
P_1
P_2
q
q''
Q
r
r_v
R
s
S
t
T
U
V
$\dot{V}$
W
$\dot{W}$
x
z

NOMENCLATURE
GREEK SYMBOLS

α	absorptance
γ	multiplication factor
ϵ	emissivity; effectiveness
η	efficiency
θ	angle
ρ	density
σ	Stefan-Boltzmann constant; standard deviation
τ	transmittance
$\overline{\tau\alpha}$	average transmittance-absorptance product

SUBSCRIPTS

a	instantaneous ambient; air
abs	absorbed
act	actual
ae	aerodynamic
app	appliances
at	attic
aux	auxiliary
ave	average
A	ambient air; area
b	boiler
bf	basement floor
bt	basement
bw	basement wall
B	basement
Bal	balance
c	ceiling; combustion; collector
ch	chimney
cl	cooling
comp	compressor
cond	conduction; condensor
conv	convection
C	combustion chamber; cold
d	daytime; delivered; destroyed
des	design

dr	door
e	exit
elec	electric
ev	evaporator
ex	exfiltration
f	flue; flow; frontal
fix	fixed costs
fl	floor
fr	friction
fuel	fuel
fur	furnace
g	gas
gain	gain
gr	gravitational; garage
G	ground
h	hood; hot
hp	heat pump
htg	heating, heating system
hx	heat exchanger
hydro	hydroelectric
in	inflow
inf	infiltration
int	internal
ir	inertial
lat	latent
lig	lights
loss	loss, total heat loss
m	month, monthly
max	maximum
min	minimum
mix	mixed
n	nighttime period
ns	night setback period
nuc	nuclear
o	overall; reference; environment; base
off	off time
on	on time
opt	optimum
out	outflow

p	products
peo	people
pl	plenum
pp	power plant
pump	pump
q	heat
r	reactants
rad	radiation
rec	recirculation flow
rej	rejected
res	resource
rf	roof
rh	reheat
R	room
s	supply air; source; surface
sa	sol-air
seas	seasonal
sens	sensible
sl	slab
sol	solar
ss	steady state
sys	system
th	thermal
total	total
turb	turbine
useful	useful
vent	ventilation
w	work
wd	window; wind
wind	wind
wl	wall
work	work
yr	annual

SUPERSCRIPTS

—	average value
*	breakeven value
°	optimum value
+	plus values only

■ 1 ■

INTRODUCTION

It is obvious that there are problems with energy both in the United States and in the world today. There is confusion over the magnitude of the available supplies, the rate of depletion of the resources, and the feasibility of shifting from one source to another. Gasoline, fuel oil, and natural gas supplies periodically have become critically short. The uncertainty over the fuel needed for comfort and business has produced a variety of reactions in suppliers, consumers, and governmental organizations. Suppliers have raised prices of some scarce fuels, and have been unable to supply other types. Consumers have responded to these price rises and fuel shortages by reducing consumption and shifting to other fuels. Governmental organizations have implemented a sometimes conflicting variety of actions. Prices of some fuels have been regulated at low values in an attempt to avoid hardship, which then acts to reduce the incentive to change consumption. Laws and codes regulating usage have been introduced, such as the 55-mph speed limit and residential building codes. Tax incentives for improvements such as home insulation and switching to an alternate renewable fuel have been passed. Fuels are alternatively declared plentiful and scarce by government. There has been no consistent governmental response to the energy situation.

Out of this morass have come a few clear messages. There is an energy problem. There is a scarcity of the nonrenewable sources such as petroleum and natural gas. Eventually these fuel sources will be depleted, and other forms will have to be used. As the readily obtained supplies decrease, the cost of extracting fuel increases. The laws of engineering and the market place mean that prices of depleting fuels must rise. At best, consumers appear to reduce demand only if prices rise since a financial incentive has to be available to effect a change. Conservation has the effect of generating new supplies of depletable sources since energy not used today is available for later use. However, it can only postpone the inevitable. Thus, increased prices and reduced demand appear to be the trend for the future.

The two sides of the energy problem are supply and demand. The size of the reserve of a given fuel and the cost of its production depend mainly on the supplier. There is little a given consumer can do to alter the type, quantity, or price of fuel available to him. His only option is to change his demand for energy. He can determine the amount consumed and the ways to reduce consumption. The supply side is generally under the control of the large energy corporations, while the demand side is under the control of the individual users.

The difference in the nature of the supplier and the consumer has a strong effect on actions. The supply industry is a small number of large companies that deliver energy to many consumers. The depletion of a source such as an oil field has a large effect on their output, while the change by any one consumer has little effect. In contrast, consumers are a very diverse group of individuals, small commercial establishments, and large companies. They are very sensitive to the change in price of fuel or the availability of their supplies. They are strongly motivated to conserve.

There are few technical problems involved in reducing energy usage. Most conservation schemes, such as insulation, lowered driving speeds, and waste heat recovery, are established. Redesign of equipment, such as more efficient air conditioners or furnaces, are technically feasible. The major constraint on changing energy usage is economic. Few, if any, users will undertake an energy conservation measure at an economic penalty. Although consumers are concerned about fuel availability and long-term energy supplies, they are motivated to save money, not energy. This is true for both individuals and institutions.

The orientation of this book is toward the demand side of the energy situation. The energy consumption for a given process will be evaluated. Alternatives to current usage will be determined and energy savings calculated. The objective here will be to determine the technical means of using less energy by maintaining the same output. An equally important objective will be to determine the economics of energy use for both the current and alternative options. The costs of additional equipment or materials will be determined and balanced against the value of the energy savings. Alternatives will be evaluated in terms of both energy and economic savings. The goal will be to determine cost-effective measures for saving energy.

1.1. HISTORICAL ENERGY CONSUMPTION PATTERNS

Energy is essential to our way of life. It provides us with comfort, transportation, and the ability to produce food and material goods. Historically, energy consumption has been directly related to the gross national product (GNP), which is a measure of the market value of the total national output of goods and services. In the United States and in the world, increases and decreases in GNP over the years appear to be directly correlated with changes in energy consumption. Those countries with higher values of both GNP and energy consumption such as the United States, Canada, Germany, and England are those that most Americans would judge to have desirable life styles. It appears that our utilization of energy has produced our current high standard of living. However, energy usage patterns must change as supplies become depleted and more expensive alternatives appear. The challenge to the engi-

CONTENTS   xi

7.5. House Heating Requirements, 168
7.6. Example Calculation, 168
7.7. Comparison Between Systems, 171
7.8. Water Source Heat Pumps, 173
7.9. Heat Pump and Storage Systems, 175
7.10. Commercial Applications, 177
7.11. Summary, 179
Suggested Reading, 179
Problems, 179

8. ELECTRICAL POWER PRODUCTION   181

8.1. Thermodynamic Principles, 181
8.2. Availability of Energy Sources, 184
8.3. Availability Destruction due to Combustion, 187
8.4. Cogeneration of Electricity and Heat, 191
8.5. Price of Delivered Electricity and Heat from Cogeneration Plants, 199
8.6. Load Management, 200
8.7. Industrial Cogeneration, 204
8.8. Summary, 206
Suggested Reading, 206
Problems, 206

9. AVAILABILITY OF FUEL FOR HEATING   209

9.1. Thermodynamic Concepts, 209
9.2. System Performance, 212
9.3. Gas-Fired Heat Pumps, 215
9.4. Summary, 220
Suggested Reading, 220
Problems, 221

10. ENERGY USE IN INDUSTRY   223

10.1. Overview, 223
10.2. Economic Optimum Insulation Levels, 225
10.3. Heat Exchangers for Waste Heat Reclamation, 228
10.4. Heat Pumps for Waste Heat Recovery, 242
10.5. Industrial Refrigeration, 248
10.6. Recycling, 253
10.7. Summary, 256
Suggested Reading, 256
Problems, 256

11. TRANSPORTATION ENERGY USE 259

11.1. Overview, 259
11.2. Automobile Transportation, 261
11.3. Vehicle Power Requirements, 263
11.4. Engine Characteristics, 272
11.5. Distribution of Energy Use for Automobiles, 277
11.6. The Effect of Speed Limits, 279
11.7. Summary, 280
Suggested Reading, 280
Problems, 280

12. RENEWABLE ENERGY SOURCES 283

12.1. Bioconversion, 283
12.2. Wood, 285
12.3. Wind, 287
12.4. Solar Energy, 291
12.5. Summary, 301
Suggested Reading, 301

APPENDIXES 303

Appendix A Weather Data for Selected Cities, 304
Appendix B Energy Content of Fuels, 306

INDEX 307

NOMENCLATURE

a	availability per unit mass; coefficient
a_{rp}	availability of reaction of fuel
A	area; availability
$\dot{A}$	rate of availability flow
b	coefficient
c	indicator for taxes for commercial buildings; specific heat
c_p	specific heat at constant pressure
C	cost; capacitance rate
C_d	discharge or drag coefficient
C_E	equipment cost per unit installed capacity
C_F	fuel cost per unit of energy
C_I	insulation cost per unit of R value
C_P	plant cost
COP	coefficient of performance
d	market discount rate or return on investment
D	fraction of first cost
D_m	days in a month
DD_m	monthly degree days
E	equipment or investment related costs; energy
$\dot{E}$	energy flow rate
f	fuel inflation rate; time fraction
F	first year fuel cost; force
g	rate of energy gain into a space; gravitational acceleration
G	energy gain into a space
h	convection coefficient; enthalpy
h_{fg}	latent heat of vaporization
h_{rp}	enthalpy of reaction of fuel
H	height
H_T	daily average incident solar radiation on a tilted surface
i	annual inflation rate
I_T	instantaneous solar insolation on a tilted surface

NOMENCLATURE

k	thermal conductivity; specific heat ratio
L	distance
LCC	life cycle costs
LCS	life cycle savings
m	mortgage interest rate; mass
$\dot{m}$	mass flow rate
M	ratio of maintenance, etc. costs to investment
n	number of years
n_d	depreciation life time
n_L	term of a loan
n_1	minimum of n and n_L
n_2	minimum of n and n_d
N	number of hours
$\dot{N}$	number of air changes per hour
N_{tu}	number of transfer units
p	property tax rate
P	perimeter; pressure
PW	present worth factor for single sum
PWF	present worth factor for annual series
P_1	present worth factor for annual costs
P_2	present worth factor for first costs
q	heat flow rate
q''	heat flow per unit area, flux
Q	total heat flow
r	tax credit on initial investment
r_v	compression ratio
R	thermal resistance; gas constant
s	entropy
S	ratio of salvage cost to investment
t	effective income tax rate; time
T	temperature
U	thermal conductance
V	volume; velocity
$\dot{V}$	volume flow rate
W	work; weight; width
$\dot{W}$	work rate, power
x	distance; thickness
z	elevation; height

NOMENCLATURE

GREEK SYMBOLS

α	absorptance
γ	multiplication factor
ϵ	emissivity; effectiveness
η	efficiency
θ	angle
ρ	density
σ	Stefan-Boltzmann constant; standard deviation
τ	transmittance
$\overline{\tau\alpha}$	average transmittance-absorptance product

SUBSCRIPTS

a	instantaneous ambient; air
abs	absorbed
act	actual
ae	aerodynamic
app	appliances
at	attic
aux	auxiliary
ave	average
A	ambient air; area
b	boiler
bf	basement floor
bt	basement
bw	basement wall
B	basement
Bal	balance
c	ceiling; combustion; collector
ch	chimney
cl	cooling
comp	compressor
cond	conduction; condensor
conv	convection
C	combustion chamber; cold
d	daytime; delivered; destroyed
des	design

NOMENCLATURE

dr	door
e	exit
elec	electric
ev	evaporator
ex	exfiltration
f	flue; flow; frontal
fix	fixed costs
fl	floor
fr	friction
fuel	fuel
fur	furnace
g	gas
gain	gain
gr	gravitational; garage
G	ground
h	hood; hot
hp	heat pump
htg	heating, heating system
hx	heat exchanger
hydro	hydroelectric
in	inflow
inf	infiltration
int	internal
ir	inertial
lat	latent
lig	lights
loss	loss, total heat loss
m	month, monthly
max	maximum
min	minimum
mix	mixed
n	nighttime period
ns	night setback period
nuc	nuclear
o	overall; reference; environment; base
off	off time
on	on time
opt	optimum
out	outflow

p	products
peo	people
pl	plenum
pp	power plant
pump	pump
q	heat
r	reactants
rad	radiation
rec	recirculation flow
rej	rejected
res	resource
rf	roof
rh	reheat
R	room
s	supply air; source; surface
sa	sol-air
seas	seasonal
sens	sensible
sl	slab
sol	solar
ss	steady state
sys	system
th	thermal
total	total
turb	turbine
useful	useful
vent	ventilation
w	work
wd	window; wind
wind	wind
wl	wall
work	work
yr	annual

SUPERSCRIPTS

—	average value
*	breakeven value
°	optimum value
+	plus values only

■ 1 ■

INTRODUCTION

It is obvious that there are problems with energy both in the United States and in the world today. There is confusion over the magnitude of the available supplies, the rate of depletion of the resources, and the feasibility of shifting from one source to another. Gasoline, fuel oil, and natural gas supplies periodically have become critically short. The uncertainty over the fuel needed for comfort and business has produced a variety of reactions in suppliers, consumers, and governmental organizations. Suppliers have raised prices of some scarce fuels, and have been unable to supply other types. Consumers have responded to these price rises and fuel shortages by reducing consumption and shifting to other fuels. Governmental organizations have implemented a sometimes conflicting variety of actions. Prices of some fuels have been regulated at low values in an attempt to avoid hardship, which then acts to reduce the incentive to change consumption. Laws and codes regulating usage have been introduced, such as the 55-mph speed limit and residential building codes. Tax incentives for improvements such as home insulation and switching to an alternate renewable fuel have been passed. Fuels are alternatively declared plentiful and scarce by government. There has been no consistent governmental response to the energy situation.

Out of this morass have come a few clear messages. There is an energy problem. There is a scarcity of the nonrenewable sources such as petroleum and natural gas. Eventually these fuel sources will be depleted, and other forms will have to be used. As the readily obtained supplies decrease, the cost of extracting fuel increases. The laws of engineering and the market place mean that prices of depleting fuels must rise. At best, consumers appear to reduce demand only if prices rise since a financial incentive has to be available to effect a change. Conservation has the effect of generating new supplies of depletable sources since energy not used today is available for later use. However, it can only postpone the inevitable. Thus, increased prices and reduced demand appear to be the trend for the future.

The two sides of the energy problem are supply and demand. The size of the reserve of a given fuel and the cost of its production depend mainly on the supplier. There is little a given consumer can do to alter the type, quantity, or price of fuel available to him. His only option is to change his demand for energy. He can determine the amount consumed and the ways to reduce consumption. The supply side is generally under the control of the large energy corporations, while the demand side is under the control of the individual users.

The difference in the nature of the supplier and the consumer has a strong effect on actions. The supply industry is a small number of large companies that deliver energy to many consumers. The depletion of a source such as an oil field has a large effect on their output, while the change by any one consumer has little effect. In contrast, consumers are a very diverse group of individuals, small commercial establishments, and large companies. They are very sensitive to the change in price of fuel or the availability of their supplies. They are strongly motivated to conserve.

There are few technical problems involved in reducing energy usage. Most conservation schemes, such as insulation, lowered driving speeds, and waste heat recovery, are established. Redesign of equipment, such as more efficient air conditioners or furnaces, are technically feasible. The major constraint on changing energy usage is economic. Few, if any, users will undertake an energy conservation measure at an economic penalty. Although consumers are concerned about fuel availability and long-term energy supplies, they are motivated to save money, not energy. This is true for both individuals and institutions.

The orientation of this book is toward the demand side of the energy situation. The energy consumption for a given process will be evaluated. Alternatives to current usage will be determined and energy savings calculated. The objective here will be to determine the technical means of using less energy by maintaining the same output. An equally important objective will be to determine the economics of energy use for both the current and alternative options. The costs of additional equipment or materials will be determined and balanced against the value of the energy savings. Alternatives will be evaluated in terms of both energy and economic savings. The goal will be to determine cost-effective measures for saving energy.

1.1. HISTORICAL ENERGY CONSUMPTION PATTERNS

Energy is essential to our way of life. It provides us with comfort, transportation, and the ability to produce food and material goods. Historically, energy consumption has been directly related to the gross national product (GNP), which is a measure of the market value of the total national output of goods and services. In the United States and in the world, increases and decreases in GNP over the years appear to be directly correlated with changes in energy consumption. Those countries with higher values of both GNP and energy consumption such as the United States, Canada, Germany, and England are those that most Americans would judge to have desirable life styles. It appears that our utilization of energy has produced our current high standard of living. However, energy usage patterns must change as supplies become depleted and more expensive alternatives appear. The challenge to the engineering profession is to increase or at least maintain our physical and economic situation at reduced levels of energy use.

The obvious targets for reduced consumption are those processes and areas which currently consume the most. In order to identify these, the overall energy consumption pattern for the United States will be surveyed. This will provide a base for determining where to search for significant reductions in usage.

The use of energy in the United States is given in Table 1.1.1 for the four traditional major sectors, denoted as Industrial, Residential and commercial, Transportation, and Utilities. Each sector consumes about an equal amount of energy. The historical pattern over time is shown in Fig. 1.1.1. The numbers in parentheses denote the percentage of end use energy. The electricity produced by the utilities sector is distributed to industry (about 60 percent) and to the commercial and residential sectors (about 40 percent). The total use including electricity is given at the top of each figure. There is predominantly more use in the industrial sector and less in transportation, but all sectors have a significant usage.

The energy use by fuel type is also shown in Fig. 1.1.1. The current reliance on liquid petroleum is readily apparent. The transportation sector is almost wholly dependent on oil. Natural gas supplies a large proportion of space and industrial heating. Electrical generation is primarily by coal. The primary energy use by fuel source is given in Table 1.1.2. Almost half of our energy is from oil which is currently in short supply. Natural gas is less critical, and meets about one-quarter of our need. Coal, which is relatively plentiful, supplies only about one-fifth. Our largest usage is of fuels which are most scarce.

The figures all show increasing energy consumption over time, and at a faster rate than population. Over the period 1950 to 1975, the population increase has been about 35 percent, while the total energy consumption increased about 2.3 times. On a per capita basis, we have almost doubled our energy use in 25 years for a current per capita energy consumption of about 400×10^6 Btu/yr. Typical energy costs are on the order of $\$5/10^6$ Btu, and the value of energy consumption amounts to a per capita expenditure of \$2000/yr. Energy costs are, obviously, a significant factor in our economy.

The major end uses of energy are identified in Table 1.1.3. Two-thirds of the total usage is accounted for in the first five categories. These are the obvious target

TABLE 1.1.1
Major Sectors of Energy Use (1978)

Sector	Energy Use	
	10^{15} Btu/yr	Percent
Industrial	18	23
Residential and commercial	17	21
Transportation	20	25
Utilities	24	31
Total	79	100

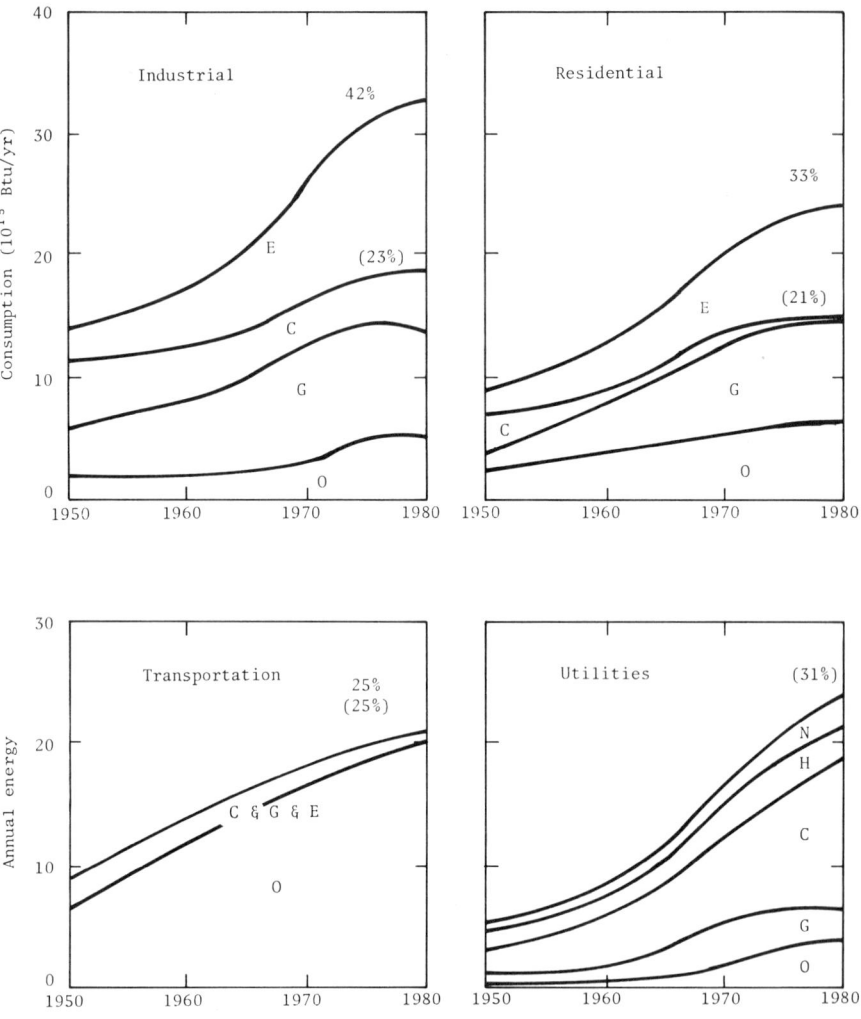

Fig. 1.1.1. Annual energy consumption by sector. The percentage of total energy consumption is shown at the top of each figure. The percentage excluding electricity is shown in parentheses.

areas in which more efficient energy use can provide significant gains. Overall, heating uses in industry and buildings comprise half of the total. Improvements in heating equipment, changes in demand, and heat recovery equipment can all reduce consumption. Automobiles use 14 percent of the total energy, but 30 percent of the petroleum supplies, and are also a major target.

Individuals control about two-thirds of the space conditioning, water heating, and lighting and are responsible for virtually all of the automotive use. Thus, about one-third of the total energy consumption results from individual actions. Consequently, such programs as thermostat turn-downs, highway speed limits, and tax

TABLE 1.1.2
Fuel Sources for Primary Energy Use

Fuel	Energy	
	10^{15} Btu/yr	Percent
Oil	38	48
Natural gas	19	24
Coal	17	21
Hydroelectric	3	4
Nuclear	2	3
Total	79	100

incentives for energy conserving options have potentially a large payoff in energy savings.

Another aspect of the historical use patterns is the time required to double energy consumption. The energy requirements of industry have doubled over a 30-year period, those of transportation and the residential-commercial sector in 20 years, while utility consumption has doubled in 10 years. If historical growth patterns were to continue, in the year 2000 we would require four times as many utilities, twice as many cars and/or roads, twice as many or twice as large houses and buildings, and so on. This would require twice the total energy supply system we currently have. Clearly, in view of our current energy situation, we cannot continue the growth rates of the past. Changes must be made to more effectively use our limited resources.

TABLE 1.1.3
Major End Uses of Energy

	Percentage of Total Consumption
Space heating	18
Process steam	16
Automotive	14
Direct heating	11
Electric motors	8
Other transportation	6
Trucking	5
Lighting	5
Water heating	4
Feedstock	4
Air conditioning	3
Refrigeration	2
Cooking	1
Other	3

1.2. FUEL PRICES

Prices of fuels have risen over the years along with energy use. The increase in price does not reflect a change in value since inflation has occurred and the value of the dollar has decreased. The current price can be adjusted using the consumer price index to reflect the true value, or cost, of energy. Representative prices of fuels are given for the period 1960 to the present in Table 1.2.1. Both the current price and the true cost (in 1960 dollars) are shown.

The current price has increased, as expected, but the trend for the true cost is somewhat surprising. Over the decade 1960 to 1970, the true cost of energy decreased slightly for all fuels. Between 1970 and 1975, gasoline and natural gas rose slightly, but still cost less than in 1960. Electrical costs remained fairly constant. Only fuel oil showed an actual cost increase. The real rise in energy costs occurred between 1975 and 1980 for all fuels. However, the percentage changes in real costs are not as large as generally perceived. Between 1970 and 1980, gasoline has increased 47 percent, fuel oil 52 percent, natural gas 56 percent, and electricity 6 percent.

The annual rate of price increase is essential to the economic analysis of energy utilizing equipment. The average rates over the last 5 years are shown in Table 1.2.2 along with the general inflation rate. It is seen that the costs of gasoline, fuel oil, and natural gas have risen about 10 percent higher than general inflation, while electricity is only slightly higher than inflation. Also shown are some projected increases for the next decade. Fuel prices are expected to rise, but not as rapidly as in the past. However, the actual cost is expected to increase faster than inflation. In light on the past history, the accuracy of the projections is extremely uncertain.

TABLE 1.2.1
Price of Fuels

Fuel	Unit Cost	Current Price				
		1960	1965	1970	1975	1980
Gasoline	¢/gal	31	32	36	55	130
Fuel oil	¢/gal	15	16	18	38	95
Natural gas	¢/100 ft^3	10	10	11	17	40
Electricity	¢/kWh	2.4	2.2	2.1	3.0	5.0

Fuel	Unit Cost	True Cost				
		1960	1965	1970	1975	1980
Gasoline	¢/gal	31	30	27	30	44
Fuel oil	¢/gal	15	15	13	21	32
Natural gas	¢/100 ft^3	10	9	8	9	14
Electricity	¢/kWh	2.4	2.1	1.6	1.6	1.7
Relative consumer price index		1.00	1.06	1.34	1.82	2.95

TABLE 1.2.2
Annual Rate of Increase

	Historical 1975–1980	Projected[a] 1980–1990
General inflation	10%	10%
Gasoline	19%	14%
Fuel oil	20%	14%
Natural gas	19%	16%
Electricity	11%	12%

[a]Necessarily speculative and very uncertain.

TABLE 1.2.3
Representative Fuel Prices (1982)

Fuel	Unit Cost	Energy Content	Cost ($/10^6 Btu)
Natural gas	45¢/100 ft^3	1000 Btu/ft^3	4.50
Oil	96¢/gal	137,000 Btu/gal	7.00
LP gas	70¢/gal	96,000 Btu/gal	7.30
Wood	$80/ton	7,000 Btu/lb$_m$	5.70
Electricity	5¢/kW	3,413 Btu/kWh	14.65
Gasoline	$1.30/gal	125,000 Btu/gal	10.60
Coal	$70/ton	13,000 Btu/lb$_m$	2.70

The current (1982) price and energy content of a variety of common fuels are given in Table 1.2.3. The unit cost is on the basis of how the fuel is sold, while a more useful basis is the cost per unit energy content. A common and consistent basis is on a per million (10^6) Btu of energy. On this basis, fuel costs vary widely. Coal is the cheapest by almost a factor of two over the next lowest price fuel, natural gas. Fuel oil, liquefied petroleum, and wood are all comparable. Gasoline costs about 30 percent more than fuel oil. Electricity is two to three times the price of the other fuels.

These fuel prices are based on the cost of purchasing the fuel. The efficiency of conversion of fuel energy into a usable form (heating, electricity, motion, etc.) depends on the devices, and will generally increase the cost of delivered energy. For example, combustion furnace efficiencies are in the range of 60 percent, and so the cost of heating with fuel oil is comparable to that with electricity. In the following chapters of this book, these conversion efficiencies will be studied in detail.

1.3. PROJECTIONS OF FUTURE ENERGY USE

The historical patterns of energy use are well known. In contrast, the future patterns are naturally extremely uncertain. The current situation dictates that usage must change and shifts from one fuel type to another must be made, but the magnitude of these changes and shifts and the times at which they will occur is unknown. As a result, projections of future consumption have been made that range between a reduction in total energy use to continued exponential growth.

The Ford Foundation studied the historical uses of energy in this country in 1974. From these data, possible future growth patterns were projected which were published in the book *A Time to Choose*. Three alternative future scenarios were developed. Two were extremes of high and low use and one attempted to realistically assess the most probable consumption path. These three scenarios will be briefly summarized to serve as guides for the technical approaches discussed in this book, and to point out changes that have occurred.

The basis for all scenarios was that sufficient food and shelter would be available, full employment would be obtained, and there would be a growth in GNP. Population projections based on demographic data were made which showed that by the year 2000 the United States population would increase by 30 percent. A mechanistic approach was then taken to determine the factors by which changes in energy consumption would occur.

In the "Historical Growth" scenario, the past trends were assumed to continue. Growth in housing occurred due to bigger and better equipped single-family homes which were located in the suburbs. People would rely on the large, low gas mileage (12 mpg) automobile for personal transportation, and use mass transit very little. There would be continued growth in the plastic, paper, and aluminum industries with little recycling of materials. The energy intensiveness of industrial production increased as in the past. These mechanisms led to an energy consumption in the year 2000 that would be 2.2 times current usage.

For the "Technical Fix" scenario, the same increases in GNP were achieved as for the "Historical Growth" scenario, but energy was utilized more efficiently. Homes would be better insulated with more efficient heating systems. Automobile fuel economy would rise (ultimately to 27 mpg), and all types of transportation would be better utilized. Industry would reduce the energy required for a given output in all processes through heat recovery and cogeneration techniques. As a result, the consumption in 2000 would be 1.4 times the current value.

The "Zero Energy Growth" scenario assumed many governmental policies would be implemented to limit growth. Planned communities of small townhouses would be developed and urban, single-family developments would be phased out. Transportation would be mainly be mass transit with very efficient automobiles (33 mpg) used sparingly. Industry would shift significantly from material goods to services, with a large reduction in energy usage. The energy consumption in 2000 would be 1.2 times the present usage.

The projections were made in 1974, and the use since that time can be examined to see which scenario is most probable. Since 1974, many of the historical growth patterns have changed. Buildings are now smaller and better insulated with more efficient heating systems. The low mileage automobile has been eliminated by legislation, while gasoline and car prices have further helped the public shift toward smaller, more economical cars. Industry has reduced waste, recycled materials, installed heat reclaim equipment, and developed cogeneration plants.

Clearly, historical growth is not occurring. The mechanisms of zero energy growth are also not occurring through either legislation or voluntary action. Many of the aspects of the technical fix scenario are being implemented, and it would appear that this path, with variations, is probably close to the one we will follow. It is important to realize that even this pattern results in a per capita increase in energy usage. It is then all the more important to determine ways to effectively use energy.

The assumptions and conclusions of the Ford Foundation report were controversial at the time they were published. It is well worth reading the comments of the advisory board, which was selected to represent a wide spectrum of business and consumer groups, in order to sense the diversity of opinion surrounding the energy issue. All members of the advisory board agreed on the need to use energy more effectively. However, there was little agreement on the means to achieve this, and whether there really was a supply problem. The criticisms ranged from condeming the zero energy growth scenario as too wasteful of energy to arguing that the historical growth scenario was too restrictive of energy use to allow the United States to maintain its status. These comments regarding the widely divergent projections of energy use only accentuate the uncertainty about the energy future.

1.4. CONVENTIONAL ENERGY SUPPLIES FOR THE FUTURE

The magnitude of the conventional, nonrenewable U.S. energy reserves provides a scale for the time span by which changes in usage must occur. Table 1.4.1 gives an estimate of the amount of the reserve remaining in both the United States and the world and also the time to deplete it. It is apparent that coal is the most plentiful source. Another way to visualize the magnitude of the resource is through the number of years at which the fuel will last at current rates of use. The column for the United States is for our use of proven reserves, while that for the world is for world use of estimated reserves.

Petroleum supplies are definitely limited, and significant changes in use must occur in the next 5 to 10 years. Oil exploration has gone on for a long time, and there are probably not major undiscovered oil fields in the continental United States. The situation with natural gas is different in that little direct exploration has occurred. Thus, the time span for this resource may increase. Coal is plentiful as a

TABLE 1.4.1
Conventional Fuel Reserves

Fuel	United States		World	
	Energy (10^{18} Btu)	Time (yr)	Energy (10^{18} Btu)	Time (yr)
Petroleum	0.2	5	4	30
Natural gas	0.2	10	13	50
Coal	40	2000	400	4000

source, and although the environmental costs may be high, it will be used heavily in the future.

The magnitude of these supplies should not allow us to become complacent. The physical nature of these supplies is that they are not renewable and will eventually be depleted. Even before that occurs, the cost of fuel will increase drastically as the reserves diminish. This will force shifts from one fuel type to other more available and less costly fuels. The economics of converting equipment from one type to another can be quite costly. Further, large shifts may place a burden on the supply facilities of the more available fuel.

These factors emphasize the need for more effective energy use. Fuel saved through nonuse is just as valuable as a new supply since it extends the life of that source. Better utilization may even be better than finding new reserves in that the attendant transportation, environmental, and handling costs are eliminated. Finally, more effective utilization saves the consumer fuel dollars which may be economically advantageous.

These conventional fuel sources will eventually be depleted though. At that time, the United States and the world will have to turn to renewable sources. These appear to be either fusion or solar energy. Fusion is a very speculative source and, at best, would not contribute for 30 years. It may be the energy supply of the future, but it is too early in its history to plan for it. For this reason, fusion is not discussed further in this book.

Solar energy forms such as agricultural crops for fuel, wind, solar heating, and electricity produced by solar are in various stages of development. Much is known in all areas, pilot projects have been built in many, and commercially available systems are available in others. The technical and economic issues surrounding these various solar forms are discussed in Chapter 12. Here, also, it is too early to assess the ultimate contribution of each form to our energy needs. At present, it appears that wind systems will eventually augment conventional power plants. Solar-photovoltaic plants may reduce or all but eliminate our reliance on fossil fuel for electricity. Solar heating of homes, domestic water, and swimming pools is technically and economically feasible at present. Growth of these sources along with penetration into the commercial market will undoubtedly occur. However, it is apparent that present concepts for solar energy cannot directly fill the needs for transportation and for heavy industry. This, again, underscores the need for better utilization of current energy supplies.

1.5. SUMMARY

The historical uses of energy, their sources, and their prices have been summarized in this chapter. Many sectors contribute to energy usage, the uses are diverse, and there are several sources. At present, oil accounts for almost one-half of our consumption, and natural gas one-quarter. These forms are the most limited, and better utilization of these two sources will bring the greatest rewards.

The major end uses of energy are for heating in residential and commercial buildings, for heating in industrial processes, for transportation, and for electric motors

SUGGESTED READING   11

...et areas where technical changes may produce great-
... by both fuel availability and price. Up to
...al dollars had decreased and there was little
...ent price rises are expected to be 2-6 percent
...this will accelerate implementation of conservation

...jections indicate that we may grow slightly in per capita energy use, which means a significant overall increase in consumption. To meet these demands, changes in supplies must occur. In the next 30 years, oil reserves will become critically short, and natural gas supplies will become low. Renewable sources such as solar will probably not be able to meet these demands. Better utilization of current supplies is an essential ingredient in meeting our future energy needs.

SUGGESTED READING

A Time to Choose, Energy Policy Project of the Ford Foundation, S. David Freeman, Director, Ballinger Pub. Co., Cambridge, MA, 1974.

R. Stobaugh and D. Yergin, Eds. *Energy Future*, Ballantine Books, New York, 1979.

Energy in Focus: Basic Data, Federal Energy Administration, Washington DC, 1980.

Efficient Utilization Research Status Report, Gas Research Institute, Chicago, IL, 1979.

R. C. Dorf, *The Energy Fact Book*, McGraw-Hill Book Co., New York, 1981.

Energy Information Administration Publications Directory – A Users Guide, USDOE Dept. DOE/EIA-0149; Annual.

Patterns of Energy Consumption in the United States, Office of Science and Technology, January 1972.

▪ 2 ▪

ECONOMIC CONSIDERATIONS

Economics play a large part in the consideration of alternatives in energy usage. Very few people acting either as individuals or on behalf of a business will undertake measures to conserve energy at a personal financial sacrifice. There usually must be some financial incentive to alter one's energy usage.

A measure that changes energy consumption usually involves an expenditure of money, such as that required to purchase and install equipment. These first costs are an expenditure of money at the present time. In contrast, the money saved through reduced fuel usage is available at times in the future. For a person considering making improvements to reduce energy use, the value of the energy saved must exceed the first costs in order for the improvement to be cost effective. The evaluation of the actual first costs and fuel savings are complicated by the differences in time that money is spent, by interest rates, by changes in prices, and by state and federal taxes.

The first cost of an improvement represents either an outlay of a sum currently in hand or a loan that must be paid back over time. Since the savings in energy costs occur in the future, they are not of as high a value as the money currently in hand. The time value of money is thus important, and the interest either that must be paid on borrowed money or that could have been earned from an alternate investment must be considered in balancing first costs against savings. Inflation occurs over time which makes a future sum even less valuable. However, energy prices may rise dramatically in the face of diminishing supplies, and so the future energy savings may actually increase. Interest payments and, for businesses, fuel payments may be deducted from income taxes, while tax credits may be available for the purchase of the improvement. These tax considerations may alter a decision. Energy conserving equipment will only be economically feasible if all of the actual costs and future savings are considered. It is important to consider all of these factors in making a decision regarding energy related investments.

ECONOMIC CONSIDERATIONS

In this chapter, the basic ideas relating to economic decisions will be reviewed first. Life cycle costing techniques will be developed to allow a ready evaluation of the true costs and benefits of energy saving investments. This formulation will prove especially helpful in optimization and in comparing several different alternatives. Several examples of the approach will be carried through in detail. This will provide a basis for all economic evaluations over a wide range of situations.

2.1. BASIC ECONOMIC CALCULATIONS

The basic definitions and relations between some of the more common economic terms are given in this section.

2.1.1. Interest Rate or Market Discount Rate

The interest rate reflects the "time value of money." A sum of money in hand today is worth more than that same sum of money in the future. The interest rate is the money paid for the use of borrowed money. An investor receives interest on the capital he invests, while a borrower pays interest to the supplier of money. The two interest rates are different. In general, a single investor receives a lower interest rate than a single borrower pays on his loan.

For a person or company considering energy conserving equipment investments, the market discount rate d stands for the interest rate that could be obtained from an alternate investment. For example, the market discount rate for a home owner considering ceiling insulation is probably equal to the interest paid on a savings account, and is on the order of 5-10 percent. For a company considering pipe insulation, however, the alternative could be to invest in a machine that would increase the plant output. A typical improvement in a process might return 20-30 percent on the investment. For the company, then, the appropriate market discount rate would be 20 or 30 percent. The interest rate to be used in an economic analysis depends on the options available to the investor.

2.1.2. Present Worth of a Single Sum

The present worth of money is its value in today's dollars. An amount of money in the future is not worth as much as the same sum today since today's money could be invested and would be worth more in the future. The present worth of an amount n years in the future equals the amount needed now that could be invested to yield the desired amount after n years. This relation between the present and future values of money is given by

$$\text{(present worth)} = \text{(future sum)} \text{(present worth factor)} \qquad (2.1.1)$$

BASIC ECONOMIC CALCULATIONS

The present worth factor for a single payment n years in the future, termed $PW(n, d)$ is given in terms of the market discount rate d by

$$PW(n, d) = \frac{1}{(1 + d)^n} \qquad (2.1.2)$$

2.1.3. Present Worth of a Series of Payments

The evaluation of the present worth of a series of payments occurs quite often in economic analyses. An annual or monthly mortgage payment may be required to pay off a given sum that is due in the future. Alternatively, the annual return from an investment may be of interest. In an energy analysis, the annual fuel payments required need to be determined; even if the fuel used remains constant over the years, the price may inflate. This relation between the present and future values is similar to that for a single sum, Eq. (2.1.1).

$$\text{(present worth)} = \text{(annual sum)} \text{(present worth factor)} \qquad (2.1.3)$$

where now the present worth factor must be evaluated for an annual series of payments.

The situation of a uniform annual series of payments that increase with time is shown schematically in Fig. 2.1.1. The first year fuel price is F, and the payment is assumed to be made at the end of the year. The present worth of the first year fuel cost is $F/(1 + d)$.

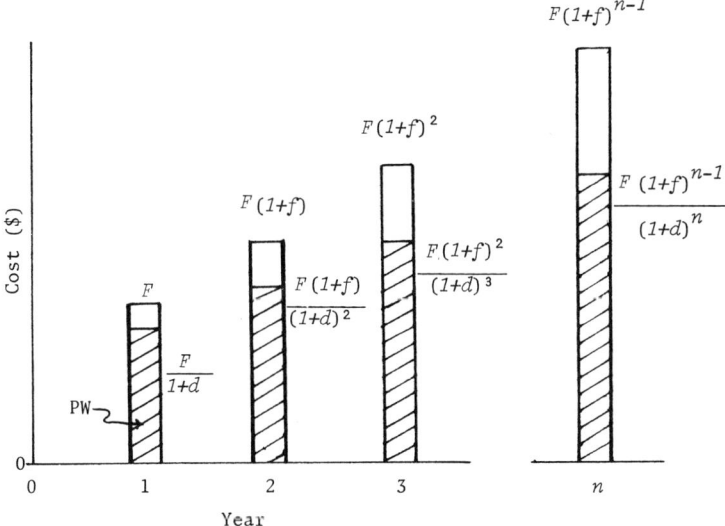

Fig. 2.1.1. Uniform annual series of increasing costs.

For the remaining years, the fuel price escalates at an annual rate f. For year n, the price in that year is then $F(1+f)^{n-1}$. However, the present worth of that fuel payment decreases at interest rate d, and in year n is worth only $F(1+f)^{n-1}/(1+d)^n$. The sum of the present worth of these series of payments is given by

$$\text{present worth factor} = \text{PWF}(n, f, d) \tag{2.1.4}$$

The present worth factor for the series is evaluated as follows.

For $f \neq d$:
$$\text{PWF}(n, f, d) = \frac{1}{(d-f)}\left[1 - \left(\frac{1+f}{1+d}\right)^n\right] \tag{2.1.5}$$

For $f = d$:
$$\text{PWF}(n, f, d) = \frac{n}{1+d}$$

For a fuel inflation rate f of 0 percent, the function $\text{PWF}(n, 0, d)$ equals the present worth factor for a uniform annual series which is commonly used in economic studies.

$$\text{PWF}(n, 0, d) = \frac{1}{d}\left[1 - \frac{1}{(1+d)^n}\right] \tag{2.1.6}$$

2.1.4. Life Cycle Costs

The life cycle costs are the present worth of all owning and operating costs for an investment. For a person or company considering an investment in energy conserving equipment, many factors have to be considered. There are first costs associated with purchasing and installing the equipment. Federal and state tax credits may be available on the equipment, which then offset the cost. There are annual costs of the fuel used, maintenance, and property taxes on the investment. There are annual tax savings due to deductions on interest payments, property tax, and, in the case of a business, fuel used and depreciation. Fuel costs, maintenance, taxes, and so on, are all not constant with time, but will generally inflate in the future. The life cycle cost approach evaluates the present worth of all of these costs. The life cycle costs of the proposed investment can then be compared to the life cycle costs of the current operation to determine if the investment is truly a good one.

2.2. ECONOMIC EXAMPLES

In order to illustrate the basic economic ideas, an example of an energy investment will be carried through on a life cycle cost basis, an annual basis, and a life cycle savings basis. These different approaches will bring out the important concepts and ideas used in evaluating the economics of different energy alternatives. The calculations will first assume constant fuel prices over time, and then the impact of fuel price changes will be considered. In this section only first costs, fuel costs, and salvage value will be included. A more complete economic analysis that includes taxes will be developed in Section 2.4.

Two alternate heating systems, A and B, will be considered. The initial investment of the first (A) is higher than that of the second (B), but the yearly energy costs are less. These energy costs are constant over the life of the systems. In addition, the first investment has a salvage value at the end of its useful life. It will be assumed that the investor has the money in hand. A discount rate of 8 percent and a 20-year life span for the economic analysis will be assumed. The two alternatives are

	A	B
First Cost	$10,000	$7,000
Annual Fuel Cost	1,000	1,500
Salvage Value	1,000	0

If the time value of money were not important, the total costs of each alternative would be the sum of first and fuel costs less the salvage value. On that basis, alternative A would cost $8000 less than B over the 20-year period. However, this calculation ignores the time value of the money and overestimates the cost of future fuel expenditures.

2.2.1. Present Worth or Life Cycle Cost Basis

The present worth of all of the costs and of the salvage value are computed and compared for the two alternatives for this approach. The present worth factor for the series of annual fuel payments from Eq. (2.1.6) is 9.818. The present worth factor for the salvage value from Eq. (2.1.2) is 0.215. The total cost is the sum of the costs less the salvage value.

	A	B
Present Worth of First Cost	$10,000	$ 7,000
Present Worth of Annual Costs	9,818	14,727
Present Worth of Salvage Value	-215	0
Life Cycle Cost	$19,603	$21,727

On a present worth basis, alternative A is better than B by $2124. This is considerably less than the previous estimate of $8000, and shows the need to accurately assess the value of future expenditures.

2.2.2. Annual Cost Basis

The annualized life cycle costs associated with the two alternatives are computed in this approach. The first costs are annualized by dividing by the present worth factor for the fuel payments (9.818). The annual credit of the salvage value is the present worth value annualized over the 20-year life span ($215/9.818).

	A	B
Annual Cost of First Cost	$1018	$ 713
Annual Fuel Cost	1000	1500
Annual Credit of Salvage	−22	0
Annualized Life Cycle Cost	$1996	$2213

On an annual cost basis, alternative A is better than B by $217. Note that the present worth of the series of annual differences equals the life cycle cost difference of $2124.

2.2.3. Life Cycle Savings Basis

In this approach, the comparison is made on the basis of the savings obtained by investing in A instead of B. This is a common approach in energy system economics when an additional investment is required to produce reduced energy costs. The life cycle savings are computed as

$$\text{LCS} = (\text{first cost of } B - \text{first cost of } A)$$
$$+ (\text{annual cost of } B - \text{annual cost of } A) \, \text{PWF}(n, 0, d)$$
$$- (\text{salvage value of } B - \text{salvage value of } A) \, \text{PW}(n, d) \quad (2.2.1)$$

For the example, the savings are

Savings in First Costs	−3000
Savings in Energy Costs	4909
Savings in Salvage Value	215
Life Cycle Savings	2124

The net savings of A over B is $2124, which is the same as the difference in the present worths or life cycle costs.

A common question associated with an economic analysis is what is the length of time required to pay off the additional investment in A. This may be computed from an expression for the savings. The payback period is the time necessary for the annual savings to pay for the added investment. This is determined as the length of time n at which life cycle savings equal zero. Setting the savings equal to zero in Eq. (2.2.1) yields an implicit expression for n in terms of the $\text{PWF}(n, 0, d)$ from which the number of years can be determined.

ECONOMIC EXAMPLES

$$\text{PWF}(n, 0, d) = \frac{[\text{first cost of } (B - A)] - [\text{salvage value of } (B - A)] \text{ PW}(n, d)}{[\text{annual cost of } (A - B)]}$$

(2.2.2)

For the example here, the payoff period at an 8 percent interest rate is between 7 and 8 years. If the life of the equipment is less than this, investment B is a better investment, while beyond this period, alternative A is economically better.

Another economic indicator is the "return on capital." This is the interest or discount rate which would have to prevail in order for alternative A not to be a better investment. This is determined by setting the life cycle savings equal to zero in Eq. (2.2.1), and determining the value of d that would result. The relation is the same as Eq. (2.2.2), except that d is unknown and the life span of the analysis n is 20 years.

For the example here, the value of the discount rate that would result is 15.5 percent and so the return on capital is then 15.5 percent. The interpretation is that if money can be invested to yield a return greater than 15.5 percent, alternative B is preferable. If not, the alternative A is more economical.

2.2.4. Life Cycle Costs with Increasing Fuel Costs

The examples in Sections 2.2.1 to 2.2.3 have assumed that fuel costs are constant over time. Historically, fuel price increases were usually at about the rate of general inflation, and the assumption of constant prices over the period of the analysis was satisfactory. The recent history shows a dramatic rise in fuel prices, and the projections for the future show further significant increases as fuels become more scarce.

It is especially important that an economic evaluation of energy conserving equipment take fuel price increases into account. Equipment such as furnaces, refrigeration systems, and insulation have a long life, and consequently produce economic and energy savings far into the future when the dollar savings are quite large. The investment in such equipment occurs at the present, and the effect of general and fuel inflation is to make a present investment more attractive than one at some later time. Life cycle cost methods can be extended to allow a determination of these effects of changes in the future.

As an example, it will be assumed that fuel prices in the previous example inflate at 10 percent per year. The value of PWF(20, 0.10, 0.08) is 22.17. The life cycle costs become

	A	B
Present Worth of First Cost	$10,000	$ 7,000
Present Worth of Annual Costs	22,170	35,255
Salvage Value	-215	0
Life Cycle Cost	$31,955	$40,255

The life cycle costs are now about double those of the constant fuel price situation. Alternative A is now a better investment by $8300, which is about four times the savings computed for constant fuel costs. Clearly, the increase in fuel prices plays a significant role in the choice of alternatives.

2.3. GENERAL LIFE CYCLE COST FORMULATION

In this section, a general life cycle cost approach will be developed. The method includes evaluation of the present worth of an annual series of payments that change over time, such as fuel costs and maintenance costs. The tax structure will be included: taxes can influence the decisions for both home owners and businessmen. The formulation follows the approach for the economic evaluation of solar energy systems as given in *Solar Engineering of Thermal Processes* by Duffie and Beckman (John Wiley & Sons, New York, 1980).

The life cycle cost of an energy conserving investment can be written as:

$$\text{LCC} = (\text{investment cost}) + (\text{fuel cost}) + (\text{miscellaneous costs})$$
$$+ (\text{property tax}) - (\text{tax savings}) \qquad (2.3.1)$$

Investment costs are those associated with the initial down payment, loan payments, if any, and installation costs. Fuel costs, property taxes, and miscellaneous expenses such as maintenance and insurance occur annually, and their present worth must be determined. Tax savings are the deductions from income tax due to interest on the loan and property tax and tax credits. For a business, additional expenses are allowed. These include the fuel costs and depreciation of equipment.

These costs can be divided into two categories: one associated with the investment cost and the other with fuel costs. The actual first cost includes both purchase and installation; generally, installation costs are proportional to the size and complexity (and thus cost) of the equipment. Similarly, maintenance costs are dependent on system cost. Mortgage payments and tax deductions for interest are dependent on the investment. State property taxes are a percent of investment, and so is depreciation. The salvage value and depreciation are a fraction of first cost. Finally, tax credits are a direct reduction in income tax based on investment. All of these fixed costs then can be combined into a multiplier of the equipment investment.

The costs associated with fuel costs are the actual fuel costs and the fuel cost deductions allowed from taxes. The distinction between investment related and fuel related costs allows Eq. (2.3.1) to be rewritten as

$$\text{LCC} = P_1 F + P_2 E \qquad (2.3.2)$$

where F is the first year fuel price and E is the equipment investment. The parameters P_1 and P_2 are present worth factors that incorporate all of the economic terms discussed above.

2.3.1. Evaluation of Economic Parameters

The evaluation of P_1 and P_2 for use in Eq. (2.3.2) depends on the following economic factors:

- c a multiplier equal to unity for a commercial investment and zero for a residential investment.
- D ratio of down payment to initial investment
- n number of years for the economic analysis
- n_L term of the loan
- n_1 minimum of n and n_L
- n_d depreciation life time
- n_2 minimum of n and n_d
- d interest or market discount rate
- m mortgage interest rate
- i general inflation rate
- f fuel price inflation rate
- p property tax rate
- t combined or effective income tax rate
- r tax credit on initial investment
- M ratio of maintenance, insurance, and so on, costs to investment
- S ratio of salvage value to investment

For a residential application, the entire fuel costs are an expense. The life cycle fuel cost equals the present worth of all fuel payments. For a commercial application, fuel expenses are deductible from income tax. Only a portion of the actual fuel expenses are incurred over the life of the device. Thus P_1 becomes

$$P_1 = (1 - ct)\, \text{PWF}(n, f, d) \qquad (2.3.3)$$

The numerical value of P_1 ranges from about 5 to 20 for residential applications, depending on the number of years. The numerical values for commercial applications are about one-half those for residential situations.

In evaluating P_2, the down payment at the present time and the future principal payments are expenses and their life cycle costs need to be evaluated. The interest paid is tax deductible and subtracts from the total cost. Property taxes increase the cost and are also deductible from income taxes. Expenses for insurance, maintenance, and so forth are only deductible from income tax for a business. These latter two will inflate at the general inflation rate. A business can depreciate the equipment, and deduct this expense from the income tax. The salvage value at the end of the life reduces the actual cost. Finally, tax credits reduce the initial cost of equipment, and are available for both residential and commercial applications. These factors are reflected by the following relation for P_2:

$$P_2 = \underbrace{D}_{\substack{\text{down} \\ \text{payment}}} + \underbrace{(1-D)\frac{\text{PWF}(n_1,0,d)}{\text{PWF}(n_L,0,m)}}_{\text{payments on principal}}$$

$$\underbrace{-t(1-D)\left[\text{PWF}(n_1,m,d)\left(m - \frac{1}{\text{PWF}(n_L,0,m)}\right) + \frac{\text{PWF}(n_1,0,d)}{\text{PWF}(n_L,0,m)}\right]}_{\text{tax deductions for interest payments}}$$

$$+ \underbrace{[p(1-t)}_{\text{property tax}} + \underbrace{M(1-ct)]}_{\text{maintenance}} \text{PWF}(n,i,d)$$

$$\underbrace{-\frac{ct}{n_d}\text{PWF}(n_2,0,d)}_{\text{depreciation}} - \underbrace{S}_{\substack{\text{salvage} \\ \text{value}}} - \underbrace{r}_{\substack{\text{tax} \\ \text{credit}}} \qquad (2.3.4)$$

The numerical value of P_2 ranges from about 0.6 to 1.5 depending mainly on the down payment and tax credits.

The salvage value term S in Eq. (2.3.4) may be greater or less than unity. For example, insulation in a house might not deteriorate, but retain the same insulative value. It has the same or even greater value at the end of the period of analysis than at the start since the price may inflate at the general inflation rate. The present value would be discounted at the market discount rate and inflated at the inflation rate, and the value of S would be calculated as

$$S = \frac{(1+i)^n}{(1+d)^n} \qquad (2.3.5)$$

It is important to be accurate in evaluating the salvage value, and different investments may have different values.

The relations for P_1 and P_2 are complex, and including many factors. However, Eqs. (2.3.3) and (2.3.4) for P_1 and P_2, along with Eqs. (2.1.5) and (2.1.2) for the present worth factors, may easily be programmed for use on a programmable calculator or computer. This is especially convenient for doing evaluations that involve different estimates of fuel inflation rate, market discount rate, and so forth.

In many applications P_1 and P_2 need be evaluated only once. This is especially true in industrial applications when many of the parameters (e.g., D, t, d, etc.) are fixed by company policy or history. In general, although the evaluation of P_2 may be tedious, it does not need to be evaluated often.

2.3.2. Examples of the General Life Cycle Cost Formulation

The example of Section 2.3 will again be considered to include the general economic factors. It will be assumed that a business is considering the two alternatives.

GENERAL LIFE CYCLE COST FORMULATION

First, the investment will be made with money at hand ($D = 1$). A property tax rate (p) of 3 percent, combined state and federal income tax rate (t) of 40 percent, a general inflation rate (i) of 6 percent, a discount rate (d) of 8 percent, and a fuel inflation rate (f) of 10 percent are assumed. The maintenance, and so on, costs for each are assumed equal, and will be omitted. The salvage value for A is again $1000 ($S = 10$ percent). The combined state and federal tax credits (r) are 20 percent. For a 20-year period of analysis, from Eq. (2.3.3) the factor P_1 is evaluated as

$$P_1 = (1 - 0.40)(22.17)$$
$$P_1 = 13.30$$

From Eq. (2.3.4), the calculation of P_2 for A is

$$P_2 = 1 + [0.03(1 - 0.4)] \, 15.60 - \frac{0.40}{20} \, 9.818 - \frac{0.10}{4.66} - 0.20$$

and

$$P_2 = 0.863$$

and similarly for B, which has no salvage value, $P_2 = 0.885$. The comparison between alternatives A and B is now

	A	B
Life Cycle Costs of Equipment	$ 8,630	$ 6,195
Life Cycle Cost of Fuel	13,300	19,950
Life Cycle Cost	$21,930	$26,145
Life Cycle Savings of A over B	$4,215	

The effect of accounting for taxes is to reduce the economic difference between A and B. The savings with no tax considerations were $8300, and this difference is reduced to about one-half.

If, instead of a 100 percent down payment, all of the investment had been borrowed ($D = 0$) at a 9 percent interest rate (m) over a 20-year term (n_L), then P_2 would be different. For A, $P_2 = 0.666$, while for B, $P_2 = 0.688$. The value of P_1 is unchanged. The economic comparison between A and B becomes

	A	B
Life Cycle Costs of Equipment	$ 6,660	$ 4,816
Life Cycle Cost of Fuel	13,300	19,950
Life Cycle Cost	$19,960	$24,766
Life Cycle Savings of A over B	$4,806	

The total costs are reduced since the first costs are now paid in the future and are worth less. However, the relative economics of each alternative remain the same.

If the two choices had been made by a home owner, the tax benefits for fuel costs, maintenance, and depreciation would not have been available. The factor P_1 is the actual present worth of all fuel payments, and equals 22.17. The factor P_2 equals the first cost plus property taxes and less salvage value and tax credits. With a 100 percent down payment, the value for A is

$$P_2 = 1 + [0.03(1 - 0.4)] \, 15.60 - \frac{0.10}{4.66} - 0.20 = 1.059$$

and $P_2 = 1.081$ for B.

The corresponding life cycle costs become

	A	B
Life Cycle Costs of Equipment	$10,590	$ 7,567
Life Cycle Costs of Fuel	22,170	33,255
Life Cycle Cost	$32,760	$40,822
Savings of A over B		$8,062

For the case where the home owner borrows the entire sum ($D = 0$), the economic factor P_2 equals 0.862 for A and 0.884 for B. The corresponding life cycle costs are

	A	B
Life Cycle Costs of Equipment	$ 8,620	$ 6,188
Life Cycle Costs of Fuel	22,170	33,255
Life Cycle Cost	$30,790	$39,443
Savings of A over B		$8,653

The economic decision for a home owner is really not different from that for a business. However, the effect of the tax structure is to increase the costs that a home owner would pay over the life of the equipment. The main impact of taxes is to allow the business to deduct fuel expenditures. Not only does this reduce the costs for fuel, but it acts to make energy conserving options less cost effective than for a home owner.

As shown by the equations for P_1 and P_2, many factors enter into the economic analysis. However, some are more important than others. In two studies, one on building insulation and the other on industrial heat exchangers, the sensitivity of the results to the economic variables was determined. It was found that the three most important parameters were the fuel inflation rate, the discount rate, and the number of years of the analysis. For an incremental increase of 1 percent in fuel inflation rate (e.g., an increase in f from 9 to 10 percent), the savings were found to increase 10 percent (e.g., from $100 to $110). For an incremental increase of 1 percent in discount rate, the savings were found to decrease 10 percent. For an increase in life span of 1 year, the savings were found to increase 8 percent. All other variables had a much smaller effect than these three.

The economic analysis is very sensitive to three parameters that may be difficult to determine. The future fuel price increases are, obviously, unknown but can be estimated to some degree. Market discount rates are better known, but there are still significant fluctuations and changes. The actual life span of equipment is also unknown. In making an economic analysis, it is important to realize the uncertainties in these parameters and in the ensuing results. It is probably best to try to estimate limits for these three parameters, and determine the economics for each set. In this way, the engineer will be better prepared to attempt to assess the cost effectiveness of proposed changes.

2.4. ECONOMIC OPTIMUM LEVEL OF INSULATION

As another example of the application of the life cycle cost approach, the situation with insulation in a wall will be considered. Two questions will be asked. The first is whether any insulation will save money, and the second is what is the optimum level of insulation to install to save the most money. As will be discussed in Sections 3.2.1 and 6.2, the heat loss through a wall decreases inversely as the insulation thickness increases, while the cost of insulation materials and labor to install insulation rises linearly with thickness. This means that the addition of a small amount of insulation to a bare wall will reduce heat loss and fuel costs significantly. However, as the thickness increases, the cost of fuel saved decreases but the cost of added insulation is the same.

The cost of fuel, the first cost of insulation, and the total cost are shown schematically in Fig. 2.4.1 as a function of insulation thickness. Fuel costs decrease and insulation costs rise, and so the total costs first decrease, go through a minimum, and then rise. The minimum cost specifies the optimum level of insulation at which the added cost of insulation produces an equal reduction in fuel costs. The savings relative to no added insulation are also shown. If costs of insulation are high or if fuel costs are low, added insulation will save energy but cost money.

As an example, the fuel and insulation costs as functions of insulation thickness in inches (x) for a given building wall are taken to be

$$\text{Annual fuel costs per square foot} = F = \frac{\$0.2}{1 + 0.75x}$$

$$\text{First cost of insulation per square foot} = E = \$0.2(1 + x)$$

The economic optimum insulation thickness will be found. The life cycle costs are given by Eq. (2.3.2).

$$\text{LCC} = P_1 \left[\frac{0.2}{1 + 0.75x} \right] + P_2 [0.2(1 + x)] \qquad (2.4.1)$$

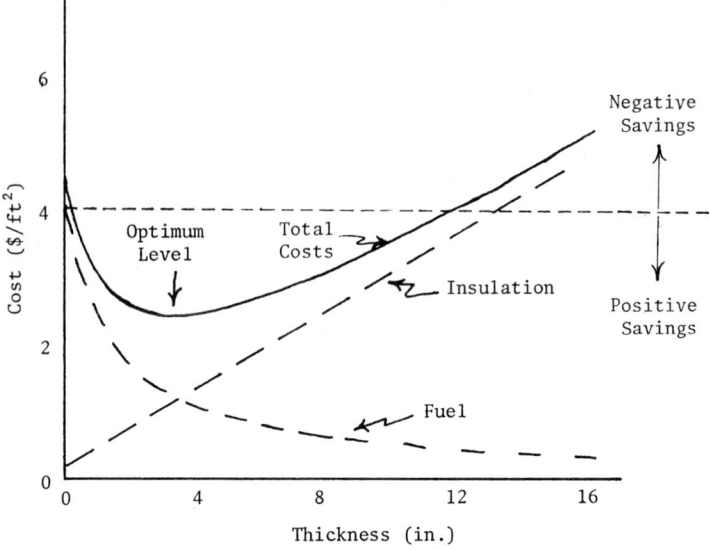

Fig. 2.4.1. Costs of insulation as functions of thickness.

The value of the minimum total cost is found by differentiating Eq. (2.4.1) with respect to thickness and equating the derivative to zero:

$$\frac{\partial LCC}{\partial x} = \frac{0.2 P_1}{(1 + 0.75x)^2} (-0.75) + 0.2 P_2 = 0 \qquad (2.4.2)$$

The value of x for minimum cost is

$$x = 1.154 \sqrt{\frac{P_1}{P_2}} - 1.333 \qquad (2.4.3)$$

For the economic parameters of Section 2.3.2 in which P_1 is 13.30 and P_2 is 0.863, the optimum thickness becomes

$$x = 3.2 \text{ in.}$$

The life cycle costs at this thickness are evaluated from Eq. (2.4.1) to be $1.51/ft^2$. For no insulation, the costs are only the fuel costs and equal $2.67/ft^2$. The life cycle savings are $1.16/ft^2$ which are positive. The insulation saves both energy and money.

The sensitivity of the optimum level to economic variables can readily be evaluated through the effect of changes in P_1 and P_2. Various combinations of economic factors and the corresponding optimum insulation levels are shown in Table 2.4.1 for residential and commercial applications. For this table, general inflation is 6 per-

TABLE 2.4.1
Economic Optimum Level of Insulation

	f	n = 10 yr			n = 20 yr		
		P_1	P_2	x (in.)	P_1	P_2	x (in.)
Residential	0	6.71	1.005	1.7	9.82	1.174	2.0
	0.10	10.07	1.005	2.3	22.17	1.174	3.7
Commercial	0	4.03	0.685	1.5	5.89	0.884	1.6
	0.10	6.04	0.685	2.1	13.30	0.884	3.1

cent, the market discount rate is 8 percent, the property tax rate is 3 percent, and the income tax rate is 40 percent for a business and 20 percent for the home owner. The depreciation life is equal to the term of the analysis (n). All investments are made from money that is currently available ($D = 1$). There is no salvage value or maintenance costs. The tax credit is 20 percent. The fuel inflation rate is either 0 or 10 percent.

Although only a specific example is considered, several general conclusions can be drawn from the results presented in Table 2.4.1. First, considering that the price of fuel will increase with time leads one to install more insulation than if price were constant. A longer term of the analysis also leads to higher insulation levels. These effects are especially important since much energy conserving equipment may have a relatively long life, and a realistic life span and estimate of fuel price increase must be made.

The effect of the tax structure is to encourage a business to install a smaller insulation thickness than a home owner would. In general, it is true that energy conserving equipment does not pay off as well for commercial applications as for residential ones. This reduces the attractiveness of commercial investments in energy conserving devices. The current income tax structure does not promote energy conservation on a nationwide basis. However, tax credits do help; for the commercial application the economic optimum thickness is about 15 percent greater with tax credits than without.

Figure 2.4.1 also demonstrates a result generally found for energy conserving devices. The curve is usually fairly flat near the minimum and the life cycle costs are close to those at the minimum over a broad range. For this example, costs are within 10 percent of the minimum over the range of 2-6 inches. This means that installing any value of insulation thickness between 2 and 6 inches will yield about the same life cycle savings. This has implications regarding the effect of a different fuel price increase than was assumed. If one installs the optimum level of insulation for a given assumed fuel price increase, and the actual fuel price increase is greater, the optimum level of savings will not be achieved. However, the savings are not too sensitive to thickness in the vicinity of the optimum, and so a value close to the optimum will still yield nearly the optimum level of savings. The main effect of fuel price changes is to change the level of savings.

2.5. SUMMARY

The basic ideas necessary for performing economic analysis of energy conserving equipment have been introduced in this section. Life cycle cost techniques are necessary in order to accurately evaluate energy conserving improvements. These methods allow an accurate determination of the true costs of fuel and of owning the device over its life span. The total life cycle costs include taxes, interest payments, and changes over time. A simple formulation as given by Eq. (2.3.2) allows a breakdown of costs into fuel and owning costs.

$$\text{LCC} = P_1 F + P_2 E \qquad (2.3.2)$$

The economic present worth factors P_1 and P_2 are given by Eqs. (2.3.3) and (2.3.4).

The examples demonstrate the strong effect of the tax structure on decisions. A business is allowed certain deductions which make energy conserving alternatives less attractive than for residences. Tax credits reduce first costs substantially and provide a large incentive to consider the alternative.

One of the most important parameters in the evaluation of costs is the fuel price increase. This parameter, which is one of the most difficult to estimate, strongly affects the life cycle savings. However, the choice between different alternatives is not similarly affected. As a result, the choice of which alternative to implement can be made rather confidently, but the value of the savings that accrue is uncertain.

SUGGESTED READING

J. A. Duffie and W. A. Beckman, *Solar Engineering of Thermal Processes*, John Wiley & Sons, New York, 1980.

E. L. Grant, W. G. Ireson, and R. S. Leavenworth, *Principles of Engineering Economy*, Ronald Press, New York, 1976.

J. A. White, *Principles of Economic Analysis*, John Wiley & Sons, New York, 1977.

PROBLEMS

2.1. A home owner is considering investing $600 in attic insulation. He calculates that he would save $40/yr at current fuel prices. His property tax rate is 3.3 percent and his combined state and federal income tax rate is 18 percent. He can earn 8 percent of his money at a savings and loan institution. A federal tax credit of 15 percent is allowed on this investment. Determine whether this is a good investment on a 20-year life for the situation of:

(a) Constant fuel prices.

(b) Fuel prices that increase 14 percent annually.

The home owner is not certain whether he should attach a salvage value to the insulation. Does whether he accounts for salvage value affect his decision? Would insulation have a salvage value, and, if so, what would be a reasonable value?

2.2. An owner of a commercial building is contemplating adding storm windows to his single-pane windows to reduce heat loss. The choices are:

(a) Leave windows as is, with current heating costs of $2000/yr.

(b) Add a single glass storm window at a cost of $2000, which reduces heating costs to $1200/yr.

(c) Add a single plastic storm window at a cost of $1000 which reduces heating costs to $1600/yr. These windows need to be replaced every 5 years.

Determine which alternative is the most economically feasible using a 20-year life. Property taxes are 3 percent, income taxes are 45 percent, fuel increases at 15 percent per year, inflation is 9 percent and a 15 percent tax credit is available. His alternate rate of return is 15 percent. State any other assumptions.

2.3. A home owner building his home wants to decide on the optimum insulation level in his ceiling. An uninsulated ceiling has an R value of 2 and would cost $0.40/ft^2 to heat for a year. Each added inch of insulation adds an R value of 3. The heating cost is inversely proportional to the total R value. The cost of insulation per square foot is $0.5 fixed charge plus $0.12/in.

(a) Determine the optimum insulation level and monetary savings using the following economic values. The market discount rate is 10 percent, general inflation is 8 percent, fuel inflates at 11 percent, property taxes are 2.5 percent, and federal income taxes are 21 percent. The cost of insulation is part of his home loan, which is for 30 years with an interest rate of 12 percent for 20 percent down payment. The analysis is for a 20-year life.

(b) If your assumption of fuel inflation rate is low by a factor of two, what is the effect on the actual economic savings compared to those for the optimum?

▪ 3 ▪

RESIDENTIAL AND COMMERCIAL BUILDING HEATING REQUIREMENTS

3.1. OVERVIEW

The total energy use in the residential and commercial sectors comprises about 33 percent of the total national energy use. Reduction of usage in these sectors can significantly affect the total consumption. The partitioning of energy use in these two areas is given in Table 3.1.1. From this, it is seen that the two major uses are for space and water heating. Nationally, space heating accounts for about 18 percent of the total energy usage, and water heating for about 4 percent. Thus, almost one-quarter of our energy goes into space and water heating, and this becomes a prime target for more effective utilization.

Approximately one-fifth of the energy use occurs in residential dwellings. The uses here are the result of individual as opposed to institutional decisions. Conservation programs such as recommendations for thermostat settings and loans for insulation which are aimed at home dwellers have a large potential for energy savings. The same programs are also applicable for small commercial establishments.

For a home owner, space and water heating costs can be a large fraction of his yearly income. Depending on his location in the country and the type of fuel available to him, the cost of energy to heat his home can range up to $2000 annually. The highest costs are usually borne by a home owner in a northern area using fuel oil or electricity, although a poorly constructed building in a southern location may also have high heating costs. Water heating costs are usually in the range of $200–$500 annually. The current cost of energy and its probable increase in the future are motivation for a home or building owner to consider techniques to reduce usage.

TABLE 3.1.1
Energy Use in Residential and Commercial Sectors

	Residential	Commercial
Space heating	33%	21%
Water heating	8	3
Cooking	3	—
Refrigeration	4	4
Air conditioning	2	5
Clothes drying	1	—
Other (including lighting)	6	10
	57%	43%

Table 3.1.1 shows that use of energy in homes and buildings for other than space and water heating is relatively small. Consumption is mainly by devices such as stoves, clothes dryers, air conditioners, and a range of other, mostly electric, appliances. These uses amount to about one-third of the total for these two sectors, or less than one-tenth of the total U.S. consumption. Since there are many different devices which contribute to the usage, it is not possible to single out one change that will significantly reduce consumption. However, for each individual home owner, there are many changes he can make which will reduce energy consumption somewhat, and so the breakdown of energy consumption by these appliances will be briefly discussed.

A significant feature of these minor uses listed in Table 3.1.1 is that electricity is the predominant source. This has a direct impact on the utilities, which must provide central power stations to meet electrical demand at any time. A breakdown of the electrical use by appliances in a typical home is given in Table 3.1.2. The annual costs are estimated for an electrical rate of 5¢/kWh. The typical U.S. household, with a typical complement of appliances, consumes about 6000 kWh of electricity per year at an annual cost of $300. It is not uncommon, though, for a home with many appliances to use 12,000 kWh annually.

The four single appliances using the largest amounts of electricity are the refrigerator, stove, freezer, and clothes dryer. Their consumption equals about two-thirds of the total and technical changes could have a significant effect. Improvements in motor design and wall insulation could reduce refrigerator and freezer consumption by 35 percent. The surface units of stoves are 60 to 70 percent efficient in terms of input energy into the food, and there is not much potential for reduction here. In contrast, conventional ovens are 10–15 percent, and so new designs such as convection ovens may be an improvement. Microwave ovens transfer energy into the food at nearly 100 percent efficiency, but the conversion of electricity into microwaves is only about 35 percent efficient. The high energy use of clothes dryers is due to the requirement to evaporate the water. Mechanical ways to remove water and solar clothes dryers are an improvement.

Appliances with a large wattage rating (e.g., toaster, dishwasher) are not necessarily large energy users since they are used much less frequently than the major users. Excluding air conditioners, these other appliances consume 1000–2000

TABLE 3.1.2
Electrical Consumption by Residential Appliances

	Annual Use (kWh)	Annual Cost
Refrigerator		
standard	1100	55
frost free	1800	90
Range and oven–regular	1200	60
Freezer	1100	55
Clothes dryer	1000	50
Air conditioner		
room	500	25
central	1500	75
Television		
color	550	26
black and white	250	12
Dehumidifier	380	19
Dishwasher	360	18
Lights	200–800	10–40
Iron	140	7
Electric blanket	100	5
Coffee maker	100	5
Washing machine	100	5
Vacuum cleaner	50	2.50
Toaster	40	2
Disposer	30	1.50
Clock	15	0.75
Can opener	10	0.50
Sewing machine	10	0.50

kWh/yr. For each appliance, there are some energy saving options. For example, not using the drying cycle of the dishwasher reduces electrical use to that required for pump and controls only. There are also trade-offs between uses; undoubtedly less energy is used in maintaining comfort at night by an electric blanket than by heating the entire house. On the other hand, reduced electrical use reduces the heating effect in the house and requires more furnace heating.

From the viewpoint of national energy use, the number of appliances in homes is important. Virtually all homes have refrigerators, stoves, and television sets and an increase in number of these appliances is only due to new housing. However, approximately only one-third of the homes have freezers, clothes dryers, air conditioners, and dishwashers, and thus there is a large potential for increased electrical consumption as new units are purchased. It is here that efficiency improvements are important. The electrical use of appliances also significantly affects the utility. Many of the appliances in a home use electricity during the peak electrical hours of the day. This means that although the energy consumption of a particular appliance is small, its rate of energy use (watts) may be high and at a time when the demand for electricity as a whole is large. For appliances then, improvements in efficiency are also valuable in reducing the demand for electrical power.

In summary, the major use of energy in the residential and commercial sectors is for space conditioning. This is true for both the individual building owner and for the nation as a whole. The use of energy for space conditioning can be thought of as having two aspects. The first is the demand side, and is represented by the transfer of energy into and out of the interior space. The second is the supply side, and is represented by the devices used to supply heating and cooling. This and the following two chapters will focus on the demand side, and Chapters 6 and 7 will cover the devices used to produce heating and cooling.

The energy flow picture for a building is complicated, and Chapter 3 is divided into a number of sections. First, the heat loss and the annual heating requirements will be considered. Section 3.2 is concerned with the mechanisms of heat loss; Section 3.3 with energy gains to the building; Section 3.4 with the design heat loss; and Section 3.5 with annual energy requirements. A detailed example of energy use in a residence and the effects of different climates will be presented in Sections 3.6 and 3.7.

The use for cooling will be discussed in a parallel manner in Chapter 4. Section 4.1 will be concerned with building heat gains; Section 4.2 with the design cooling load; Section 4.3 with annual cooling requirements; and Section 4.4 with an example calculation. Chapter 5 will conclude with the consideration of simultaneous heating and cooling loads in a multizone office building.

Historically, the calculations presented in this chapter have been used to size the heating or cooling plant and the related equipment. The emphasis has been on providing a large enough system so that the occupants do not complain of discomfort. As a result, the accuracy of the calculations has not been of particular concern. The recent need to estimate energy use and utility bills for buildings has made these calculations more important and has created a need for more precise estimates. The calculations are also essential for determining the major avenues of heat loss or gain and for evaluating cost-effective measures for reducing them. Thus, it is necessary to consider in detail the energy flows through building structures.

3.2. BUILDING HEAT LOSSES

The mechanisms by which heat losses occur from a building are depicted schematically in Fig. 3.2.1. There are combined conduction–convection–radiation heat flows through the roof, window, door, and wall surfaces from the room to the outside air. The driving potential for these flows is the difference between the room and ambient temperatures. There is a similar heat flow through the floor and into the ground due to the temperature difference between the house and ground. Energy is carried by air flows up through the chimney and cracks around windows, doors, and the foundation. These air flows enter the house at ambient temperature, are heated to room temperature, leave the heated space, and thus are energy losses from the house. They are termed the infiltration loss.

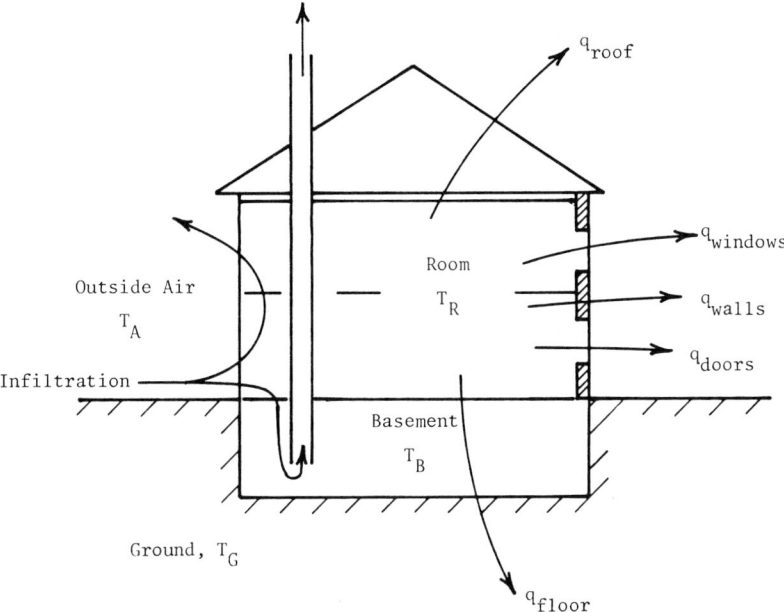

Fig. 3.2.1. Schematic of heat flows from a building.

The heat loss from the building equals the sum of these energy flows out of the space and is given by

$$q_{loss} = q_{walls} + q_{windows} + q_{door} + q_{floor} + q_{roof} + \dot{E}_{infiltration} \qquad (3.2.1)$$

In the following sections the components of the heat loss and the governing temperatures will be evaluated. These will then be used to determine the total heat loss and the energy supplied to the building.

3.2.1. Wall, Roof, and Door Heat Flows

The heat flow through these elements of a building structure is governed by a combination of the conduction, convection, and radiation mechanisms. A typical frame wall such as used in residential construction is shown in Fig. 3.2.2a. The mechanisms of heat transfer through each section are shown schematically. There are many parallel and series heat flow paths through the wall. It will be convenient to combine these individual heat flows into an overall conduction coefficient using steady-state heat transfer expressions. First, the governing equations for conduction, convection, and radiation will be presented. They will then be combined to yield an expression for the total heat flow through the wall. [For the development of the basic heat transfer equations, see a standard heat transfer text such as Holman's *Heat Transfer* (McGraw-Hill Book Co., New York, 1976) or the ASHRAE guide, *ASHRAE Handbook of Fundamentals*.]

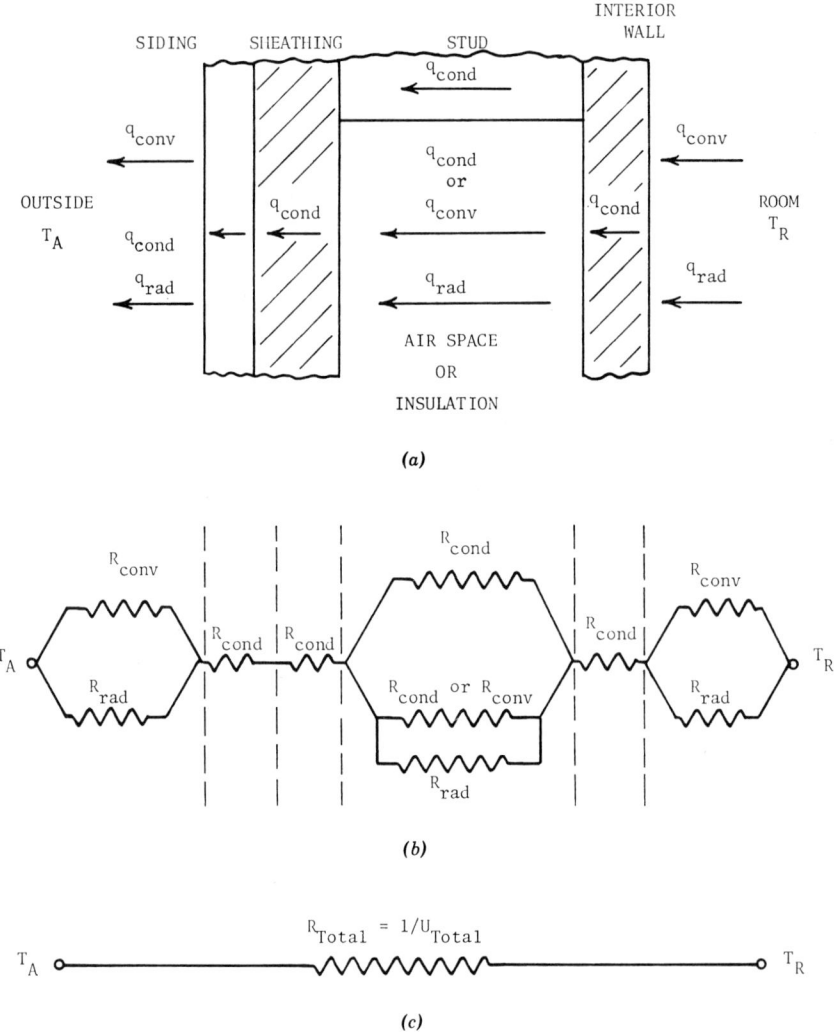

Fig. 3.2.2. Heat transfer analysis for a frame wall. (*a*) Schematic of heat flows in a wall. (*b*) Thermal circuit for heat flow. (*c*) Equivalent thermal resistance.

The conduction of heat through a plane wall with temperatures T_1 and T_2 at each face under steady conditions is given by

$$q_{\text{cond}} = \frac{kA(T_1 - T_2)}{L} \tag{3.2.2}$$

The convection of heat from a surface at T_1 to surrounding air at T_2 is given by

$$q_{\text{conv}} = hA(T_1 - T_2) \tag{3.2.3}$$

The radiation heat flow between two plane surfaces at T_1 and T_2 is given by

$$q_{rad} = \frac{A\epsilon_1\epsilon_2\sigma(T_1^4 - T_2^4)}{[1 - (1 - \epsilon_1)(1 - \epsilon_2)]} \qquad (3.2.4)$$

Each of the heat flow expressions can be put in the form of a temperature difference and thermal resistance

$$q = \frac{(T_1 - T_2)A}{R} \qquad (3.2.5)$$

Alternatively, the expressions can be written in terms of a thermal conductance as

$$q = UA(T_1 - T_2) \qquad (3.2.6)$$

The conductance is the reciprocal of the thermal resistance and both of these quantities will be used in determining heat flows. For the conduction and convection heat flows, the thermal resistance and conductance are readily determined from Eqs. (3.2.2) and (3.2.3). For conduction,

$$R_{cond} = \frac{1}{U_{cond}} = \frac{L}{k} \qquad (3.2.7)$$

and for convection,

$$R_{conv} = \frac{1}{U_{conv}} = \frac{1}{h} \qquad (3.2.8)$$

The radiation resistance and conductance are more complicated, and are found by expanding the temperature difference in Eq. (3.2.4).

$$(T_1^4 - T_2^4) = (T_1^2 + T_2^2)(T_1 + T_2)(T_1 - T_2)$$

or

$$(T_1^4 - T_2^4) = (T_1^3 + T_1^2 T_2 + T_1 T_2^2 + T_2^3)(T_1 - T_2)$$

The first term in parentheses is approximately equal to four times the average temperature cubed:

$$(T_1^3 + T_1^2 T_2 + T_1 T_2^2 + T_2^3) \simeq 4T_{ave}^3$$

where

$$T_{ave} = \frac{T_1 + T_2}{2}$$

38 RESIDENTIAL AND COMMERCIAL BUILDING HEATING REQUIREMENTS

Equation (3.2.4) can then be rewritten as

$$q_{rad} = A \frac{4\epsilon_1 \epsilon_2 \sigma T_{ave}^3}{[1 - (1 - \epsilon_1)(1 - \epsilon_2)]} (T_1 - T_2)$$

The thermal resistance and conductance become

$$R_{rad} = \frac{1}{U_{rad}} = \frac{[1 - (1 - \epsilon_1)(1 - \epsilon_2)]}{4\epsilon_1 \epsilon_2 \sigma T_{ave}^3} \qquad (3.2.9)$$

The formulation of heat flows in terms of thermal resistances allows the heat flow through the wall to be represented as a thermal circuit. Figure 3.2.2b shows the circuit corresponding to the wall of Fig. 3.2.2a. The parallel and series resistances of Fig. 3.2.2b can be combined and replaced by an equivalent thermal resistance as shown in Fig. 3.2.2c. The heat flow through the wall is then expressed either in terms of an overall wall resistance R_{wl} as

$$q_{wl} = \frac{(T_R - T_A) A}{R_{wl}} \qquad (3.2.10)$$

or in terms of an overall conductance and wall area as

$$q_{wl} = UA_{wl}(T_R - T_A) \qquad (3.2.11)$$

where the resistance and conductance are related by

$$U_{wl} = \frac{1}{R_{wl}} \qquad (3.2.12)$$

In order to evaluate the resistances and conductances, the thermal properties of the materials used to construct the wall are needed. Property values for common building materials are given in Table 3.2.1. The entries are in terms of the "R value" of the material which is its thermal resistance in English units. The thermal conductivity can be obtained from the R value and thickness using Eq. (3.2.7).

As an example of the calculation procedure, the resistance and conductance of the frame wall of Fig. 3.2.2 will be determined. The building materials are assumed to be nonreflective ($\epsilon = 0.9$). The stud is a nominal 2 × 4 ($1\frac{5}{8} \times 3\frac{5}{8}$ inches) and the studs are placed 16 inches apart. The heat flow area of the stud is 10 percent of the total wall area. The conductance of the parallel heat flow paths through the studs and air space will be determined first on a per square foot of total wall area basis:

$$U_{stud} = \frac{k}{L} \frac{A_{stud}}{A_{wall}} = \frac{(0.10 \text{ Btu/hr ft }°F)(0.1)}{(3.625/12) \text{ ft}} = 0.03 \text{ Btu/hr ft}^2 \text{ }°F$$

$$U_{air\,space} = \frac{1}{R} \frac{A_{air\,space}}{A_{wall}} = \frac{1}{1.01} \text{ Btu/hr ft}^2 \text{ }°F \, (0.9) = 0.89 \text{ Btu/hr ft}^2 \text{ }°F$$

TABLE 3.2.1
Property Values of Structural Materials, Insulation, Air Spaces, and Surface Films

	R (hr ft^2 °F/Btu)		R (hr ft^2 °F/Btu)
Exterior Materials		*Air Spaces* ($> \frac{3}{4}$-in.)	
Wood bevel siding	0.81	Heat flow *up*	
Stucco/in.	0.20	Nonreflective	0.87
$\frac{1}{2}$-in. insulated board		Reflective—one	
sheathing	1.32	surface	2.23
$\frac{1}{4}$-in. plywood	0.31	Heat flow *down*	
$\frac{1}{4}$-in. hardwood	0.18	Nonreflective	1.02
$\frac{3}{4}$-in. softwood board	0.94	Reflective—one	
Building paper	0.06	surface	3.55
8-in. concrete blocks	2.18	Heat flow	
Common brick/in.	0.20	*horizontal*	
Face brick/in.	0.11	Nonreflective	1.01
Sand and gravel		Reflective—one	
concrete/in.	0.08	surface	3.48
Wood siding shingles	0.87		
Asbestos-cement		*Surface Air Films*	
shingles	0.03	*(Including Radiation)*	
Asphalt roof shingles	0.44	INSIDE	
Wood roof shingles	0.94	Heat flow *up*	
		Nonreflective	0.61
Interior Materials		Reflective	1.32
$\frac{1}{2}$-in. gypsum board	0.45	Heat flow *down*	
$\frac{1}{2}$-in. plaster	0.32	Nonreflective	0.92
$\frac{3}{4}$-in. hardwood finish		Reflective	2.55
flooring	0.68	Heat flow	
Floor tile	0.05	*horizontal*	
Carpet and fibrous pad	2.08	Nonreflective	0.68
Carpet and foam rubber		Reflective	1.54
pad	1.23	OUTSIDE	
		15-mph wind	
Insulation (per inch)		(winter)	0.17
Fiberglass (batts)	3.2	7.5-mph wind	
Styrofoam sheets	4.2	(summer)	0.25
Fiberglass (loose)	4.2		k
Mineral wool (loose)	3.0		(Btu/hr ft °F)
Cellulose (loose)	3.3		
Foam (blown)	4.4	Thermal	
		Conductivity	

	ϵ	α_{sol}		
			Air	0.015
Radiation Properties			Wood	0.10
Building materials	0.9	0.4–0.9	Water	0.34
Glass	0.94	0.04	Glass	0.45
Aluminum foil	0.1	0.2	Steel	26
White paint	0.95	0.25	Aluminum	110
Black paint	0.95	0.9	Steel	220
Aluminum paint	0.55	0.54		

Since the resistances are in parallel, the conductances can be added directly to determine the total conductance

$$U_{total} = 0.92 \text{ Btu/hr ft}^2 \text{ °F}$$

and the resistance is

$$R = 1.09 \text{ hr ft}^2 \text{ °F/Btu}$$

A comparison of the stud and air conductances shows that the heat flow through the studs is small compared to that through the air space and the contribution of the stud could be neglected.

It is interesting to note that a 3.5-inch-thick air layer would have a pure conduction conductance of 0.045 Btu/hr ft² °F. However, convection occurs in the air space and this increases the conductance. The conductance for radiation alone is about 0.8 Btu/hr ft² °F. This shows that the major path for heat flow through the air space between the interior and exterior walls is by radiation.

The overall wall resistance is obtained by adding up the component resistances. The wall has gypsum board on the interior and sheathing plus wood siding on the exterior. The values of the component resistances are taken from Table 3.2.1 and the previous calculations as follows:

	R (hr ft² °F/Btu)
Interior air film	0.68
½-in. gypsum board	0.45
Air space and stud	1.09
½-in. sheathing	1.32
Wood bevel siding	0.81
Outside air film (winter)	0.17
	4.52

The overall resistance is 4.52 hr ft² °F/Btu and the conductance is 0.22 Btu/hr ft² °F.

If the air space is filled with $3\frac{1}{2}$ inches of fiberglass batt insulation, the R value of the space becomes

$$R = (3.2 \text{ hr ft}^2 \text{ °F/Btu in.}) \, 3.5 \text{ in.} = 11.2 \text{ hr ft}^2 \text{ °F/Btu}$$

and the conductance of the space is reduced to

$$U_{airspace} = \frac{1}{11.2} \text{ Btu/hr ft}^2 \text{ °F} \, (0.9) = 0.080 \text{ Btu/hr ft}^2 \text{ °F}$$

The overall wall resistance is then increased to 12.52 hr ft² °F/Btu and the conductance is 0.080 Btu/hr ft² °F.

The main effect of insulation is to reduce radiation heat transfer across the space. It also eliminates convection due to air flow, but replaces it by conduction through the insulation. The resistance of the air space alone is increased by a factor of about 10, and the overall wall resistance increases by a factor of 2.8. This illustrates the value of insulation.

The calculation of resistances to heat flow through a roof, ceiling, or door follows the same procedure as for walls. For a ceiling separating a room from an attic, it is commonly assumed that the ceiling is the major resistance, and that the roof surfaces contribute little. This is realistic in that most attics are well ventilated to prevent moisture buildup. As a consequence the attic temperature is close to that of the ambient. Also, this is conservative in that the ceiling heat loss will be overestimated slightly.

3.2.2. Window Heat Flows

Heat flow through windows is treated similarly to that for walls, roofs, and doors. It is conventional to add the external and internal radiation and convection resistances to that for the glass itself and report an overall conductance. The heat losses are then calculated by

$$q_{wd} = UA_{wd}(T_R - T_A) \tag{3.2.13}$$

Typical values for the conductance for different window types are given in Table 3.2.2. The winter values assume a 15-mph wind, and the summer values a 7.5-mph wind in accordance with the ASHRAE recommendations (see Table 3.2.1).

Table 3.2.2 indicates that the heat loss through a single-pane window can be

TABLE 3.2.2
Conductance Values for Windows

	U (Btu/hr ft² °F)	
	Winter	Summer
Single glazing	1.13	1.06
Double glazing, $\frac{1}{4}$-in. space	0.65	0.61
Double glazing, $\frac{1}{2}$-in. space	0.58	0.56
Double glazing, 1–4-in. space	0.56	0.54
Triple glazing, $\frac{1}{4}$-in. space	0.47	0.45
Triple glazing, $\frac{1}{2}$-in. space	0.36	0.35

TABLE 3.2.3
Component Resistances for Double Glazed Windows

	R (hr ft^2 °F/Btu)
Inside	0.68
Glass	0.02
Air space	1.01
Glass	0.02
Outside	0.17
	1.90

$U = 0.53$ Btu/hr ft^2 °F

halved through the use of double glazed windows. However, the loss through a double glazed window is only reduced 20 to 30 percent by the addition of a third pane. There are diminishing returns here in that at some point additional panes are not cost effective. A large spacing, such as would be found using storm windows, is the best insulator, although spacings between glazings greater than about 1 inch affect the heat loss very little.

A breakdown of the component resistances for a double glazed window is given in Table 3.2.3. The property values are taken from Table 3.2.1 and the glass resistance is computed from Eq. (3.2.7). The slight difference between this computed value and that given in Table 3.2.2 is due to the higher emissivity of glass than that assumed for the air space in Table 3.2.1. It is seen that the major heat transfer resistance is that of the air space, while the glass panes themselves contribute little. As was discussed previously, the major mechanism of heat flow across the air space is by radiation. Thus, improvements in window design are those such as surface treatments and films which reduce radiation.

3.2.3. Floor and Basement Heat Flows

The heat flow from an interior space through the floor and into the ground depends strongly on the type of building construction. Crawl spaces, in which the floor is separated from the ground by an air space, are common in many U.S. locations. Slab construction, in which a concrete slab is first poured and then the house constructed on it, is common in the southern areas of the country. Heating ducts or pipes may be laid in this slab. In northern areas, full basements under the house floor are common and basements may either be heated by the furnace, or unheated.

The heat loss for each of these types of floor-basement combinations is different, and depends on whether there is insulation on the floor, slab perimeter, or basement walls. Each of these types will be considered separately. The relations will be based on the information in the ASHRAE guides.

Crawl spaces under floors are usually well ventilated, and the air temperature ap-

BUILDING HEAT LOSSES

proaches the outdoor ambient. The heat loss from the house to the floor can be computed from a relation similar to that for walls:

$$q_{fl} = UA_{fl}(T_R - T_A) \tag{3.2.14}$$

For buildings with slab floors, the heat loss depends on the building perimeter P. The values depend on the thermal insulation, and the ASHRAE guide gives relations for heat flows from concrete floors at or near grade with different insulation resistance values and for different environmental temperatures. For slab floors without heating ducts in the slab, the heat flow can be computed from

$$q_{fl} = [0.5(T_R - T_A) + (20 - 4R_{sl})] P \tag{3.2.15}$$

For floors with heating ducts in the slab, the heat flow is given by

$$q_{fl} = [1.0(T_R - T_A) + (20 - 6R_{sl})] P \tag{3.2.16}$$

In Eqs. (3.2.15) and (3.2.16) the governing temperature difference is that between the room and the outdoor ambient. The value for R_{sl} is the thermal resistance of insulation on the edge (hr ft^2 °F/Btu). For the wall insulation to be effective, it must extend 2 ft or more below the house floor.

The heat loss through full basements depends on the total resistance from the house interior to the surrounding ground. If the basement is heated at the room temperature, the total resistance is the combination in parallel of the basement wall and basement floor resistances. If the basement is unheated, the floor resistance must be added in series to this combination.

The conductance-area product for basement walls that are completely below grade and insulated over their entire surface is given by

$$UA_{bw} = \left[0.035 + \left(\frac{0.22}{R_{bw} + 0.6}\right)\right] A_{bw} \tag{3.2.17}$$

where R_{bw} is the resistance of added insulation on the wall (hr ft^2 °F/Btu). The basement floor conductance-area product is given by

$$UA_{bf} = 0.025 A_{bf} \tag{3.2.18}$$

The basement thermal conductance is then the sum of these in parallel:

$$UA_{bt} = \left[\left(0.035 + \frac{0.22}{R_{bw} + 0.6}\right) A_{bw} + 0.025 A_{bf}\right] \tag{3.2.19}$$

44 RESIDENTIAL AND COMMERCIAL BUILDING HEATING REQUIREMENTS

If the basement is heated at room temperature, the heat loss is that through the basement floor and walls only. It will be termed the floor loss and is given by

$$q_{bt} = UA_{bt}(T_R - T_G) \qquad (3.2.20)$$

For unheated basements, the house floor thermal resistance may be important and must be added to the basement wall and floor combination. The floor conductance is

$$UA_{fl} = \frac{A_{fl}}{R_{fl}} \qquad (3.2.21)$$

The heat flow through the floor goes through the basement and floor resistances in series and is given by

$$q_{bt} = \frac{T_R - T_G}{1/UA_{bt} + 1/UA_{fl}} \qquad (3.2.22)$$

The ground temperature depends on the deep soil temperature which is essentially equal to the annual average ambient temperature. It also depends on the ambient temperatures which vary each month. For the calculation of heat loss on a monthly basis, it is recommended that the average of the monthly average temperature and the yearly average temperature be used as the ground temperature.

3.2.4. Infiltration and Ventilation Heat Flow

Energy loss due to infiltration is usually a major component of house heating requirements and, unfortunately, this is often the hardest term to estimate. The air flow leaving a building is at room temperature, while the replacement air enters at the ambient temperature. The energy flow associated with this air flow is given in terms of flow rate and air enthalpy by

$$\dot{E}_{inf} = \dot{m}(h_R - h_A)$$

Using ideal gas relations, this can be expressed as

$$\dot{E}_{inf} = \rho C_p \dot{V}(T_R - T_A) \qquad (3.2.23)$$

For commercial buildings, $\dot{V}$ is usually specified in terms of the air flow per occupant. Recommended values from ASHRAE range from 7 to 25 cfm per person depending on occupant density with the higher value suggested for more densely occupied buildings. The fresh air requirement is based on that needed to overcome odor buildup since meeting the oxygen needs of people is about 0.3 cfm per person. These values are somewhat arbitrary; for example, the Wisconsin Administrative Code, in an effort to promote energy conservation, has established a maximum of 5 cfm per person as the ventilation requirement for most applications.

BUILDING HEAT LOSSES

Residences are generally not mechanically ventilated, and air flow occurs through cracks and openings in the building structure. There are two major mechanisms creating infiltration. The "stack effect" is due to the temperature difference between the house and ambient. The warmer inside air is more buoyant than that outside and tends to flow outside through cracks and the chimney flue. The "wind effect" is due to increased pressure on windward surfaces which creates infiltration through cracks.

There is considerable uncertainty as to what value to use for the infiltration flow rate since the flow areas of cracks and the wind pressure differences are not well known. The flow rate is usually expressed in terms of the number of air changes per unit time ($\dot{N}$), where one air change per hour represents the entire volume of air in the house being replaced with fresh air in a 1-hour period. It has been conventional to use one air change per hour as the flow rate for a typical house, although the actual value depends both on house construction and ambient temperatures. A poorly weather stripped house in a cold climate might have 1.5 air changes per hour, while an electrically heated home or a tightly weather stripped house in a moderate climate might have 0.6 air changes per hour.

The air infiltration rates recommended by ASHRAE are of the form

$$\dot{N} = a + b(T_R - T_A) \qquad (3.2.24)$$

The values of the coefficients a and b are a function of the "tightness" of the house; this depends on the degree of weather stripping. House construction is denoted by "loose," which is typical of old or poorly constructed homes, "medium," which is representative of conventional construction, and "tight," which represents new buildings with good workmanship and special precautions taken to reduce infiltration. The values of the coefficients are given in Table 3.2.4 for these three levels of construction. The infiltration is then calculated from

$$\dot{E}_{inf} = \rho C_p V \dot{N}(T_R - T_A) \qquad (3.2.25)$$

where V is the heated volume of the house. The value of ρC_P for air is 0.018 Btu/ft^3 °F. For houses with combustion furnaces, it is usually assumed that the infiltration loss does not include the effect of air flow up the flue. That loss will be accounted for in the furnace efficiency.

Equation (3.2.25) shows that infiltration is proportional to the temperature dif-

TABLE 3.2.4
Values of Coefficients for Winter
Infiltration Rates [Eq. (3.2.24)]

Construction Level	a	b
Tight	0.280	0.0063
Medium	0.408	0.00873
Loose	0.483	0.01224

ference between the room and the ambient. It is convenient to define an infiltration conductance-area product as

$$UA_{\text{inf}} = \rho C_p V \dot{N} \qquad (3.2.26)$$

3.2.5. Attached Garages and Porches

Unheated garages and porches attached to houses increase the thermal resistance from the interior to the surroundings. The procedure for calculating the effect of these structures on heat loss follows that for unheated basements. The various thermal resistances between the house interior, the unheated space, and the ambient are calculated. One major resistance, which is difficult to assess, is the infiltration resistance. Garages and porches are probably represented by "loose" construction, and a value of one air change per hour is probably representative. The resulting thermal circuit can then be solved to find the heat loss from the house interior.

3.2.6. Total Heat Loss

The total house heat loss is obtained summing the individual heat losses as determined in the previous sections. The individual losses, with the exception of basement losses for slab and full basements, are all dependent on the temperature difference between the room and the ambient. The resulting expression for heat loss can be written as

$$q_{\text{loss}} = (UA_{wl} + UA_{dr} + UA_{rf} + UA_{wd} + UA_{fl} + UA_{\text{inf}})(T_R - T_A) + Q_{bt}$$
$$(3.2.27)$$

The term in parentheses is called the overall conductance-area product for the house, UA_o. Equation (3.2.27) can then be written as

$$q_{\text{loss}} = UA_o(T_R - T_A) + q_{bt} \qquad (3.2.28)$$

This is a convenient way to represent building heat losses.

3.2.7. Room and Ambient Temperature

The heat loss calculations depend on values for the room and ambient temperatures. In the evaluation of building annual energy requirements, calculations are performed on a monthly basis. The correct temperatures to use in the calculations are the monthly average temperatures. For the ambient temperatures, this is the average of the daily temperatures, denoted $\bar{T}_A$. It is available for many locations and data for selected locations are given in Appendix A.

The monthly average room temperature depends on the thermostat setting, and whether it is set back at night or not. For the situation in which the daytime temperature is one value $(T_{R,d})$ and the nighttime setting another value $(T_{R,ns})$, the monthly average room temperature is

$$T_R = T_{R,d}(1 - f_{ns}) + T_{R,ns} f_{ns} \qquad (3.2.29)$$

where f_{ns} is the fractional number of hours for night setback.

The relations developed in this section are also used to determine the design heating load, which is useful for sizing equipment such as furnaces and heat pumps. For this calculation, the relevant temperatures are the maximum room temperature and the minimum ambient temperature that the house could experience. This calculation will yield the maximum heat loss from the structure.

3.3. ENERGY GAINS

The heating requirements of a building are met by heat from the furnace and from other energy sources in the interior. These latter sources, termed heat gains, are due to people, appliances, equipment, and solar heat through windows. These gains will be discussed in the following sections.

3.3.1. Heat Generation due to People

People generate heat through the process of metabolism, and this energy is added to the interior space. There are two components of this energy. The heating effect due to heat transfer from the relatively warm person to the cooler room is called sensible heating. There is also moisture added due to breathing and perspiration, and this is termed latent heating. During winter heating conditions, the sensible heat addition offsets a portion of the heat loss. The latent heating adds moisture to the room, which may be a desirable feature during periods of low ambient humidity.

Representative values of the sensible heating effect for different activities are given in Table 3.3.1. A typical value for use in home heating calculations is 250 Btu/hr, or about 75 W per person. This is a relatively small term compared to house heating requirements, and extreme accuracy is not needed.

TABLE 3.3.1
Heat Generation for People

Activity	Sensible Heat Flow (Btu/hr)
Seated	225
Office work	250
Walking	250
Light work	275
Heavy work	580

3.3.2. Heat Generation due to Equipment and Appliances

The heat generated from these sources can be estimated from the electrical appliances and equipment that are used in the heated spaces. For both residential and commercial buildings, electrical consumption records are the most accurate source. For new construction, estimates may have to be made based on the expected appliances that will be installed. The nationwide average electrical usage is 6000 kWh/yr, although a newly constructed well-equipped home might consume 12,000 kWh. These values correspond to an average rate of 2300 to 4600 Btu/hr (700 to 1400 W). During winter, a consumption rate higher than average might be expected due to shorter daylight hours, more cooking, electric blankets, and so on. A value of 3000 Btu/hr (880 W) may be representative for heating calculations.

The amount of electricity used in residences varies over the year. The greatest consumption occurs in winter due to, presumably, lights, cooking, TV, electric blankets, and so on, being used a lot. The consumption decreases during the year and, for a nonairconditioned house, is lowest in summer. It is recommended that the nonairconditioning electrical use be distributed over the year inversely proportional to the average number of daylight hours for each month. The average day length is tabulated in Appendix A.

For example, if the annual consumption were 6000 kWh, the average hourly rate would be 2338 Btu/hr. During January in Madison, the day length is 9.2 hours, and the average rate would be

$$2338 \text{ Btu/hr } \frac{12}{9.2} = 3049 \text{ Btu/hr}$$

In contrast, in July the day length is 15.1 hours and the gain would be 1858 Btu/hr.

3.3.3. Heat Gain Through Windows

Solar energy entering through windows often is a significant energy gain during winter. The average amount that enters during a month is given by the product of the daily average incident solar radiation, the transmittance–absorptance product of the window, the window area, and the number of days per month.

$$G_{sol} = \bar{H}_T (\overline{\tau \alpha}) A_{wd} D_m \qquad (3.3.1)$$

The term $\bar{H}_T$ is the daily average energy flux over the month (Btu/day ft^2) incident on a window. It depends on location, time of year, and orientation of the surface. Values for selected cities are given in Appendix A for vertical surfaces and different compass directions.

The average transmittance–absorptance product for the room and window $(\overline{\tau \alpha})$ is mainly a function of the number of glazings. Virtually all of the energy entering a room is absorbed, and so the room is essentially black (a perfect absorber). Values of $(\overline{\tau \alpha})$ for conventional window constructions are given in Table 3.3.2.

TABLE 3.3.2
$(\overline{\tau\alpha})$ Values for Windows

Number of Glazings	$(\overline{\tau\alpha})$
1	0.83
2	0.71
3	0.64
4	0.57

3.4. DESIGN HEATING REQUIREMENTS

The design heating requirement is the heat rate required to maintain the building at the desired room temperature (daytime thermostat setting) on the coldest day of the year. It is assumed that there are no internal or window gains, and that all heat is supplied by the furnace. The heating load computed in this manner is used to size the heating equipment in the building. It represents the largest heating demand that could be expected. There are several considerations which provide a margin of safety in this calculation. Solar energy gains through windows, internal heat generation by lights and people, and the thermal capacitance of the building mean that this design load is rarely experienced. Further, a factor of safety is used in furnace selection in that usually the nearest larger size is installed. As will be discussed in Chapter 6, oversized equipment increases energy consumption, but the practice is to be conservative.

The design heating load is calculated from

$$q_{des} = UA_{o,des}(T_R - T_{des}) + q_{bt,des} \tag{3.4.1}$$

The design outdoor air temperature is that below which temperatures occur 1 percent of the time in the design, or average, winter; this corresponds to about 40 hours during the year. Values of the design temperatures are taken from the *ASHRAE Handbook of Fundamentals* and are given in Appendix A for various locations. The overall conductance–area product and the basement losses are calculated using this design temperature as the ambient temperature in the appropriate infiltration and basement relations.

A worksheet for the calculation of the design heat loss rate has been developed. It is given in Table 3.4.1. An example calculation is carried out in Section 3.6.

3.5. ANNUAL ENERGY CONSUMPTION

The annual energy consumption is only indirectly related to the design load. The outdoor design temperature is a low value chosen so that it rarely occurs, while the outdoor temperature that determines the annual energy consumption is an average

TABLE 3.4.1
Design Heat Loss Rate

House daytime temperature _____ °F
Ambient design temperature _____ °F

	Area	R	UA
Wall	_____	_____	_____
Door	_____	_____	_____
Ceiling	_____	_____	_____
Crawl space	_____	_____	_____
Garage	_____	_____	_____
Porch	_____	_____	_____
Window	_____	_____	_____

Infiltration _____
Total UA_o _____ Btu/hr °F
Basement: Maximum heat loss rate _____ Btu/hr
q_{des} = _____ Btu/hr

Infiltration:
  Heated volume _____ ft³
  Construction type _____

Basement:
  Slab: Perimeter _____ ft
  Full: Heated basement _____ yes _____ no
    House floor area _____ ft², R value _____ hr ft² °F/Btu
    Basement wall area _____ ft², R value _____ hr ft² °F/Btu
    Basement floor area _____ ft², R value _____ hr ft² °F/Btu
    Basement UA _____ Btu/hr °F

over the course of the year. The appropriate indoor temperature is not the thermostat setting since internal gains maintain the house temperature at comfortable levels at moderately low ambient temperatures. A house "balance temperature" exists which means heating from the furnace is necessary only when the ambient is below this value. This balance temperature depends on the house conductance-area product and the gains. It is conventional to characterize the average temperature difference between the balance and ambient temperatures as the "degree days" for the month.

The energy consumption for each month depends on the monthly average conditions for each month. The annual consumption is the sum of these monthly values. The procedure for calculating the annual consumption will be described in the following sections.

3.5.1. Balance Temperature

The balance temperature for the building is the outdoor ambient temperature at which the internal and solar gains exactly offset the losses from the structure. The furnace is off under these conditions, and an instantaneous steady-state energy balance on the building yields

$$g - UA_o(T_R - T_{\text{Bal}}) - q_{bt} = 0 \qquad (3.5.1)$$

The balance temperature is then given by

$$T_{\text{Bal}} = T_R - \frac{g - q_{bt}}{UA_o} \qquad (3.5.2)$$

The instantaneous balance temperature evaluated from Eq. (3.5.2) varies over the month since room temperature, gains, and some components of the conductance–area product vary. It is convenient to use a monthly basis, and Eq. (3.5.2) can be integrated over the month and then divided by the number of hours in the month. The monthly average balance temperature is given by

$$T_{\text{Bal}} = T_R - \frac{\bar{g} - \bar{q}_{bt}}{UA_o} \qquad (3.5.3)$$

where $\bar{g}$ and $\bar{q}_{bt}$ are the average hourly rates of heat gain and basement loss, respectively, over the month. The monthly average conductance–area product is evaluated as described in Section 3.2. T_R is the building average temperature as given by Eq. (3.2.28).

The actual balance temperature for the house also depends on the time period in the day, and thus the monthly average balance temperature should really be an average for these different periods. For example, during daytime, solar gains occur and the thermostat setting is higher than during the night. The balance temperature for these two periods may be significantly different.

The appropriate values to use for gains depend on the period during the day and on house construction. It is reasonable to assume that the gains due to electrical appliances and people are distributed uniformly over each hour of the month, both day and night. Most electrical appliances such as refrigerators, freezers, and so on, are on during both periods. If information is available on the actual time distribution, this could be used. The monthly average gains for these two internal sources are given by

$$\bar{g}_{\text{int}} = \frac{G_{\text{peo}} + G_{\text{app}}}{N_m} \qquad (3.5.4)$$

where the value of G is the total for the month, and N_m is the number of hours in the month. Similarly, it is assumed that the same basement loss occurs each hour of the month, and the average basement loss is

$$\bar{q}_{bt} = \frac{Q_{bt}}{N_m} \tag{3.5.5}$$

For houses of light construction, such as frame houses, the solar absorbed during daylight hours is not stored for use during nighttime. These solar gains are available only during the day, and the average rate is

$$\bar{g}_{sol} = \frac{G_{sol}}{N_d D_m} \tag{3.5.6}$$

during the daytime period, and zero during the night. N_d is the number of daylight hours per day. For houses of heavy construction, such as those with thick concrete floors and concrete partitions, these gains may be available during the night. In this situation, the assumption is that the solar gains are distributed uniformly over each hour of the month, and the gains in Eq. (3.5.4) would include G_{sol}.

A common situation is that of a house of light construction with solar gains through windows. The thermostat is turned down during the night and night insulation (drapes) is added at the end of day when the sun goes down. The corresponding three balance temperatures are:

Daytime:

$$T_{Bal,d} = T_{R,d} - \frac{\bar{g}_{sol} + \bar{g}_{int} - \bar{q}_{bt}}{UA_d} \tag{3.5.7}$$

Night (before setback):

$$T_{Bal,n} = T_{R,d} - \frac{\bar{g}_{int} - \bar{q}_{bt}}{UA_n} \tag{3.5.8}$$

Night setback:

$$T_{Bal,ns} = T_{R,ns} - \frac{\bar{g}_{int} - \bar{q}_{bt}}{UA_n} \tag{3.5.9}$$

These balance temperatures will be used in calculating the heat loss for each of these periods.

3.5.2. Heating System Requirements

The instantaneous rate of heat loss from a structure depends on the difference in temperature between the interior and the environment that exists at that time. The total energy required for heating is the integrated value of the instantaneous heat losses. Some of the heat loss is met by solar gains, while the rest must be supplied by the heating system. The breakdown between these two energy forms and the determination of the energy required from the heating system will be discussed in this section.

The instantaneous heat loss from the building is given by Eq. (3.2.28) as the sum of envelope and basement losses:

$$q_{loss} = UA_o(T_R - T_a) + q_{bt} \tag{3.5.10}$$

where T_a is the instantaneous ambient temperature. This heat loss is met by energy supplied from the heating system and from gains:

$$q_{loss} = q_{htg\,sys} + g \tag{3.5.11}$$

The energy supplied by the heating system is evaluated from combining Eqs. (3.5.10) and (3.5.11) as

$$q_{htg\,sys} = UA_o(T_R - T_a) + q_{bt} - g \tag{3.5.12}$$

The daily energy supplied is given by integrating Eq. (3.5.12) over the day

$$q_{htg\,sys,d} = \int_0^{24} [UA_o(T_R - T_a) + q_{bt} - g]\, dt \tag{3.5.13}$$

or, in terms of average quantities

$$Q_{htg\,sys,d} = [UA_o(T_R - T_A) + \bar{q}_{bt} - \bar{g}]\,24 \tag{3.5.14}$$

The daily average temperature T_A is usually evaluated as the average of the minimum and maximum temperatures for that day. The daily temperature excursions are usually relatively small, and the temperature variation is more-or-less sinusoidal, and so this is not a bad approximation. The term $(T_R - T_A)$ is called the "heating degree days" for that day. For example, if the room temperature is 75°F and the average air temperature is 45°F, there are 30 heating degree days for that day.

The evaluation of the heating requirement can be made in terms of the balance temperature rather than the room temperature. Equations (3.5.7)-(3.5.9) can be combined with Eq. (3.5.14) and the daily heating requirement becomes

$$Q_{htg\,sys,d} = \sum_i UA_{o,i}(T_{Bal,i} - T_A) N_i \tag{3.5.15}$$

where $i = 1 - 3$ and denotes the daytime, nighttime, and night setback periods, respectively.

The total heating energy required over a month is obtained by summing Eq. (3.5.15) over the number of days in the month.

$$Q_{htg\,sys,m} = 24 \sum_i \sum_{D_m} UA_{o,i}(T_{Bal,i} - T_A) \frac{N_i}{24} \tag{3.5.16}$$

Here, the term $\Sigma_{Dm}(T_{Bal} - T_A)$ is the number of heating degree days for that month and period, $DD_{m,i}$. The quantity $(N_i/24)$ is the time fraction for the period, f_i. This allows Eq. (3.5.16) to be written compactly as

$$Q_{htg\ sys, m} = 24 \sum_i UA_{o,i} f_i DD_{m,i} \qquad (3.5.17)$$

Monthly values of degree days have been evaluated for a wide variety of locations, and are tabulated in the ASHRAE guides. These tabulated values were established in the 1950s and assume that the building interior is kept at 75°F, and that the internal and solar gains provide enough heat to maintain the temperature 10°F higher than the environment. The building balance temperature is thus 65°F, and the tabulated values are for a 65°F base.

There are several changes that have occurred to make the 65°F base obsolete. House insulation has increased significantly since the 1950s, and as a result the same internal heat production maintains the house at a temperature greater than 10°F above the ambient. Internal gains have also increased due to greater use of electrical appliances. Government recommendations and mandates and fuel price increases have all served to reduce thermostat settings in public and private buildings. For example, a survey of home thermostat settings in Madison, Wisconsin showed an average value of about 69°F rather than 75°F. Finally, it has become common to turn down thermostats at night either manually or automatically. All of these changes mean that new values of degree days need to be calculated. This will be discussed in Section 3.5.3.

3.5.3. Degree-Day Evaluation

The number of degree days in a given month is the sum of the daily difference between the balance temperature and the daily average temperature. When each of the daily average temperatures is below the balance temperature, the calculation is straight forward. From Eq. (3.5.16) the degree days for each balance temperature are defined as

$$DD_{m,i} = \sum_{D_m} (T_{Bal,i} - T_A) \qquad (3.5.18)$$

When each daily value of T_A is less than $T_{Bal,i}$, the term in parentheses is positive for each day, and all terms contribute to the sum. The degree-day expression becomes simply

$$DD_{m,i} = (T_{Bal,i} - \bar{T}_A) D_m \qquad (3.5.19)$$

where $\bar{T}_A$ is the monthly average temperature.

When the balance and monthly average temperatures are close in value, the difference between the balance and ambient temperatures is not positive every day due to the day-to-day variation in ambient temperature. Only the positive values contribute to the degree-day calculation; when T_A is greater than T_{Bal}, heating is not required. Equation (3.5.19) must then be modified.

The situation is indicated schematically in Figure 3.5.1 for Madison, Wisconsin in March. The hourly occurrences of ambient temperature in 5°F increments (bins) are given as a function of the daily average temperature. The monthly average temperature $\bar{T}_A$ is 32°F. Two typical house balance temperatures are shown, one at 65°F and the other at 38°F. The degree days for a balance temperature of 65°F can be calculated directly from Eq. (3.5.18) since all temperatures are less than 65°F. The number of degree days are 1023°F day.

If Eq. (3.5.19) is used to compute degree days for the 38°F balance temperature, a value of 186°F day would result. However, this result would be in error since this calculation would take credit for temperatures above 38°F. The heat gain from these temperatures, shown as shaded in Fig. 3.5.1, cannot be used to offset the heating requirements due to the lower ambient temperatures. The degree days are greater than 186°F day.

A method has been developed to account for this effect. First, the difference between the balance and monthly average temperatures is calculated and normalized with the standard deviation of the daily average temperatures. A parameter F has been statistically correlated with this difference. The normalized temperature dif-

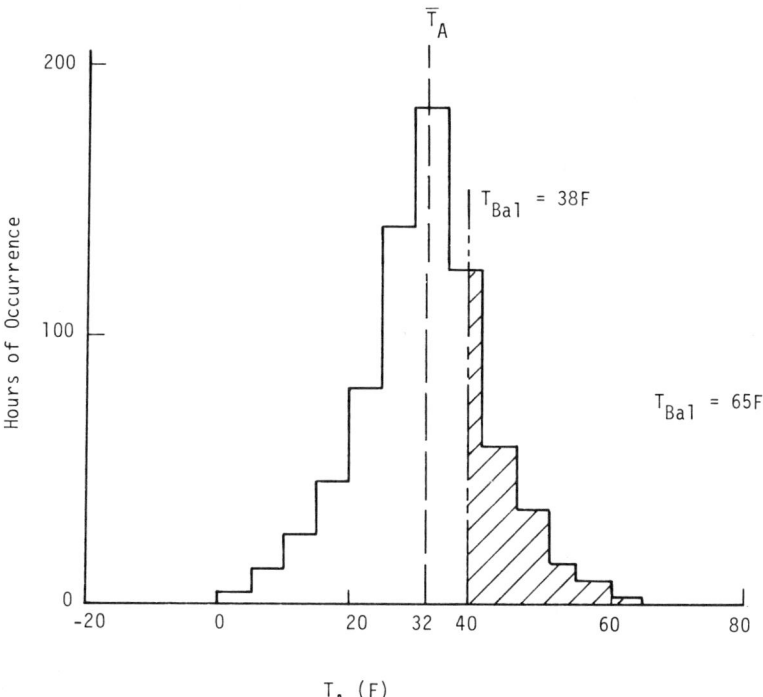

Fig. 3.5.1. Hours of occurrence as a function of ambient temperature (March, Madison, WI).

ference is

$$H = \frac{T_{Bal} - \bar{T}_A}{\sigma} \tag{3.5.20}$$

The parameter F is calculated as:
For $H > 0$:

$$F = 0.34e^{-4.7H} - 0.15e^{-7.8H} \tag{3.5.21}$$

For $H < 0$:

$$F = 0.34e^{4.7H} - 0.15e^{7.8H} - H \tag{3.5.22}$$

The standard deviation σ is calculated as

$$\sigma = (4.79 - 0.0337\bar{T}_A)\sqrt{D_m} \tag{3.5.23}$$

The actual degree days are then computed from

$$DD_{m,i} = (T_{Bal,i} - \bar{T}_A + \sigma F) D_m \tag{3.5.24}$$

For the ambient temperature of 32°F, the value of σ is 20.7°F. For the selected balance point of 38°F, the value for H is 0.29 and the value of F from Eq. (3.5.21) is 0.071. The corresponding degree days from Eq. (3.5.24) is 232°F day. This value is 25 percent higher than that calculated from Eq. (3.5.19).

The calculation of degree days following Eqs. (3.5.21) through (3.5.24) is important only during those months in which the ambient temperature is within 10°F of the balance temperature. These are, typically, spring and fall months. The heating loads in these months are usually small, and extreme accuracy is not required.

3.5.4. Calculation of Annual Heating Requirement

The accurate estimation of annual energy consumption incorporates the various factors discussed in the previous sections. The basic steps are given in this section followed by a worksheet (Table 3.5.1) to facilitate the calculations. An example for a dwelling is given in Section 3.5.5. The steps are:

1. Calculate the house average temperature based on the daytime and nighttime thermostat settings from Eq. (3.2.29).
2. Calculate the UA products for the walls, doors, ceilings, crawl space, garage, porch, and windows for daytime and nighttime periods following the procedures in Section 3.2. These values are constant over the course of the year.
3. From the house construction index (tight, medium, loose) and the monthly average temperature, compute the infiltration UA from Eq. (3.2.26).

TABLE 3.5.1
Heat Loss Calculation (Constant Loss Coefficients)

Daytime temperature _____ °F, Nighttime temperature _____ °F, Hours for setback _____ hr,
Average house temperature _____ °F

	Daytime			Nighttime	
	Area	R	UA	R	UA
Walls	___	___	___	___	___
Doors	___	___	___	___	___
Ceiling	___	___	___	___	___
Crawl space	___	___	___	___	___
Garage	___	___	___	___	___
Windows	___	___	___	___	___
Total			___ Btu/hr °F		___ Btu/hr °F

Garage: Common wall area _____ ft², R value _____
Common floor area _____ ft², R value _____
Garage wall area _____ ft², R value _____
Garage volume _____ ft³, Infiltration UA _____
Garage UA _____ Btu/hr °F

TABLE 3.5.1 (Continued)
Heat Loss Calculation (Variable Loss Coefficients)

Infiltration:
 Heated volume _____ ft^3
 Construction type _____

Basements
 Slab: perimeter _____ ft Yearly average temperature _____ °F
 Full: Basement heated _____ yes _____ no R value _____ hr ft^2 °F/Btu
    House floor area _____ ft^2 R value _____ hr ft^2 °F/Btu
    Basement wall area _____ ft^2 R value _____ hr ft^2 °F/Btu
    Basement floor area _____ ft^2
    Basement UA _____ Btu/hr °F

	J	F	M	A	M	J	J	A	S	O	N	D
Monthly average temperature (°F)												
Infiltration (Btu/hr °F)												
Daytime UA_o (Btu/hr °F)												
Nighttime UA_o (Btu/hr °F)												
Average ground temperature (°F)												
Basement loss (Btu/hr)												

Heat Loss Calculation (Monthly Gains)

	J	F	M	A	M	J	J	A	S	O	N	D
Daylength (hr)	—	—	—	—	—	—	—	—	—	—	—	—

Average Gains (Btu/hr)

Windows

Orientation	A	$\tau\alpha$	J	F	M	A	M	J	J	A	S	O	N	D
S			—	—	—	—	—	—	—	—	—	—	—	—
E			—	—	—	—	—	—	—	—	—	—	—	—
W			—	—	—	—	—	—	—	—	—	—	—	—
N			—	—	—	—	—	—	—	—	—	—	—	—
Total window gain			—	—	—	—	—	—	—	—	—	—	—	—
Internal gains			—	—	—	—	—	—	—	—	—	—	—	—
Net daytime gain (Btu/hr) (including basement)			—	—	—	—	—	—	—	—	—	—	—	—
Net nighttime gain (Btu/hr) (including basement)			—	—	—	—	—	—	—	—	—	—	—	—

TABLE 3.5.1 (*Continued*)
Heat Loss Calculation (Monthly and Annual Losses)

	J	F	M	A	M	J	J	A	S	O	N	D
Balance temperatures (°F)												
Daytime	—	—	—	—	—	—	—	—	—	—	—	—
Night (without setback)	—	—	—	—	—	—	—	—	—	—	—	—
Night (with setback)	—	—	—	—	—	—	—	—	—	—	—	—
Degree days (°F day)												
Daytime	—	—	—	—	—	—	—	—	—	—	—	—
Night (without setback)	—	—	—	—	—	—	—	—	—	—	—	—
Night (with setback)	—	—	—	—	—	—	—	—	—	—	—	—
Monthly heat loss (10^6 Btu)												
Daytime	—	—	—	—	—	—	—	—	—	—	—	—
Night (without setback)	—	—	—	—	—	—	—	—	—	—	—	—
Night (with setback)	—	—	—	—	—	—	—	—	—	—	—	—
Total (10^6 Btu/mo)	—	—	—	—	—	—	—	—	—	—	—	—
Annual heat loss	———— Btu/yr											

4. Add up all component UA values to get the house UA for each month.
5. From the average ground temperature and basement type, compute the basement heat loss rate from equations in Section 3.2.3.
6. From the window area, construction, and orientation together with the radiation data, compute the solar gain for each month using Eq. (3.3.1).
7. Compute the balance temperature for each month and time period using Eqs. (3.5.7)–(3.5.9).
8. Compute the degree days for each period during each month using equations in Section 3.5.3.
9. Compute the monthly heating supplied by the furnace during each time period using Eq. (3.5.17). Add up the fractions for each time period to get the total monthly load. Add up the monthly totals to get the yearly heating requirement.

3.6. EXAMPLE CALCULATION FOR ANNUAL ENERGY CONSUMPTION

As an example of the procedure for calculating annual energy consumption, a typical house will be studied. The energy use for the structure insulated in a conventional manner will be calculated first and then compared to the use for the structure in an uninsulated condition. The largest contributors to the overall energy use will be determined, and the cost effectiveness of various insulation techniques will be evaluated.

The house will be a typical midwestern two-story house of frame construction. It has a pitched roof and unheated attic, a full unheated basement, and an attached two-car garage. The foundation size is 25 × 30 ft, and the interior wall height is 8 ft. The window area is 12 percent of the wall area with the house oriented so that 60 ft^2 of window face south and 50 ft^2 face the other three directions. There are two outside doors. The annual electrical consumption is 7700 kWh, corresponding to an average rate of 3000 Btu/hr. Four people contribute an additional 1000 Btu/hr. The relevant heat transfer areas of these surfaces are given in Table 3.6.1.

Heat loss calculations will be done for both insulated and uninsulated houses. Frame wall construction similar to that described in Section 3.2.1 is used, with the addition of fiberglass batt insulation in the insulated house. Single glazed and double glazed windows are installed in the respective houses. The doors for the uninsulated case are hollow-core wood doors, while those for the insulated house are metal foam-filled doors. The insulated doors and double-pane windows also provide better weatherstripping and reduce infiltration. Drapes are pulled at sunset in both houses and add an R value of 0.5 to the windows. The insulation for the ceiling in the insulated structure is 8 inches of blown mineral wool on top of the wallboard ceiling. The basement walls are poured concrete, and for the insulated house, are finished with wool paneling with the 1-inch gap between wall and paneling filled with fiberglass batts. The thermal property values are taken from Table 3.2.1 and the corresponding R values are given in Table 3.6.2.

TABLE 3.6.1
Characteristics of Example House

Surface	Area (ft^2)
Walls	1508
Windows	210
Doors	42
Ceiling	750
Basement walls	880
Basement floor	750

Heated volume is 12,000 ft^3

Location: Madison, Wisconsin
Thermostat setting:
    Daytime   72°F
    Nighttime   66°F
Hours for night setback: 8 hr

TABLE 3.6.2
Insulation Levels for Example House

	R Values (hr ft^2 °F/Btu)	
	Uninsulated	Insulated
Walls	4.5	12.5
Windows—Day	0.88	1.85
Night	1.38	2.35
Doors	1.9	7.3
Ceiling	1.8	25.8
Basement walls	0	4
Construction quality	Medium	Tight

The garage calculations assume that a standard two-car garage has a volume of 4000 ft^3. The common wall area is 160 ft^2, and the R value of the remaining 1600 ft^2 of garage wall and roof area is 2 hr ft^2 °F/Btu.

The calculations on the worksheet (Table 3.6.3) are for the insulated house. They are straightforward, but tedious. A computer program (FLOAD) has been developed to carry out these calculations.

The calculations below show how the entries for January are obtained. The circled numbers refer to the corresponding lines in Table 3.6.3.

① $T_{ave} = (72 \times 16 + 66 \times 8)/24 = 70°F$

② $UA = 1348/12.5 = 108$ Btu/hr °F

EXAMPLE CALCULATION FOR ANNUAL ENERGY CONSUMPTION

③ $UA = 42/7.3 = 6$ Btu/hr °F

④ $UA = 750/25.8 = 29$ Btu/hr °F

⑤ Garage: $R_{gr,wl} = 12.5/160 = 0.078$ hr °F/Btu

$$UA_{inf} = 0.018 \times 4000 \times 1 = 72 \text{ Btu/hr °F}$$

$$UA_{wl} = 1600/2 = 800 \text{ Btu/hr °F}$$

$$UA = (72 + 800) = 872 \text{ Btu/hr °F}$$

$$UA_{gr} \; 1/(1/872 + 0.078) = 13 \text{ Btu/hr °F}$$

⑥ $UA_{wd,d} = 210/1.85 = 114$ Btu/hr °F

$UA_{wd,n} = 210/2.35 = 89$ Btu/hr °F

⑦ $UA_d = (108 + 6 + 29 + 0 + 13 + 114) = 270$ Btu/hr °F

$UA_n = (108 + 6 + 29 + 0 + 13 + 89) = 245$ Btu/hr °F

⑧ Basement: $R_{fl} = 2/750 = 0.00267$ hr ft² °F/Btu

$$UA_{wl} = 880 \, [0.035 + 0.22/(4 + 0.6)] = 72.9 \text{ Btu/hr °F}$$

$$UA_{fl} = 750/40 = 18.7 \text{ Btu/hr °F}$$

$$UA = (72.9 + 18.7) = 91.6 \text{ Btu/hr °F}$$

$$UA_{bt} = 1/(0.00267 + 1/91.6) = 74 \text{ Btu/hr °F}$$

⑨ $UA_{inf} = 0.018 \times 12000 \, [0.28 + 0.0063(70 - 19.4)] = 129$ Btu/hr °F

⑩ $UA_{o,d} = (270 + 129) = 399$ Btu/hr °F

⑪ $UA_{o,n} = (245 + 129) = 374$ Btu/hr °F

⑫ $T_G = (19.4 + 45.2)/2 = 32.2$ °F

⑬ $q_{bt} = 74(70 - 32.2) = 2797$ Btu/hr

⑭ $q_{wd} = 60 \times 0.71 \times 906/9.2 = 4195$ Btu/hr

⑮ $q_{wd} = 50 \times 0.71 \times 372/9.2 = 1435$ Btu/hr

⑯ $q_{wd} = 50 \times 0.71 \times 372/9.2 = 1435$ Btu/hr

⑰ $q_{wd} = 50 \times 0.71 \times 159/9.2 = 614$ Btu/hr

(18) $q_G = 1000 + 3000 \times 12/9.2 = 4910$ Btu/hr

(19) $q_{G,d} = (7649 + 4910 - 2797) = 9792$ Btu/hr

(20) $q_{G,d} = (4910 - 2797) = 2113$ Btu/hr

(21) $T_{Bal,d} = 72 - 9792/399 = 47.4°F$

(22) $T_{Bal,n} = 72 - 2113/374 = 66.4°F$

(23) $T_{Bal,ns} = 66 - 2113/374 = 60.4°F$

(24) $DD_d = 31(47.4 - 19.4) = 868°F$ day

(25) $DD_n = 31(66.4 - 19.4) = 1457°F$ day

(26) $DD_{ns} = 31(60.4 - 19.4) = 1271°F$ day

(27) $Q_d = 399 \times 868 \times 9.2$ hr/day $= 3.18 \times 10^6$ Btu

(28) $Q_n = 374 \times 1457 \times 6.8$ hr/day $= 3.70 \times 10^6$ Btu

(29) $Q_{ns} = 374 \times 1271 \times 8$ hr/day $= 3.80 \times 10^6$ Btu

(30) $Q = (3.18 + 3.70 + 3.80) \times 10^6 = 10.7 \times 10^6$ Btu/mo

There are some initial conclusions that can be drawn from the conductance–area products on pages 1 and 2 of the worksheets. Windows comprise only 12 percent of the surface area, and yet are 40 percent of the envelope conductance. It is difficult to insulate windows, and the greatest reductions in heat loss come with reduced window area. The ceiling is the easiest surface to insulate, and with moderate levels of insulation it can contribute very little to heat loss. Frame walls, on the other hand, are relatively difficult to insulate to the same level. It has become common to add a styrofoam sheet (R value of 4.2 hr ft^2 °F/Btu) to the outside of the wall before the sheathing is applied. This reduces the wall UA by 30 percent but the loss would still be considerably higher than that for the ceiling. Increased thicknesses of wall insulation would probably require custom-built door and window frames to accommodate the thicker walls. This will also require significant changes in the traditions of our construction industry with increased costs until new ways are accepted. Finally, infiltration, which is the most difficult component to determine, is one of the most significant heat loss paths.

The components of the monthly gains are plotted over the course of the year in Fig. 3.6.1. Solar gains are the greatest contributors to the gains and are fairly constant over much of the year. They drop considerably in November and December due to a combination of lower radiation levels and the orientation of the windows. A greater window area on the south walls could produce an increase in the gains in

TABLE 3.6.3
Heat Loss Calculations (Constant Loss Coefficients)

Description Insulated house
Daytime temperature 72°F, Nighttime temperature 66°F, Hours for Setback 8 hr,

① Average house temperature 70°F

		Daytime			Nighttime	
	Area	R	UA		R	UA
② Walls	1348	12.5	108			108
③ Doors	42	7.3	6			6
④ Ceiling	750	25.8	29			29
⑤ Crawl space	0	—	0			0
⑥ Garage			13			13
⑥ Windows	210	1.85	114		2.35	89
⑦ Total			270 Btu/hr °F			245 Btu/hr °F

Garage: Common wall area 160 ft², R value 12.5
Common floor area 0 ft², R value 0
Garage wall area 1600 ft², R value 2
Garage volume 4000 ft³, Infiltration UA 72
Garage UA 13 Btu/hr °F

TABLE 3.6.3 (Continued)
Heat Loss Calculation (Variable Loss Coefficients)

Infiltration:

Heated volume 12,000 ft^3

Construction type tight

⑧ Basements:

Slab: Perimeter 0 ft   Yearly average temperature 45.2°F

Full: House floor area 750 ft^2 R value 2 hr ft^2 °F/Btu

Basement wall area 880 ft^2 R value 4 hr ft^2 °F/Btu

Basement floor area 750 ft^2 R value 40 hr ft^2 °F/Btu

Basement UA 74 Btu/hr °F

	J	F	M	A	M	J	J	A	S	O	N	D
⑨ Infiltration (Btu/hr °F)	129	127	112	95	80	66	61	63	75	88	110	124
⑩ Daytime UA_o (Btu/hr °F)	399	395	381	364	349	334	329	332	344	356	378	393
⑪ Nighttime UA_o (Btu/hr °F)	375	371	357	339	325	310	305	308	320	332	354	369
⑫ Average ground temperature (°F)	32.2	33.1	38.5	44.8	50.3	55.7	57.5	56.6	52.0	47.5	39.4	34.0
⑬ Basement loss (Btu/hr)	2797	2731	2331	1865	1458	1058	925	992	1332	1665	2264	2664

Heat Loss Calculation (Monthly Gains)

			J	F	M	A	M	J	J	A	S	O	N	D
Daylength (hr)			9.2	10.3	11.8	13.1	14.5	15.1	14.8	13.2	12.2	10.9	9.5	8.9

Average Gains (Btu/hr)

Windows

Orientation	A	$\tau\alpha$	J	F	M	A	M	J	J	A	S	O	N	D
⑭ S	60	0.71	4195	4517	3888	2933	2491	2332	2467	3127	3704	4303	3453	3264
⑮ E	50	0.71	1435	1899	2196	2295	2512	2671	2720	2770	2383	1989	1300	1104
⑯ W	50	0.71	1435	1899	2196	2295	2512	2671	2720	2770	2383	1989	1300	1104
⑰ N	50	0.71	614	796	969	1168	1442	1646	1576	1358	1048	840	628	534
Total window gain			7679	9111	9249	8691	8957	9320	9483	10,025	9518	9121	6681	6006
⑱ Internal gains			4910	4495	4050	3748	3482	3384	3432	3727	3950	4364	4790	5040
⑲ Net daytime gain (Btu/hr) (including basement)			9792	10,875	10,968	10,574	10,981	11,646	11,990	12,760	12,136	11,820	9207	8382
⑳ Net nighttime gain (Btu/hr) (including basement)			2113	1764	1719	1883	2024	2326	2507	2735	2618	2699	2526	2376

TABLE 3.6.3 (*Continued*)
Heat Loss Calculation (Monthly and Annual Losses)

	J	F	M	A	M	J	J	A	S	O	N	D
Balance temperature (°F)												
㉑ Daytime	47.4	44.5	43.2	43.0	40.5	37.1	36.1	33.6	36.7	38.8	47.6	50.7
㉒ Night (without setback)	66.4	67.2	67.2	66.4	65.8	64.5	63.8	63.1	63.8	63.9	64.9	65.6
㉓ Night (with setback)	60.4	61.2	61.2	60.4	59.8	58.5	57.8	57.1	57.8	57.9	58.9	59.6
Degree days (°F day)												
㉔ Daytime	868	652	347	85	3	0	0	0	0	10	414	859
㉕ Night (without setback)	1457	1288	1091	654	322	50	0	0	150	417	933	1320
㉖ Night (with setback)	1271	1120	905	474	136	10	0	0	89	245	753	1135
Monthly heat loss (10^6 Btu)												
㉗ Daytime	3.2	2.7	1.6	0.4	0	0	0	0	0	0.1	1.5	3.0
㉘ Night (without setback)	3.7	2.7	1.7	0.6	0.2	0	0	0	0.2	0.7	2.2	3.5
㉙ Night (with setback)	3.8	3.5	2.7	1.3	0.4	0	0	0	0.2	0.8	2.2	3.4
㉚ Total (10^6 Btu/mo)	10.7	8.9	6.0	2.3	0.6	0	0	0	0.4	1.6	5.9	9.9
Annual heat loss	46.3×10^6 Btu/yr											

EXAMPLE CALCULATION FOR ANNUAL ENERGY CONSUMPTION

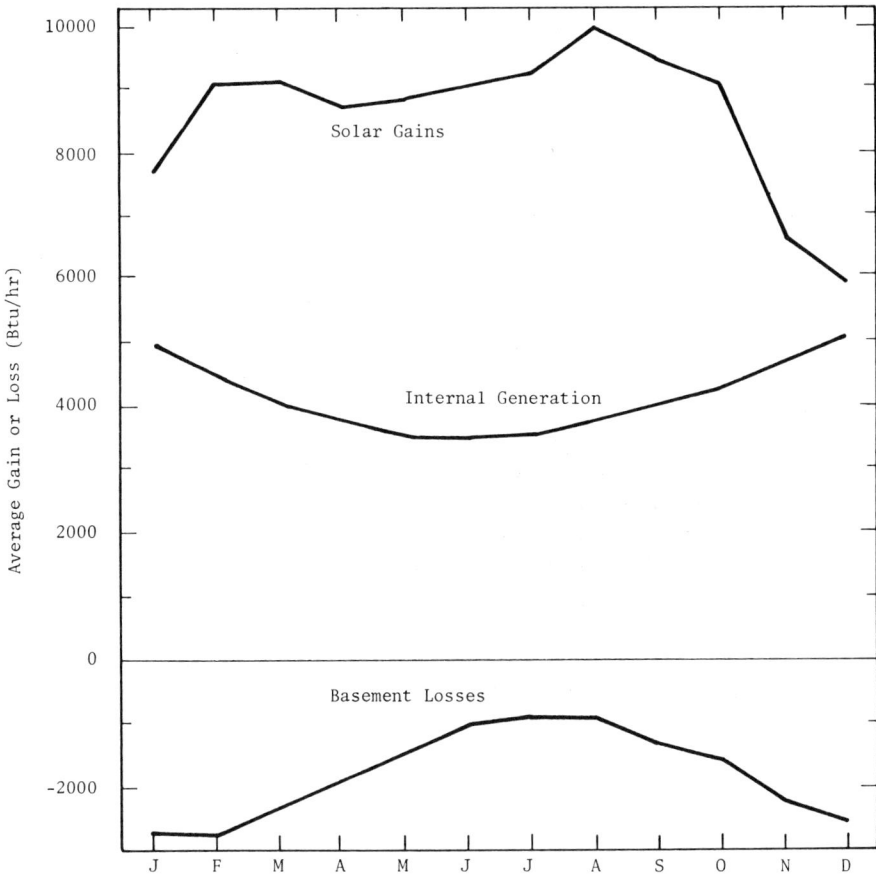

Fig. 3.6.1. Components of heat gains for example house.

these months. This is indicated by line 19 of the worksheet, page 3, which shows that the largest solar gains through the south windows occur in the winter months.

The generation from appliances and people contributes about one-third to the total gains. Since the electrical use of appliances is also used for heating, reduced electrical consumption increases heating requirements which may be an unexpected result of conservation. Basement losses are relatively minor compared to the gains.

The monthly values for the heat loss during different time periods are given on lines 27, 28, and 29. The period of greatest loss occurs during the nighttime periods. There are no solar gains to offset losses and so the loss is higher than during the daytime.

There are significant heat losses only during the months of October through April. A heating load analysis for these 7 months only would be sufficient. For more southern locations, the heating season would be even shorter and the calculations could be reduced accordingly.

The cost of heating can be estimated and compared to the costs of insulation.

For natural gas at $3.50/10^6$ Btu and a furnace efficiency of 70 percent, the annual fuel cost is $230.

The heat loss was also calculated for the uninsulated building. It was assumed that the house is the same size and construction, but without insulation in walls, doors, ceiling, and basement. Single-pane windows are installed, which means that there is less weatherstripping and more infiltration loss. The resulting annual heat loss is 233×10^6 Btu. The corresponding cost is $1041, which is four times that of the insulated dwelling.

The purchased energy flows through the various components of the insulated and uninsulated houses are given in Table 3.6.4. These flows are less than the total heat transferred and are determined by subtracting the proportionate gains from the total heat loss. These entries reflect, then, the energy that must be purchased to maintain comfort.

The energy savings through the use of insulation techniques are also shown. The reduction in infiltration loss is due to better door and window weatherstripping, and so the infiltration savings are divided equally between these two components. Increasing insulation levels has two effects. The first is the direct effect of reducing heat loss. The second is that increased insulation reduces the balance temperature, which means that the gains are more effectively utilized.

The greatest savings occur due to ceiling insulation, with walls next. Double-pane windows and basement insulation save about equally. Insulated doors save the least, and are mainly valuable in reducing infiltration.

The prices of insulation techniques are given in Table 3.6.5. These prices were obtained from a survey of home insulation suppliers in the Midwest in 1980. The prices quoted for wall and attic insulation include installation, while those for doors and windows do not. However, an insulated door or window costs about as much to install as an uninsulated one, and so installation costs cancel out of the savings computation. These prices are averages from a number of suppliers, and undoubtedly vary between regions and suppliers.

The economic costs and savings for each of the insulation techniques are given in Table 3.6.6. The life cycle savings for each component are positive, which means

TABLE 3.6.4
Purchased Energy Flows

	Uninsulated	Insulated	Savings (10^6 Btu/yr)
Walls	57.4	13.7	43.7
Doors	3.5	0.5	9.6[a]
Ceiling	64.7	3.0	61.7
Basement	28.1	7.3	20.8
Windows	29.8	10.3	26.1[a]
Infiltration	24.7	11.5	—[a]
	208.2	46.3	161.9

[a]Infiltration savings divided equally between doors and windows.

TABLE 3.6.5
Prices of Insulation

Wall insulation (includes installation)	
Fiberglass batts	(15¢ + 4.5¢/in.)/ft²
Styrofoam sheets ($\frac{1}{2}$–2 in.)	(10¢ + 3¢/in.)/ft²
Attic insulation (includes installation)	
Fiberglass	(15¢ + 5¢/in.)/ft²
Cellulose	(10¢ + 4¢/in.)/ft²
Windows (does not include installation)	
Single-pane	$6/ft²
Double-pane	$10/ft²
Triple-pane	$14/ft²
Vinyl sheets	$1.50/ft²
Doors (does not include installation)	
Ordinary	$100
Insulated	$300
Patio storm door—Glass	$350
Plastic	$100

that each technique is cost effective. Attic insulation is the most cost effective, with walls next. Basement insulation and double-pane windows follow. Insulated doors are not cost effective unless they reduce infiltration also. Each component has a relatively short payback period, which is on the order of 1–5 years. These insulation techniques are definitely cost effective.

If the example house had been built as a single-story ranch-style house of the same floor area, the envelope area would be greater. The single-story house would have twice the ceiling area and about three-quarters the wall area. As a result its total surface area would be 10 to 15 percent more than that of the two-story home. However, it is possible to insulate the ceiling to a much greater degree than the walls. For the levels of Table 3.6.2 the annual heat loss of the ranch home would be about 10 to 15 percent lower than that of the two-story home. The envelope area is not a guide to energy use.

TABLE 3.6.6
Cost Effectiveness of Insulation Techniques

Surface	Insulation Costs ($)	Annual Savings ($)	Life Cycle Savings ($)
Wall	603	218	4,140
Doors	400	48	590
Ceiling	150	308	6,660
Basement	352	104	1,890
Windows	420	130	2,400
Total	1,925	808	15,680

72 RESIDENTIAL AND COMMERCIAL BUILDING HEATING REQUIREMENTS

The proper orientation of windows can significantly increase solar gains and reduce heating requirements. If all of the windows for the example dwelling had been placed on the south wall, the annual heating requirements would drop from 46.3 to 37.0×10^6 Btu/yr for an annual savings of $47. This demonstrates how orientation on a lot and attention to window placement can reduce heating bills. There are practical limits to the amount of window area possible on a wall. Further, too much area can cause overheating in winter and increase air conditioning requirements in summer. With care, then, windows can be used to advantage.

These calculations were based on a daytime thermostat setting of 72°F. For a northern location, a drop of 1°F in the thermostat setting over the entire day would reduce heat loss about 3 percent. A drop of 1°F for the nighttime setback period would reduce heating requirements by about 1 percent. Night setback is achieved either manually or with a clock thermostat. A clock thermostat costs

TABLE 3.6.7
Design Heat Loss Rate

House daytime temperature 72°F
Ambient design temperature −11°F

		Area	R	UA
(1)	Wall	1348	12.5	108
(2)	Door	42	7.3	6
(3)	Ceiling	750	25.8	29
	Crawl space	0	—	0
(4)	Garage	160	12.5	13
(5)	Window	210	1.85	114
(6)	Infiltration			173
(7)	Total UA_o			443 Btu/hr °F

(8) Basement: Maximum heat loss rate 6142 Btu/hr
(9) q_{des} = 42,900 Btu/hr

Infiltration:
    Heated volume   12,000 ft³
    Construction type tight

Basement:
    Slab: Perimeter 0 ft
    Full: House floor area   750 ft², R value 2 hr ft² °F/Btu
           Basement wall area 880 ft², R value 4 hr ft² °F/Btu
           Basement floor area 750 ft², R value 40 hr ft² °F/Btu
              Basement UA 40 Btu/hr °F

EXAMPLE CALCULATION FOR ANNUAL ENERGY CONSUMPTION

$100-$200 installed. The value of the energy savings are about $15 annually, or $340 for a 20-year life, and so a clock thermostat is cost effective.

The design heating load is important for sizing the heating system. The calculations are shown on the worksheet (Table 3.6.7). The design heating load is about 43,000 Btu/hr. The installed furnace or heat pump should be big enough to meet this load; however, it is unlikely that this load will ever be felt. As will be discussed in Chapter 6, this sizing technique contributes to low seasonal efficiencies. The presence of people, appliances, and solar radiation will reduce the actual maximum heat loss rate.

For the uninsulated house, the design heat loss rate is about 122,000 Btu/hr. This means that the installed furnace or heat pump in the uninsulated case must be three times bigger than that for the well-insulated house. This reduction in size would probably save $100 to $200 for the furnace and up to $2000 for a heat pump. Thus, insulation is important in reducing the heating system costs as well as fuel costs.

This example illustrates the approach used in evaluating energy conservation in residential buildings. It is important to note that savings in fuel costs occur for a relatively small investment. Current construction prices are about $50/ft^2 of floor area, and the example house could cost about $75,000 to build. Insulation techniques represent only 3 percent of this cost yet save $15,000-$20,000 over the life of the analysis. It appears that home owners can ill-afford not to insulate sufficiently.

The calculations for design heat loss rate worksheet (Table 3.6.7) are

① $UA = 1348/12.5 = 108$ Btu/hr °F

② $UA = 42/7.3 = 6$ Btu/hr °F

③ $UA = 750/25.8 = 29$ Btu/hr °F

④ Assume garage is at ambient temperature and that garage walls have little resistance.

$UA = 160/12.5 = 13$ Btu/hr °F

⑤ $UA = 210/1.85 = 114$ Btu/hr °F

⑥ $UA_{inf} = 0.018 \times 12,000 [0.28 + 0.0063(72 - -11)] = 173$ Btu/hr °F

⑦ $UA_o = (108 + 6 + 29 + 13 + 114 + 173) = 443$ Btu/hr °F

⑧ Basement:

$R_{fl} = 2/750 = 0.00267$ hr °F/Btu

$UA_{wl} = 880 [0.035 + 0.22/(4 + 0.6)] = 72.9$ Btu/hr °F

$UA_{fl} = 750/40 = 18.7$ Btu/hr °F

74 RESIDENTIAL AND COMMERCIAL BUILDING HEATING REQUIREMENTS

$$UA = (18.7 + 72.9) = 91.6 \text{ Btu/hr }°F$$

$$UA_{bt} = 1/(0.00267 + 1/91.6) = 74 \text{ Btu/hr }°F$$

$$q_{bt, \max} = 74(72 - -11) = 6142 \text{ Btu/hr}$$

(9) $\quad q_{des} = 443(72 - -11) + 6142 = 42,900 \text{ Btu/hr}$

3.7. REGIONAL VARIATIONS IN ENERGY USE

3.7.1. Annual Degree Days

The detailed calculations presented in this section are needed in order to accurately estimate the heating loads of buildings. As was discussed in Section 3.5.4, it had become conventional to calculate degree days for a 65°F base, and use these values to compute heat loss. The house heating load was then computed with annual degree days as

$$q_{\text{loss}} = UA_o DD_y 24 \qquad (3.7.1)$$

where DD_y are the annual degree days. This relation is, obviously, oversimplified and cannot account for the many conservation methods and changes that have been implemented in recent years. Nevertheless, this relation has been employed for calculating heat loss.

The annual degree days are useful in describing the severity of the winter heating season. Table 3.7.1 lists annual degree days for four different locations in the United States. These vary from what would be termed mild winters (Los Angeles) to severe (Madison, Wisconsin).

Another aspect of the annual degree days is the variation from year to year. Table 3.7.2 presents results for the last 8 years in Madison, Wisconsin. The average is fairly close to the tabulated value, but there is a considerable spread. In 1972, the winter was 6 percent colder, based on degree days, than the average, while in 1973 it was 13 percent warmer. It is apparent that the heat loss calculations presented in this chapter are useful for predicting average values, not the loss for any one year.

TABLE 3.7.1
Annual Heating Degree Days

Madison, WI	7863
Chicago, IL	6155
Washington, DC	4224
Los Angeles, CA	2061

REGIONAL VARIATIONS IN ENERGY USE

TABLE 3.7.2
Variation of Degree Days for Madison

Year	Actual Degree Days
1971	7600
1972	8360
1973	6840
1974	7366
1975	7235
1976	7935
1977	7593
1978	8310
1979	7726
Average	7663

3.7.2. Influence of Climate on Heating Requirements

In order to illustrate the effect of climate on heating requirements, calculations were carried out for the four locations in Table 3.7.1 for the example house of Section 3.6. The heating requirements and costs for both insulated and uninsulated houses were determined, and are shown in Table 3.7.3. The heating loads decrease with the decrease in degree days tabulated in Table 3.7.1, but there is not the strict proportionate relation of Eq. (3.7.1). Conservation techniques are quite successful in reducing energy costs and usage for all climates. The greatest percent reductions come in the warmer climates, while the greatest dollar reductions occur for colder climates.

The life cycle savings due to the added insulation measures are given in Table 3.7.4. It is assumed that the insulation costs given in Table 3.6.5 are representative for all cities, and that natural gas is available as the fuel for heating. The life cycle savings are positive for all locations except Los Angeles, indicating that the measures are cost effective in northern areas. These savings would be even greater if fuel

TABLE 3.7.3
Influence of Climate on Heating Requirements, Annual Heating Requirements and Cost

	Insulated		Uninsulated	
Location	10^6 Btu	$	10^6 Btu	$
Madison, WI	46	230	208	1041
Chicago, IL	33	165	165	825
Washington, DC	20	100	98	490
Los Angeles, CA	4.3	22	21	105

TABLE 3.7.4
Life Cycle Savings for
Different Climates

Location	LCS
Madison, WI	15,680
Chicago, IL	12,370
Washington, DC	6,390
Los Angeles, CA	-490

oil or electricity were used for heating, and would then prove cost effective for Los Angeles. It is beneficial to both the home owner individually and the nation as a whole that conservation techniques be applied in all locations.

3.8. SUMMARY

This chapter has covered the basic concepts of heat transfer and has presented the methods necessary for calculating the heat loss or gain through building structures. A detailed description of heating requirements for residences has been carried out. A worksheet (Table 3.5.1) has been developed which aids this calculation. Calculation procedures for larger, commercial buildings are similar in nature. The significant different feature is that it is common in commercial buildings to have simultaneous heating and cooling, which requires different systems. The mechanisms of heat loss are the same, however, and the techniques in this chapter are applicable.

The chapter concludes with an evaluation of insulation techniques in various climates. Insulation to some degree is effective in most locations. It becomes even more cost effective as fuel prices rise or as shifts to more expensive forms occur. It is beneficial to individuals and the community to insulate sufficiently.

SUGGESTED READING

ASHRAE Handbook of Fundamentals, American Society of Heating, Refrigeration, and Air Conditioning Engineers, Atlanta, GA, 1977, 1981, Chapters 2, 21, 22, 23, 24.

ASHRAE Cooling and Heating Load Calculation Manual, GRP 158, American Society of Heating, Refrigeration, and Air Conditioning Engineers, New York, 1979.

J. P. Holman, *Heat Transfer*, McGraw-Hill Book Co., New York, 1976.

W. A. Beckman, J. A. Duffie, S. A. Klein, and J. W. Mitchell, F-Load, A Building Heating Load Calculation Program, to be published in *ASHRAE Transactions*.

T. Kusuda and T. Saitoh, *Simplified Heating and Cooling Energy Analysis Calculations for Residential Applications*, NBSIR 80-1961, National Bureau of Standards, July 1980.

Energy Conservation in Building Design, ASHRAE Standard 90-75, American Society of Heating, Refrigeration, and Air Conditioning Engineers, Atlanta, GA, 1975.

PROBLEMS

3.1. Obtain a price for one insulation technique or component from a local vendor. The technique should be suitable for use on either a commercial or residential building such as wall insulation, triple-pane windows, and so on. The price should include installation by the vendor.

3.2. A house under design has 1800 ft^2 of floor area, a total window area equal to 15 percent of the gross wall area, and a full, unheated basement. It may be either a single- or two-story home, and is located in Madison, WI.

Estimate the annual heating requirements, the design heating load, and the life cycle costs for this house under two conditions below:

(a) The house is built using standard frame construction but has no wall, ceiling, door, and so on, insulation and has single-pane windows.

(b) The same house is insulated appropriately. This means that you select the values of insulation you would use if you were building the house.

In estimating the life cycle costs, use reasonable economic parameters. Assume that the furnace uses natural gas for heating at a cost of \$4.6/10^6 Btu and an efficiency of 70 percent. Show your calculations in neat, tabular form.

3.3. For Problem 3.2, change the location to Chicago, IL.

3.4. For Problem 3.2, change the location to Washington, DC.

3.5. For Problem 3.2, change the location to Los Angeles, CA.

3.6. For the house of Problems 3.2–3.5, estimate the annual heating requirements and life cycle costs if the house is built to conform to the ASHRAE recommendations. These are specified as the overall conductance of walls plus windows and doors of 0.19 Btu/hr ft^2 °F, a ceiling conductance of 0.05 Btu/hr ft^2 °F, and a floor conductance of 0.08 Btu/hr ft^2 °F.

3.7. For the house of Problems 3.2–3.5, find the economic optimum level of one aspect of insulation (e.g., ceiling insulation, thermal drapes).

3.8. For the house of Problems 3.2–3.5, determine the influence of daytime and nighttime thermostat settings on heating requirements and fuel cost.

3.9. For the house of Problems 3.2–3.5, determine the effect on annual heating requirement and cost of increasing the window area on the south, holding total window area the same.

3.10. (a) A frame wall is built with 2 × 4's on 16-inch centers. It is then insulated with urea-formaldehyde foam. Determine the overall R value of the wall.

(b) After 3 months, the foam has shrunk 4 percent in all directions. Determine the overall R value of the wall.

(c) Foam costs are about $1.50/ft^2 of wall area. For a location with 5000°F days, determine whether the foam insulation is cost effective both before and after shrinkage.

4

RESIDENTIAL AND COMMERCIAL AIR CONDITIONING REQUIREMENTS

4.1. OVERVIEW

Under warm, humid conditions, heat and moisture enter buildings from the environment. In most areas of the country, air conditioning is needed to remove the energy and water vapor and maintain comfort in the interior space. Air conditioning currently accounts for about 8 percent of the energy used in the residential and commercial sectors, and this usage will increase since air conditioning is becoming widely accepted in both commercial buildings and residences. Virtually all air conditioners are electrically driven, which affects the electrical generation requirements of utilities. The electrical consumption due to air conditioning is large, and, for most utilities, the peak electrical loads occur in summer. The electrical generating system must be able to meet the maximum load imposed on it. Air conditioning usage is thus a major factor in determining both the installed capacity of the power plant and its total electrical output.

The energy used for air conditioning depends on the climate, the interior conditions, and the building structure. Historically, the interest has been on sizing of equipment, and the ASHRAE guides present detailed methods for evaluating peak loads. These calculations have been extended to estimate annual consumption. The maximum cooling loads are based on design wet and dry bulb temperatures for each location. These are the 1 percent values, which means that the environmental conditions are exceeded only 1 percent of the time, or 30 hours during the summer.

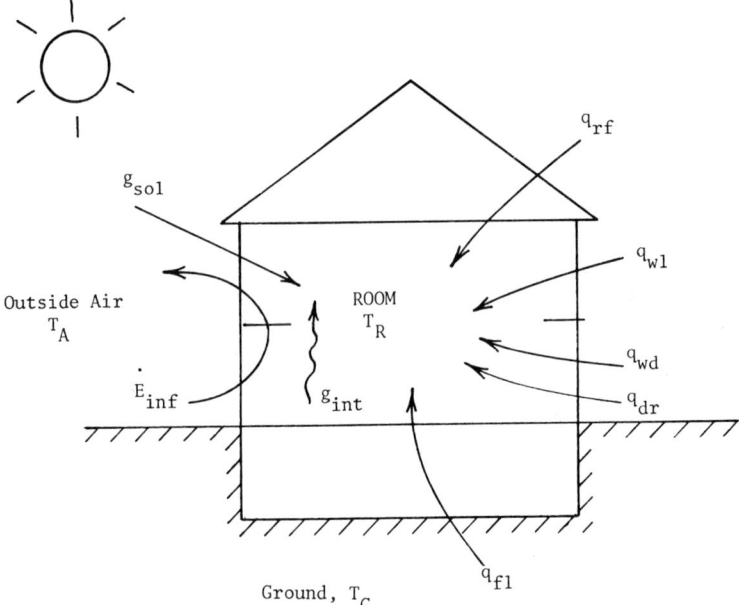

Fig. 4.1.1. Schematic of heat gains in a building.

The room conditions are between 73 to 78°F and 30 to 70 percent relative humidity. Recent government legislation has attempted to set 80°F as a design room temperature for air conditioning, but it appears that 78°F is a more realistic, and acceptable, upper temperature.

The seasonal energy requirements depend on the heat and moisture flows into the building. These are basically the same as for heating, and are shown schematically in Fig. 4.1.1. The solar gain and the internal heat and moisture generation become major contributions in the cooling load. The total energy gain is determined from an energy balance on the building as

$$\dot{E}_{gain} = q_{wl} + q_{rf} + q_{fl} + q_{wd} + q_{dr} + \dot{E}_{inf} + g_{int} + g_{sol} \tag{4.1.1}$$

The calculation of the major components of the energy gain will be discussed first in the following sections. Then the design cooling load will be determined. This load together with modified degree days will be used in estimating seasonal energy requirements. Finally, some examples of energy conservation will be given.

4.2. BUILDING ENERGY GAINS

4.2.1. Heat Gains Through Walls and Doors

The building structure is the same for summer cooling as for winter heating, and thus the overall conductance is essentially unchanged. Although the summer design

BUILDING ENERGY GAINS

wind speed (7.5 mph) is lower than that in winter (15 mph), this lowers the external surface conductance only a small amount.

There are, though, significant differences between summer and winter conditions. Winter temperatures are usually well below room temperature and daily fluctuations in ambient temperature do not affect the total heat flow since the heat flow is mostly out of the structure. In contrast, summer temperatures often fluctuate around room temperature, and heat flows into and out of the house. Energy storage in the structure itself can be significant in that the peak heating effect at midday is attenuated by the time it reaches the interior. This storage effect is accentuated by the very large changes in solar insolation over the day. Due to these storage effects, thermal energy transferred into the wall during the day may actually never reach the interior, but be transferred back out at night. This storage of energy in the walls and interior space needs to be accounted for in the calculation of heat gain.

The exterior of the building is exposed to both the ambient temperature and solar insolation. There are heat gains due to both of these effects. The heat flow situation with solar radiation impinging on a wall is shown in Fig. 4.2.1. The corresponding thermal circuit is also shown. With absorption of solar energy on the wall surface, the exterior wall temperature is raised and may be considerably higher than the air temperature. Without solar energy absorption on the wall surface, the heat flow through the wall would have been given by

$$q_{wl} = UA_{wl}(T_A - T_R) \tag{4.2.1}$$

where the overall conductance includes both the inside and outside convection conductances. With solar radiation, there is an additional heating effect. The calculation of heat gain will be of the same form as Eq. (4.2.1), but with a different temperature for T_A when solar radiation is present.

The heat flow through a wall can then be computed from an energy balance on the surface of the wall. For the left-hand surface in Fig. 4.2.1 the energy balance is

$$\alpha I_T A = q_A + q_{wl}$$

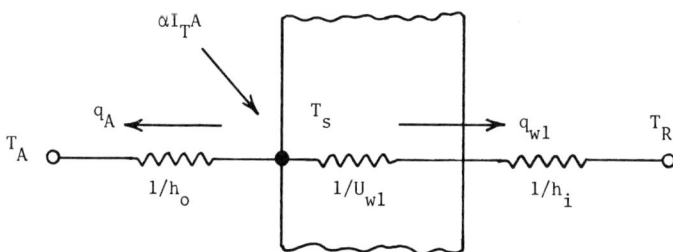

Fig. 4.2.1. Actual heat flows.

or

$$q_{wl} = \alpha I_T A - q_A \tag{4.2.2}$$

where α is the solar absorptance of the wall and I_T is the incident solar radiation flux. Introducing the convection mechanism equation for the heat flow from the surface to the air allows Eq. (4.2.2) to be written as

$$q_{wl} = \alpha I_T A - h_o(T_s - T_A) \tag{4.2.3}$$

It is convenient to define a fictitious air temperature T_{sa} that gives the same heat flow as that given by Eq. (4.2.3). This temperature is called the sol-air temperature, and is defined so that the heat flow through the wall is given by

$$q_{wl} = h_o A(T_{sa} - T_s) \tag{4.2.4}$$

Equating Eq. (4.2.3) and (4.2.4) allows T_{sa} to be calculated as

$$T_{sa} = T_A + \frac{\alpha I_T}{h_o} \tag{4.2.5}$$

Since T_{sa} is the effective air temperature that gives the same heat flow that actually occurs, the actual thermal circuit can be represented as in Fig. 4.2.2. The sol-air temperature replaces the air temperature in Eq. (4.2.1) and accounts for solar absorption at the wall surface. The rate of heat gain can then be calculated as

$$q_{wl} = UA_{wl}(T_{sa} - T_R) \tag{4.2.6}$$

where U is again the overall wall conductance. Equation (4.2.6) could also be used for heat loss calculations. However, in winter the higher outside convection coefficient reduces the difference between sol-air and air temperatures, and the overall temperature difference is quite large. Further, neglecting the solar effects in winter is conservative for sizing equipment and estimating energy consumption.

The use of the instantaneous values of the sol-air temperature in calculating heat

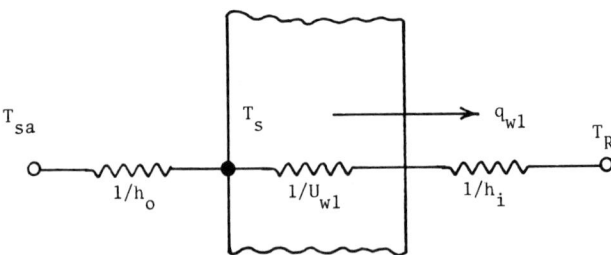

Fig. 4.2.2. Equivalent circuit using sol-air temperature.

gains leads to values that are higher than actually experienced. The reason lies in that there is a large variation in solar flux, and consequently sol-air temperature, over the day. As a result, the walls store energy, and heat flows into the interior are attenuated. If the air temperature drops below room temperature at night, heat may flow back out of the building without ever reaching the interior. The sol-air temperature concept must be modified to account for the variation in solar radiation and effects of storage on the daily air conditioning requirement.

The design heat flow through walls and doors will be calculated using room temperature and a design value for the sol-air temperature. This sol-air temperature will be calculated using the design ambient temperature and a daytime-average solar flux for the hottest month. This heat flow will be more representative of the actual heat flow into the structure than the value computed on an instantaneous basis. This method is similar to the total equivalent temperature difference (TETD) approach of the ASHRAE guides. However, it is more useful in including those mechanisms that contribute to air conditioning loads.

The design value of the sol-air temperature is then given by

$$T_{sa} = T_{des} + \frac{\alpha \bar{I}_T}{h_o} \qquad (4.2.7)$$

The average solar radiation flux is given by dividing the daily average solar insolation H_T by the day length.

As an example of this calculation, consider the insulated frame wall of Section 3.2 which has a U value of 0.11 Btu/hr ft² °F. The wall will be facing east. The solar absorptivity of the outside surface of dark painted or stained walls is about 0.9. Under summer wind conditions, the outside heat transfer coefficient is 4 Btu/hr ft² °F. The house will be located in Madison, Wisconsin where July is the warmest month, and the interior will be maintained at 78°F.

For Madison, the summer design temperature is 91°F. The average solar radiation flux on the east wall is

$$\bar{I}_T = (1134 \text{ Btu/ft}^2 \text{ day})/(14.8 \text{ hr/day}) = 76.6 \text{ Btu/hr ft}^2$$

The design sol-air temperature is

$$\bar{T}_E = 91°F + 0.9 \frac{76.6}{4} = 108.2°F$$

The average heat flux into the structure for this wall is then

$$q''_{wl} = 0.11 \text{ Btu/hr ft}^2 \text{ °F } (108.2 - 78) \text{ °F} = 3.3 \text{ Btu/hr ft}^2$$

The solar gain can be reduced by changing the wall properties. If the wall were painted white, the solar absorbtance would drop to 0.25. The sol-air temperature would become 95.8°F, and the average heat flux would be 2.0 Btu/hr ft². This illustrates the importance of surface properties on heat gains into the building.

84 RESIDENTIAL AND COMMERCIAL AIR CONDITIONING REQUIREMENTS

This method is similar to the TETD approach in that the TETD essentially equals the difference between the sol-air and room temperatures. The main advantage of this approach is that the effect of wall and weather parameters can be calculated directly.

4.2.2. Heat Gains Through Roofs and Attics

For buildings with flat roofs, the same procedures as for walls are followed. The incident radiation is the value on a horizontal surface.

The situation for a roof with an attic space is more complicated. There are two heat flow paths into the attic as shown schematically in Fig. 4.2.3. The heat flow into the room through the ceiling is given by

$$q_c = UA_c(T_{at} - T_R) \qquad (4.2.8)$$

The attic temperature is evaluated from an energy balance on the attic. It is assumed that there is no energy stored in the attic itself. This yields

$$UA_{rf}(T_{sa} - T_{at}) + UA_{\inf}(T_{des} - T_{at}) - UA_c(T_{at} - T_R) = 0 \qquad (4.2.9)$$

The attic temperature is evaluated as

$$T_{at} = \frac{(UA_{rf}T_{sa} + UA_{\inf}T_{des} + UA_c T_R)}{(UA_{rf} + UA_{\inf} + UA_c)} \qquad (4.2.10)$$

In order to evaluate the various terms, some assumptions have to be made. These are not critical since the attic heat flow is small in a well-insulated house. The sol-

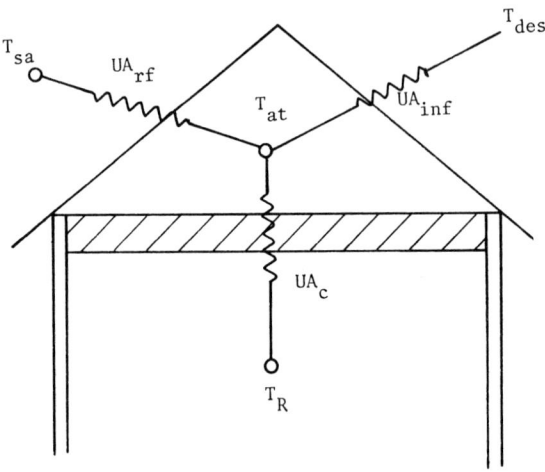

Fig. 4.2.3. Thermal circuit for a roof–attic combination.

BUILDING ENERGY GAINS

air temperature will be evaluated using Eq. (4.2.7) and the value of horizontal solar radiation.

The volume of the attic and the area of the roof will be estimated assuming that the house floor plan is square and that the roof pitch is 30°. This will yield the following relations

$$A_{rf} = 1.15 A_c \qquad (4.2.11)$$

$$V_{at} = 0.145(A_c)^{3/2} \qquad (4.2.12)$$

For the conductance due to infiltration, it will be assumed that there is one air change per hour in the attic space. The resulting infiltration UA becomes

$$UA_{inf} = 0.0027(A_c)^{3/2} \text{ Btu/hr °F}$$

The roof R value is that due to the roof structure (rafters, plywood, shingles) and is approximately 2 hr ft² °F/Btu.

These relations are only approximate. If more accuracy is desired, a more refined calculation of the terms may be performed.

4.2.3. Heat Gain Through Basements

During the cooling season, the heat loss through a basement reduces the cooling load. The heat loss through full and slab basements, and for crawl spaces, is calculated in the same manner as for the heating season. This heat gain, which is negative, then subtracts from the total cooling load.

4.2.4. Heat Gains Through Windows

The solar energy entering through a window in the cooling season is treated similarly to that in the heating season. These heat flows are shown schematically in Fig. 4.2.4. The heat conducted through the window is calculated in the same manner as heat conducted through walls using the summer design sol-air temperature as

$$q_{wd} = UA_{wd}(T_{sa} - T_R) \qquad (4.2.13)$$

The sol-air design temperature essentially equals the ambient design temperature since the absorptance of windows is low (about 0.04).

The solar energy which enters the interior and which is absorbed in the space is given by

$$q_{sol} = \bar{I}_T A_{wd}(\overline{\tau\alpha}) \qquad (4.2.14)$$

where $(\overline{\tau\alpha})$ is the product of the average window transmittance and the absorptance of the building interior. The average incident solar radiation rate is again the daily

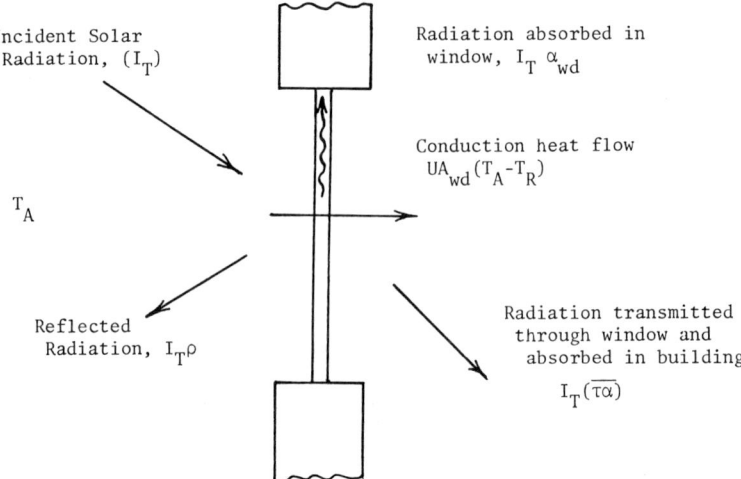

Fig. 4.2.4. Schematic of energy flows for a window.

average incident solar divided by the hours in the day. If there are significant shading effects from exterior structures such as awnings, trees, and adjacent buildings, an estimate of the reduction in average incident flux must be made. This is relatively important in that the window gains contribute significantly to the air conditioning load. It is probably best to underestimate the shading effect of these structures to be conservative. Internal shading may be obtained by surface coatings, drapes, and blinds, and are included in the $(\overline{\tau\alpha})$ product for the window. Selected values obtained from the ASHRAE shading coefficients are given in Table 4.2.1.

As an example of this calculation procedure, consider a single-pane window on the east facing surface of the wall in the example in Section 3.6.1. For the design temperature of 91°F and the average solar radiation flux of 76.6 Btu/hr ft^2, the de-

TABLE 4.2.1
Window Properties

Type	Transmittance–Absorptance $(\tau\alpha)$			U Value (Btu/hr ft^2 °F)
	No Shades	Venetian Blinds	Drapes	
Single glazing	0.83	0.45	0.3–0.5	1.06
Double glazing	0.71	0.43	0.25–0.45	0.61
Single, grey coating	0.38	0.31	0.29	1.00
Double, grey coating	0.28	0.25	0.24	0.55
Single, silver coating	0.19	0.15	0.14	0.80
Double, silver coating	0.14	0.13	0.12	0.50

sign sol–air temperature is

$$T_{sa} = 91°F + 0.04 \frac{76.6}{4} = 91.8$$

Within negligible error, this sol–air equals the design temperature. The heat flux by conduction is

$$q''_{wd} = 1.06(91 - 78) = 14 \text{ Btu/hr ft}^2$$

The solar gain is

$$q''_{sol} = 76.6(0.83) = 64 \text{ Btu/hr ft}^2$$

and the total heat gain is 78 Btu/hr ft². For a double-pane window, the solar flux becomes 54 Btu/hr ft², or about 15 percent less than the single-pane window. The corresponding conduction flux is reduced to 9 Btu/hr ft². The total heat gain is reduced by 20 percent to 63 Btu/hr ft². The main effect of insulating glass in the summer time is to reduce the direct solar gain and not the heat conducted through the window. The conducted heat flow is small compared to the direct solar gain.

Coatings can be very effective in reducing solar gains. As seen from Table 4.2.1 a coating or film on the window could reduce the heat gain to 20 percent of the uncoated value. Drapes and blinds are also effective, but do not reduce heat gains as much as coatings. The choice between these techniques depends on the application. Coatings are permanent, and reduce the solar heating during winter, while drapes and blinds can be employed selectively in the summer. For this reason coatings are preferred on commercial buildings where air conditioning loads are high, while uncoated glass together with drapes or blinds are applicable to residences.

4.2.5. Heat and Moisture Gain by Infiltration

Infiltration rates in summer are affected by the same factors of winds and the temperature difference between the interior and the ambient as in winter. However, wind speeds are generally less in summer than winter, and there is a reduced temperature difference between the room and environment. This reduces the driving potential for free convection infiltration flows. Further, during summer fuel-fired heating equipment is not operating, and there are no gains corresponding to the flue losses during the heating season. As a result, infiltration contributes much less to heat gain than to heat loss.

The values of air changes per hour recommended by the ASHRAE guides for summer cooling conditions are given by the same type of relation as for heating.

$$\dot{N} = a + b(T_{des} - T_R) \qquad (4.2.15)$$

TABLE 4.2.2
Coefficients for Summer Infiltration Rates

Construction	a	b
Tight	0.21	0.0072/°F
Medium	0.31	0.0084/°F
Loose	0.31	0.0140/°F

where $\dot{N}$ is the number of air changes per hour and a and b are functions of house construction as given in Table 4.2.2. The thermal energy carried into the space due to the air temperature is called the sensible load, and is given by

$$\dot{E}_{\text{inf,sen}} = \rho c_p V \dot{N} (T_{\text{des}} - T_R) \qquad (4.2.16)$$

For typical conditions, the value of ρc_p is again 0.018 Btu/ft³ °F.

Infiltration also brings moisture into the building which must removed by the air conditioner. The air is cooled to condense the water vapor, and the condensate drains from the house. The energy required to remove this moisture equals the product of infiltration air flow rate, the difference in humidity ratio between the environment and the room, and the latent heat of vaporization of water. This component of infiltration, called the latent load, is given by

$$\dot{E}_{\text{inf,lat}} = \rho h_{fg} \dot{N} V (w_{\text{des}} - w_R) \qquad (4.2.17)$$

For typical conditions, the product of the air density and latent heat of vaporization equals 80 Btu/ft³. The room humidity for the ASHRAE comfort conditions is 0.008 lb_m/lb_m.

4.2.6. Heat and Moisture Generation Due to People

People contribute both heat and moisture to a room, and both must be removed. The direct heating effect is again the sensible heating, while the moisture added through breathing and perspiration is latent heating. Both energy flows depend on many variables including age, sex, and activity, and representative values are given in Table 4.2.3. Moisture generation is as important as sensible heating for moderately active people.

4.2.7. Heat and Moisture Generation due to Appliances and Equipment

The sensible heat generation from these sources may be estimated in a variety of ways. For commercial buildings, the electrical power consumption of equipment, which ultimately goes into heating the building, can be determined from nameplate information. For residences, the estimation is a little more difficult. The average household electrical energy use is 6000 kWh/yr, which corresponds to a uniform hourly rate of 2300 Btu/hr. However, people usually would not have a large num-

TABLE 4.2.3
Sensible and Latent Heat Flows for People

Activity Level	Sensible Heat Flow (Btu/hr)	Latent Heat Flow (Btu/hr)	Total Heat Flow (Btu/hr)
Seated	210	140	350
Office work, light	230	190	420
Office work, medium	255	255	510
Light work	375	435	810
Medium work	345	695	1040
Heavy work	565	1035	1600

ber of lights, ovens, electric blankets, and so on, turned on on hot days and this value is too high for air conditioning calculations. The value of heat gains recommended by ASHRAE for use in design calculations for residences is 1200 Btu/hr.

The latent gains due to these sources are more difficult to estimate. For commercial or industrial calculations, the moisture from specific pieces of equipment may be estimated. For residential calculations, it is suggested that between $\frac{1}{2}$ and 1 pound of water per hour may evaporate due to cooking, showers, water used in sinks, and so on. The corresponding latent load is 500-1000 Btu/hr.

4.3. DESIGN COOLING LOAD

The cooling load used to specify the size of the air conditioner is obtained from the component gains described in Section 4.2. These gains add up to yield the total energy flow rate into the structure. The air conditioner installed in the house must be large enough to meet this load and maintain comfort conditions. This design load will also be used in the estimation of annual cooling energy requirements. A worksheet for the design load is given in Table 4.3.1, and the procedure is described as follows:

1. Enter the relevant weather and house data.
2. Calculate the daytime average sol-air and attic temperatures.
3. Calculate infiltration flow rate.
4. Calculate sensible heat flow rate through walls, doors, ceiling, and by infiltration.
5. Calculate the direct solar gains through windows.
6. Estimate the internal sensible gains due to people and appliances.
7. Estimate the latent loads due to infiltration, people, and appliances.
8. Add up all gains to determine the total design heat gain rate.

An example calculation is given in Section 4.5.

TABLE 4.3.1
Design Cooling Load

Location _____ Month _____

$\bar{T}_A$ _____ °F, T_{des} _____ °F, w_{des} _____ lb$_m$/lb$_m$

T_R _____ °F, w_R _____ lb$_m$/lb$_m$

Infiltration: House volume _____ ft^3

              Construction type _____

              $\dot{N}$ _____ changes/hr

Roof-ceiling:

     Ceiling: A_c _____ ft^2, R_c _____ hr ft^2 °F/Btu, UA_c _____ Btu/hr °F

       Roof: A_{rf} _____ ft^2, R_{rf} _____ hr ft^2 °F/Btu, UA_{rf} _____ Btu/hr °F

     Infiltration: V_{inf} _____ ft^3, UA_{inf} _____ Btu/hr °F

                                                   Sum _____ Btu/hr °F

     Walls: R_{wl} _____ hr ft^2 °F/Btu

    Doors: R_{dr} _____ hr ft^2 °F/Btu

Window: R_{wd} _____ hr ft^2 °F/Btu

      Floor: R_{fl} _____ hr ft^2 °F/Btu

Sensible gains

		A (ft^2)	UA (Btu/hr °F)	$\alpha I_t/h_o$ (°F)	q (Btu/hr)
Walls	S	____	____	____	____
	E	____	____	____	____
	N	____	____	____	____
	W	____	____	____	____
Doors	S	____	____	____	____
	E	____	____	____	____
	N	____	____	____	____
	W	____	____	____	____
Windows		____	____	____	____
Ceiling		____	____	____	____
Floor		____	____		____
Infiltration			____		____
Sum			____		_____
			$(\overline{\tau\alpha})$		
Windows:	S	____	____		____
	E	____	____		____

TABLE 4.3.1 (*Continued*)

	A (ft^2)	UA (Btu/hr °F)	$\alpha I_t/h_o$ (°F)	q (Btu/hr)
N	___	___		___
W	___	___		___
People				___
Appliances				___
Basement				___
Sum of gains				___
Total sensible gain				_____
Latent gains				
Infiltration				___
People				___
Appliances				___
Total latent gain				_____
Design cooling load				_____ Btu/hr
Air conditioner size				_____ tons

4.4. ANNUAL COOLING ENERGY REQUIREMENTS

Many methods have been proposed for estimating annual energy consumption, and all are only approximate. The method suggested in the ASHRAE guides is based on the experience of the utility companies in meeting the air conditioning loads imposed on their systems. The ratio of annual cooling requirements to the rated energy input of the equipment is used to estimate the rated full hours of operation of properly sized equipment during the normal cooling season. The total energy consumption is then estimated from the design cooling load, the estimated hours of operation, and the seasonal coefficient of performance (COP) of the air conditioner.

The ASHRAE guides stress that these estimations are only approximate, and may be in error by as much as ±20 percent. Individual differences in occupant lifestyle, house orientation, shading from trees, and so on, create the uncertainty. The recommendation is that wherever possible, past utility bills be relied upon.

The method recommended here is based on one recently developed by NBS, and termed the variable-base degree-day method. In this method, the design cooling load is first calculated, and then the house balance temperature is calculated. This is the ambient temperature for which the solar and internal gains balance the losses, and is calculated in the same manner as for the heating season. The gains used,

92 RESIDENTIAL AND COMMERCIAL AIR CONDITIONING REQUIREMENTS

though, are for the design conditions. The balance temperature is

$$T_{Bal} = T_R - \frac{(\bar{g}_{sol} + \bar{g}_{int} - \bar{q}_{bt})}{UA_o} \quad (4.4.1)$$

where the overall conductance area product UA_o is for summertime conditions.

For a house that is tightly sealed, the seasonal cooling load would be calculated using this balance temperature and the ambient temperatures similar to that for the heating load as described in Section 3.5.2. The monthly cooling load would be

$$Q_{cl,m} = UA_o 24 \sum_{D_m} (T_A - T_{Bal})^+ \quad (4.4.2)$$

where the superscript + signifies that the sum includes terms only when T_A is greater than T_{Bal}.

The calculation in this manner using degree days assumes a relation between heat loss and ambient temperature as shown in Fig. 4.4.1. The cooling load is assumed to be a linear function of ambient temperature, with a value of zero at the balance temperature and equal to the design cooling load at the design temperature. This assumes a correlation between solar gains and air temperature, which is that the solar increases in proportion to air temperature. While this is only an approximation, it appears to be reasonably accurate.

For a well-built house, the balance temperature is quite low and Eq. (4.4.2) would indicate that air conditioning would be required under cool to cold conditions. In practice, most people would open doors and windows and use natural or forced ventilation when the outside temperature was below the desired room temperature. This is also shown schematically in Fig. 4.4.1. The degree-day calculation needs to be modified to include the use of outside air for cooling.

The use of outside air for cooling reduces the number of cooling degree days. This is shown schematically in Fig. 4.4.2, where the frequency of occurrence of ambient temperatures as a function of ambient temperature is given. The degree days

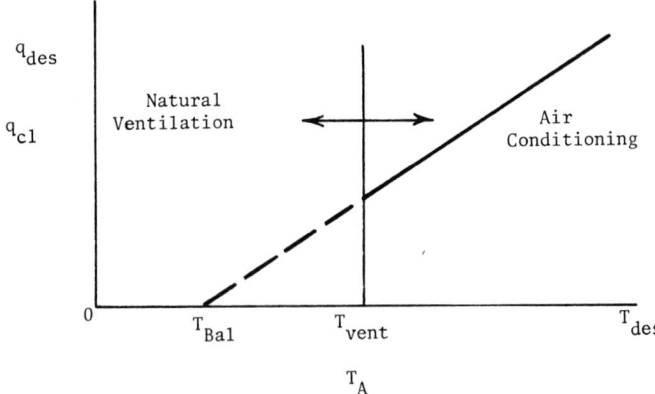

Fig. 4.4.1. Schematic of relation between cooling load and ambient temperature.

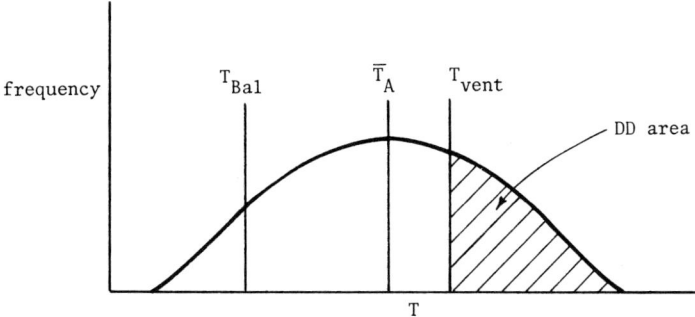

Fig. 4.4.2. Temperature distribution and degree days for air conditioning.

are based on the balance temperature, but only the area above the ventilation temperature is included in the calculation. These degree days are computed as the sum of the degree days based on the ventilation temperature and then shifted to the balance temperature. This is evaluated as

$$DD_m = DD_{T_{vent}} + D_m(T_{vent} - T_{Bal}) \frac{D_{vent}}{D_m} \qquad (4.4.3)$$

The degree days based on the balance temperature are calculated as described in Section 3.5.3. The term H is redefined as

$$H = \frac{\bar{T}_A - T_{vent}}{\sigma} \qquad (4.4.4)$$

and the degree days from Eq. (3.5.25) become

$$DD_{T_{vent}} = (\bar{T}_A - T_{vent} + \sigma F) D_m \qquad (4.4.5)$$

In Eq. (4.4.3), the term (D_{vent}/D_m) is the fraction of the month that the ambient is above the ventilation temperature. It is a function of the parameter H. The ratio (D_{vent}/D_m) is tabulated in Table 4.4.1.

As an example of the degree-day calculation procedure, a situation in July will be considered. The average ambient temperature is 73°F, the ventilation temperature is 78°F, and the balance temperature is 55°F. Following Section 3.5.3, the standard deviation is computed from Eq. (3.5.24) as

$$\sigma = [4.79 - (0.0337)73] \sqrt{31} = 13.0°F$$

H is computed from Eq. (4.4.4) as

$$H = \frac{(73 - 78)\,°F}{13°F} = -0.38$$

TABLE 4.4.1
Distribution of Days for Cooling

$\dfrac{\bar{T}_A - \bar{T}_{\text{vent}}}{\sigma}$	$\dfrac{D_{\text{vent}}}{D_m}$	$\dfrac{\bar{T}_A - \bar{T}_{\text{vent}}}{\sigma}$	$\dfrac{D_{\text{vent}}}{D_m}$
−2.0	0.001	0	0.500
−1.5	0.006	0.1	0.584
−1.0	0.032	0.2	0.664
−0.8	0.062	0.3	0.734
−0.6	0.115	0.4	0.795
−0.4	0.204	0.6	0.885
−0.3	0.266	0.8	0.938
−0.2	0.336	1.0	0.968
−0.1	0.415	1.5	0.994
0	0.500	2.0	0.991

This term is negative, and so F is computed from Eq. (3.5.23) as

$$F = 0.34 e^{4.7(-0.38)} - 0.15 e^{7.8(-0.38)} - (-0.38) = 0.43$$

The degree days based on T_{vent} are evaluated from Eq. (4.4.5) or

$$DD_{T_{\text{vent}}} = [73 - 78 + 0.43(13.0)]\ 31 = 49°\text{F day}$$

From Table 4.4.1, the fraction of the time that the temperature is above the ventilation temperature is

$$\frac{D_{\text{vent}}}{D_m} = 0.22$$

The degree days from Eq. (4.4.3) become

$$DD_m = 49 + 31(78 - 55)(0.22) = 188\ °\text{F day}$$

With degree days calculated in this manner, the sensible load is determined as

$$Q_{m,\text{sens}} = UA_o 24 DD_m \tag{4.4.6}$$

The latent load is assumed to be the same fraction of the monthly sensible load as it was for design conditions. This implies a linear relation between latent loads and ambient temperature similar to that depicted in Fig. 4.4.1. The monthly latent load becomes

$$Q_{m,\text{lat}} = Q_{m,\text{sens}} \left(\frac{q_{\text{des,lat}}}{q_{\text{des,sens}}} \right) \tag{4.4.7}$$

TABLE 4.4.2
Annual Cooling Requirements

UA_o _____ Btu/hr °F T_R _____ °F T_{Bal} _____ °F T_{vent} _____ °F

Month												
$\bar{T}_A$ (°F)												
σ(°F)												
H												
$DD_{T_{vent}}$ (°F day)												
D_{vent}/D_m												
DD_m (°F day)												
$Q_{m,sens}$ (10^6 Btu)												
$Q_{m,lat}$ (10^6 Btu)												
Q_m (10^6 Btu)												

Seasonal cooling load _____ 10^6 Btu
Seasonal COP _____
Seasonal electrical load _____ kWh

96 RESIDENTIAL AND COMMERCIAL AIR CONDITIONING REQUIREMENTS

The seasonal load is the sum of the monthly loads. The total electrical requirement is the seasonal load divided by the seasonal COP of the cooling system. The approximate value for a window unit is 2.0, and that for a central unit is 2.15.

A worksheet has been developed to aid the calculation of cooling loads, and it is given in Table 4.4.2. The procedure is:

1. Enter the house overall conductance-area product, the room temperature, the balance temperature from Eq. (4.4.1), and the ventilation temperature.
2. For each month enter the monthly average ambient temperature and compute the standard deviation from Eq. (3.5.23).
3. Compute the parameter H from Eq. (4.4.4) and degree days based on T_{vent} from Eqs. (3.5.21) or (3.5.22) and (4.4.5).
4. Compute the fraction of the time the temperature is above the ventilation temperature from Table 4.4.1.
5. Compute the total degree days from Eq. (4.4.3) and the sensible cooling load from Eq. (4.4.6).
6. Compute the latent cooling load from Eq. (4.4.7).
7. Add up all months to get the seasonal load, enter the seasonal COP of the air conditioner, and compute the seasonal electrical load.

4.5. EXAMPLE CALCULATION FOR DESIGN AND SEASONAL COOLING LOAD

The design cooling load of the house described in Section 3.6 will be evaluated as an example. It will be assumed that all windows are unshaded, and that the garage walls are treated as house walls. The house will be maintained at a 75°F inside temperature. The various quantities are on the worksheet, and an explanation of how to calculate them follows. The calculations are performed for Madison, Wisconsin using the weather data in Appendix A. July is the hottest month and used for the design calculations.

The components of the design cooling load are given on the filled-out worksheet, Table 4.5.1. The heat flow through windows are the greatest single source of heat gain and comprise about 45 percent of the total. The heat conducted through walls, doors, and ceiling contributes only about 23 percent and infiltration amounts to only 5 percent. Gains due to electrical appliances are large and equal 10 percent. The latent contribution is 15 percent of the total.

Clearly, the first area to consider in reducing the air conditioning load is the windows. Shading techniques such as trees, awnings, roof overhangs, and drapes are all effective. Most houses are not so completely exposed as this example, and these gains could be significantly lower. The second area to consider is that of internal electrical consumption. While this is only 10 percent of the air conditioning load, it is really costly in terms of electricity. This energy is paid for twice; first as it is supplied and then second, as it is removed.

TABLE 4.5.1
Design Cooling Load

Location __Madison__ Month __July__

$\bar{T}_A$ __69.8__ °F, T_{des} __91__ °F, w_{des} __0.014__ lb$_m$/lb$_m$

T_R __75__ °F, w_R __0.008__ lb$_m$/lb$_m$

Infiltration: House volume __12,000__ ft^3

Construction type __Tight__

① $\dot{N}$ __0.33__ changes/hr

Roof-ceiling:

② Ceiling: A_c __750__ ft^2, R_c __25.8__ hr ft^2 °F/Btu, UA_c __29__ Btu/hr °F

③ Roof: A_{rf} __862__ ft^2, R_{rf} __2.1__ hr ft^2 °F/Btu, UA_{rf} __410__ Btu/hr °F

④ Infiltration: V_{inf} __5956__ ft^3, UA_{inf} __107__ Btu/hr °F

Sum __546__ Btu/hr °F

Walls: R_{wl} __12.5__ hr ft^2 °F/Btu

Doors: R_{dr} __7.3__ hr ft^2 °F/Btu

Window: R_{wd} __1.85__ hr ft^2 °F/Btu

Floor: R_{fl} ___ hr ft^2 °F/Btu

Sensible gains

			A (ft^2)	UA (Btu/hr °F)	$\alpha I_t/h_o$ (°F)	q (Btu/hr)
⑤	Walls	S	399	31.9	13.0	829
		E	350	28.0	17.2	930
		N	409	32.7	10.0	850
		W	350	28.0	17.2	930
	Doors	S	21	2.9	13.0	84
		E	0			0
		N	21	2.9	10.0	75
		W	0			0
	Windows		210	113.5		1,816
⑥	Ceiling		750	29.1	21.2	1,083
	Floor		0			0
⑦	Infiltration			71.3		1,140
⑧	Sum			340		7,737
				$(\tau\alpha)$		
⑨	Windows:	S	60	0.71		2,467
		E	50	0.71		2,720
		N	50	0.71		1,576
		W	50	0.71		2,720

RESIDENTIAL AND COMMERCIAL AIR CONDITIONING REQUIREMENTS

TABLE 4.5.1 (Continued)

		A (ft²)	UA (Btu/hr °F)	$\alpha I_t/h_o$ (°F)	q (Btu/hr)
⑩	People				840
⑪	Appliances				1,200
⑫	Basement				-1,428
⑬	Sum of gains				10,095
	Total sensible gain				17,832
	Latent gains				
⑭	Infiltration				1,901
⑮	People				560
⑯	Appliances				550
	Total latent gain				3,011
	Design cooling load				20,843 Btu/hr
⑰	Air conditioner size				1.7 tons

The calculations for the design cooling load are:

① $\dot{N} = 0.21 + 0.0072(91 - 75) = 0.33$ change/hr

② $UA_c = 750/25.8 = 29$

③ $A_{rf} = 1.15(750) = 862$

For wood shingles, plywood, and surface films, R_{rf} from Table 3.2.1 is

$$R_{rf} = (0.25 + 0.94 + 0.31 + 0.61) = 2.1 \text{ hr ft}^2 \text{ °F/Btu}$$

$UA_{rf} = 862/2.1 = 410$

④ $V = 0.29(750)^{3/2} = 5956 \text{ ft}^3$

$UA_{inf} = (1 \text{ change/hr})(0.018)(5956) = 107 \text{ Btu/hr °F}$

⑤ South wall $UA = 399/12.5 = 31.9$ Btu/hr °F

$\alpha I_T/h_o = 0.9(857/14.8)/4 = 13.0°F$

$q = 31.9(91 + 13.0 - 75) = 829$ Btu/hr

⑥ $T_{sa} = 91 + 0.9(1933/14.8)/4 = 91 + 29.4 = 120.4°F$

$T_{at} = (410 \times 120.4 + 107 \times 91 + 29 \times 75)/546 = 112.2°F$

EXAMPLE CALCULATION FOR COOLING LOAD

$\alpha I_t / h_o = 112.2 - 91 = 21.2°F$ (effective value)

$q = 29.1(112.2 - 75) = 1083$ Btu/hr

⑦ $UA = (0.33)(0.018)(12{,}000) = 71.3$ Btu/hr

$q = 71.3(91 - 75) = 1140$ Btu/hr

⑧ $UA_o = 340$ Btu/hr °F

$q_{envelope} = 7737$ Btu/hr

⑨ $S: q = (60)(0.71)(857/14.8) = 2467$ Btu/hr

⑩ $q_{peo} = 4(210) = 840$ Btu/hr

⑪ $q_{app} = 1200$ Btu/hr

⑫ $q_{bt} = 74(55.7 - 75) = -1428$ Btu/hr (see Table 3.6.3)

⑬ Sum = $(2467 + 2720 + 1576 + 2720 + 840 + 1200 - 1428) = 10{,}095$

⑭ $\dot{E}_{inf} = (0.33)(12{,}000)(0.014 - 0.008)(80) = 1901$ Btu/hr

⑮ $\dot{E}_{peo} = 4(140) = 560$ Btu/hr

⑯ $\dot{E}_{app} = 550$ Btu/hr

⑰ Size = $(20{,}843$ Btu/hr$)/(12{,}000$ Btu/hr ton$) = 1.7$ tons

The seasonal air conditioning requirements are summarized in Table 4.5.2. The major loads occur in the months of June through August, and probably the calculations could be limited to these months with little error. For example, in May the air conditioner would be on only 2 percent of the time, or 15 hours for the month. Most people probably would not turn on the air conditioner under these conditions.

The same calculations were carried out for the house in an uninsulated condition. The heat flow through various components are summarized in Table 4.5.3. It can be seen that the effect of insulation is mainly to reduce heat gain through walls and ceiling. For the uninsulated dwellings the ceiling comprises about one-quarter of the total gain, and insulation reduces this gain to an almost negligible proportion. Wall and door heat gains decrease similarly. Double-pane windows reduce the solar gain by 17 percent over the single-pane value due to the second pane. Infiltration, which can amount to one-third of the heating load, adds only a small amount to the cooling load in both insulated and uninsulated houses. Insulation reduces the size of the installed air conditioner significantly. The cost of 1.5 tons of additional capacity would be about $750. The added cost of insulation (Table 3.6.6) is then partially offset by this savings.

TABLE 4.5.2
Annual Cooling Requirements

① ② UA_o __340__ Btu/hr °F T_R __75__ °F T_{Bal} __45.3__ °F T_{vent} __75__ °F

①	③ Month	A	M	J	J	A	S	O
	$\bar{T}_A$ (°F)	44.6	55.4	66.2	69.8	68.0	59.0	50.0
④	σ(°F)	18.3	16.3	14.0	13.6	13.9	15.3	16.8
⑤	H	-1.7	-1.2	-0.63	-0.38	-0.50	-1.04	-1.4
⑥	$DD_{T_{vent}}$ (°F day)	0	0	7	20	12	1	0
⑦	D_{vent}/D_m	0.004	0.02	0.11	0.22	0.16	0.03	0.010
⑧	DD_m (°F day)	4	18	105	215	159	28	9
⑨	$Q_{m,sens}$ (10^6 Btu)	0.03	0.15	0.86	1.75	1.30	0.23	0.07
⑩	$Q_{m,lat}$ (10^6 Btu)	0.00	0.03	0.14	0.29	0.22	0.04	0.01
⑪	Q_m (10^6 Btu)	0.03	0.18	1.00	2.04	1.52	0.27	0.08

Seasonal cooling load __5.12__ 10^6 Btu

Seasonal COP __2.15__

Seasonal electrical load __698__ kWh

EXAMPLE CALCULATION FOR COOLING LOAD

The calculations for the annual cooling requirements are:

(1) $$T_{Bal} = 75 - \frac{10{,}095}{340} = 45.3°F$$

(2) $T_{vent} = 75°F$ assuming wind and window fans are used.

(3) From Table 3.6.3, only May–Sept are expected to have cooling loads.

Calculations for July:

(4) $$\sigma = (4.79 - 0.0337 \times 69.8)\sqrt{31} = 13.6°F$$

(5) $$H = (69.8 - 75)/13.6 = -0.38$$

(6) $$F = 0.43$$

$$DD_{T_{vent}} = (69.8 - 75 + 13.6 \times 0.43)(31) = 20°F \text{ day}$$

(7) From Table 4.4.1 for $F = -0.38$, $D/D_m = 0.22$

(8) $$DD_m = 20 + 31(75 - 45.3)(0.22) = 215°F \text{ day}$$

(9) $$Q_{m,sa} = (340)(215)(24) = 1.75 \times 10^6 \text{ Btu/mo}$$

(10) $$Q_{m,lat} = 1.75 \times 10^6 (3011/17{,}832) = 0.29 \times 10^6 \text{ Btu/mo}$$

(11) $$Q_m = (1.75 + 0.29) \times 10^6 = 2.04 \times 10^6 \text{ Btu/mo}$$

For an already constructed house, the most effective technique for reducing air conditioning costs would be interior drapes, blinds, shades, or exterior awnings. In summer, east and west windows have the greatest gains, and these occur in morning

TABLE 4.5.3
Air Conditioning Requirements for Madison, Wisconsin

	Insulated House	Uninsulated House
Design heating load	20,843 Btu/hr	38,640 Btu/hr
Air conditioner size	1.7 tons	3.2 tons
Component heat gains		
Walls and doors	3,698 Btu/hr	10,706 Btu/hr
Ceiling	1,083	9,090
Windows (conduction)	1,816	3,818
Infiltration	1,140	1,520
Windows (solar)	9,483	11,086
Seasonal energy use	698 kWh	747 kWh
Air conditioning cost	$35	$37

and afternoon, respectively. Drapes cost in the range of $1-2/ft^2, and are definitely cost effective on east and west facing windows. The energy gain for north and south facing windows are one-quarter to one-half that of the east and west ones, but shading devices still have a benefit.

Surprisingly, the seasonal energy use of the uninsulated house is essentially equal to that of the insulated one. The reason lies in the relation between degree days, the balance temperature, and the ventilation temperature. The balance temperature for the uninsulated house is much higher than that of the insulated one (66.1°F compared to 45.3°F). As indicated in Fig. 4.4.2, the degree days are based on the same ventilation temperature and are the same, but those based on the balance temperature are much lower for the uninsulated house. Thus, even though the conductance is three times higher, the total energy in is about the same. This is not a general conclusion, though. For climates where the ambient air temperature is higher than the ventilation temperature over much of the season, insulation would significantly reduce air conditioning energy requirements.

The calculations have been done for Madison, Wisconsin, which has a relatively mild summer. If the same houses were located in Washington, DC, an area with a warmer and more humid summer, the cooling loads would increase. The same insulated house would require a 1.8-ton air conditioner, and the seasonal air conditioning bill would be $190. Without insulation, a 3.8-ton unit costing about $1000 more would be required, and the seasonal cost would be $250. In this case, the installation of insulation would be cost effective based on air conditioning alone.

4.6. SUMMARY

This chapter has presented the detailed methodology needed to estimate heat and moisture gains and air conditioning loads for buildings. Worksheets have been developed for calculating design loads (Table 4.3.1) and seasonal energy use (Table 4.4.2). These methods are based on degree-day calculations using the building balance temperature and incorporating physical mechanisms. They are expected to be reasonably accurate for evaluating loads and energy conservation measures.

SUGGESTED READING

ASHRAE Handbook of Fundamentals, American Society of Heating, Refrigeration, and Air Conditioning Engineers, Atlanta, GA, 1977, 1981, Chapters 23, 25.

T. Kusuda and T. Saitoh, *Simplified Heating and Cooling Energy Analysis Calculations for Residential Applications*, NBSIR 80-1961, National Bureau of Standards, July 1980.

PROBLEMS

4.1. A house under design has 1800 ft^2 of floor area, a total window area equal to 15 percent of the total wall area, and a full unheated basement. It may be either a single- or two-story home, and is located in Madison, Wisconsin. Estimate the annual air conditioning load and the design load for:

(a) Standard frame construction without insulation and with single-pane windows.

(b) The same house insulated appropriately.

4.2. For Problem 4.1, change the location to Chicago, Illinois.

4.3. For Problem 4.1, change the location to Washington, DC.

4.4. For Problem 4.1, change the location to Los Angeles, California.

4.5. For the house of Problems 4.1-4.4, estimate the life cycle costs of air conditioning. The air conditioner has an energy efficiency rating (EER) of 7.2 Btu of cooling per watt-hour of electrical input. Summer electrical rates are 6.0 ¢/kWh.

4.6. Determine the cost effectiveness of adding drapes to the house of Problems 4.1-4.4 during the air conditioning season.

4.7. Reflective films added to a window reduce the solar transmittance to 0.55 and the conductance by 5 percent at a cost of $0.90/ft^2. Determine if this is a cost-effective option.

■ 5 ■

HEATING AND COOLING OF COMMERCIAL BUILDINGS

Commercial buildings differ from residential structures in several ways. The larger size of a multistory building means that all rooms may not naturally be at the same temperature as is usually the case in a residence. The interior rooms do not have walls and windows separating them from the environment, and so there are no heat losses to or gains from the ambient. These interior rooms, termed the central core or zone, usually require cooling at all times of the year. In contrast, the rooms along the outside wall of the building, termed the peripheral zone, require heating in winter and cooling in summer. It is also common for exterior rooms to require heating at the same time interior rooms require cooling.

Further differences between commercial and residential structures arise due to the function of the building. Office buildings, department stores, and restaurants have a greater number of people per unit area during the occupied period than do most houses. The energy generation by people may then be enough to require cooling, while heating is needed during unoccupied times. The heating effect of lights and equipment is also important during occupied periods.

The differences between different zones and different time periods requires heating and cooling systems that are more sophisticated and complex than those for residences. There is usually a centralized system which then distributes heating and cooling to the zones as needed. The size of the heating and cooling loads, and the resulting energy bills, means that it is cost effective to carefully design and operate the system.

This chapter will briefly discuss the space conditioning of multizone buildings. The load profiles characteristic of such structures will be illustrated. A space conditioning system that can meet the loads will be analyzed and ways to improve its performance evaluated. The energy required for illumination of interior spaces is a large fraction of a building's energy budget and will be considered. Finally, the criteria for occupant comfort and their interaction with the energy requirements will be discussed.

5.1. LOAD PROFILES

The calculation of heating and cooling loads for multizone buildings follows the same procedures as for residences. It is usually necessary to include the effects of the thermal capacitance of the structure, and the methods to account for this are explained in the *ASHRAE Handbook of Fundamentals*. It is probably easiest to calculate the loads using computer programs, and several exist that are commonly used for such purposes. The basic ideas, though, are the same as in Chapters 3 and 4.

An example of the annual load profile for a multizone building will be shown. This profile was determined for a typical three-story office building that is 100 ft on a side. The total floor area is 30,000 ft^2, and the central core area is 10,000 ft^2. There are four peripheral zones (N, E, S, and W) of equal floor area. The building is occupied from 8 AM to 5 PM and the occupant density is 100 ft^2 per person. This is representative of a bank, department store, or office building. The illumination level amounts to 2 W/ft^2. This is in the range of modern practice which is 1-3 W/ft^2, depending on function.

The building characteristics are given in Table 5.1.1. The walls and ceiling are not highly insulated, as will be discussed later. Generation from lighting is seen to be a major heat source, especially in the core zone which has no exterior walls or windows.

A typical daily load profile is shown in Fig. 5.1.1. The average ambient temperature is 20°F, which is representative of winter conditions. The positive loads are for air conditioning, and the negative values are for heating.

During nighttime, the peripheral zones and the top floor of the core all require

TABLE 5.1.1
Characteristics of Example Building

Zone	Floor A_{fl} (ft^2)	Walls A_{wl} (ft^2)	U_{wl} (Btu/hr ft^2 °F)	Windows A_{wd} (ft^2)	U_{wd} (Btu/hr ft^2 °F)	Generation g_{peo} (Btu/hr)	g_{lig} (Btu/hr)
Core	10,000	0	–	0	–	25,000	68,000
Peripheral	5,000	2,000	0.1	400	0.6	12,500	34,000

Ceiling: U = 0.1 Btu/hr ft^2 °F
Occupancy: 8 AM to 5 PM
100 ft^2/person
Illumination at 2 W/ft^2

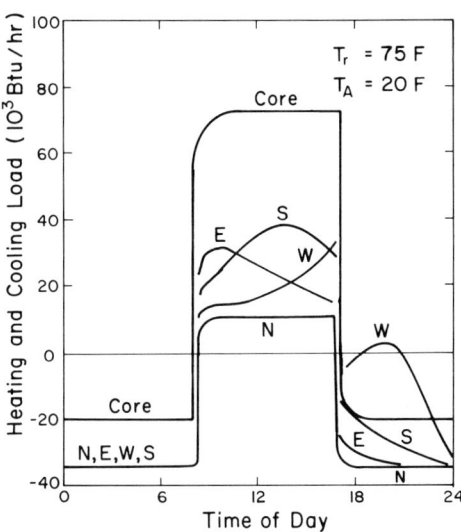

Fig. 5.1.1. Building loads for typical winter day.

heating. The effect of occupying the building is to immediately require air conditioning in all zones. Further, there are solar gains through windows which also produce an air conditioning load. These solar heating effects persist for some time after the sun sets, as evidenced by the loads for the west facing zone. This load profile demonstrates the need for a space conditioning system that can meet loads which vary over the course of the day and change from heating to cooling. The system must be able to respond rapidly to these changing conditions.

The annual load profile for the building is shown in Fig. 5.1.2. The net load is shown which accounts for the use of heat rejected from the condenser of the air conditioner to help offset the heating loads. Most modern systems incorporate this feature. The loads were calculated for Madison, Wisconsin. Even for this northern location, the major requirement is for cooling. There are also months in spring and fall in which both heating (during weekends) and cooling (during workdays) are re-

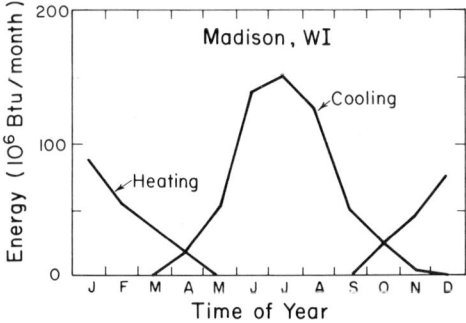

Fig. 5.1.2. Monthly space conditioning loads for a commercial building.

quired. This profile again demonstrates the need for a more complex system than for a residence.

Some general conclusions about conservation techniques can also be drawn from this example. First, cooling loads created by illumination can be large and measures that reduce this use can have a significant effect. Current levels are in the range of 1-3 W/ft^2. The lower value is for lighting systems that use zone or spot lighting in which only the local working area is highly illuminated. The higher value is for buildings that are uniformly illuminated.

Second, solar gains are significant heat sources, especially for the east and west facing zones in summer. Techniques such as coatings and overhangs which reduce these loads are cost effective. The savings from reducing air conditioning costs are much greater than those from using solar energy to offset heating. In commercial buildings, solar gains occur during the occupied period and are less effective than they are for residences.

Finally, wall and window insulation may actually increase energy costs. As indicated in Fig. 5.1.1, air conditioning of peripheral zones is required during cold weather. A reduction in wall insulation, for example, would have reduced cooling costs by allowing a greater heat flow out. There would be, however, increased heating costs at night. The optimum insulation level would be determined considering both heating and cooling costs, and is generally lower than for residences.

5.2. SPACE CONDITIONING SYSTEMS

5.2.1. System Description

The systems used to provide space conditioning must meet a variety of criteria. Both heating and cooling must be supplied, often simultaneously, but to different zones. The nature of the load in a given zone may change over the day, and is quite different during occupied and unoccupied periods. The system usually has a centrally located chiller and boiler for high efficiency. Air is commonly used as the circulating fluid and hot or cold air can then be distributed to all zones. It is also common to have a hot water system for heating only.

There are also requirements to introduce fresh air for the occupants. Usually these are set to control the buildup of odor and smoke, and recommended flow rates are in the range of 2-10 cfm per person. Some areas such as kitchens and restrooms require a continuous exhaust flow. The number of fresh air changes for the entire building, though, is less than that for residential buildings. This, plus the presence of moisture generating equipment, means that the circulating air must be dehumidified to remove moisture. The system must have a provision to introduce a controlled amount of outside air, to dehumidify the circulating air, and to further cool or to heat air as necessary to provide comfort in each room or zone.

There are many system types in use today that will meet these criteria. In this chapter, a generic system will be considered. The performance in the simplest mode of operation will be evaluated. Then, modifications that reduce energy consump-

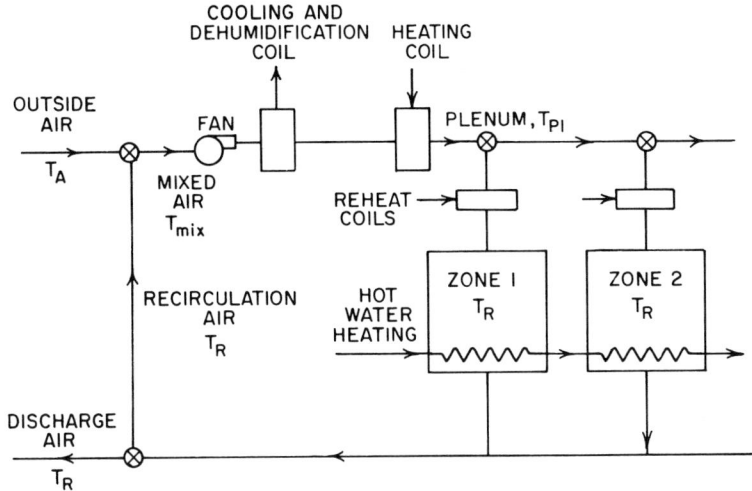

Fig. 5.2.1. Schematic of space conditioning system.

tion will be considered. The goal will be to illustrate basic ideas and provide a foundation for analyzing systems in detail.

The typical generic space conditioning system is shown in Fig. 5.2.1. Outside fresh air and recirculated room air are mixed prior to the delivery fan. The air passes through the cooling coil where it is also dehumidified. It then passes through a heating coil where heat is added as needed. The air in the plenum, which will be distributed to all zones, is usually in the range of 50-55°F.

The air flow to each zone may first pass through reheat coils. This would supply heat to the air if the zone required heating. Heat would also be supplied by a hot water heating system which has the same effect as the reheat coils. The air leaves the rooms at room temperature and mixes with that from other zones. A fraction (10-20 percent) is discharged to the atmosphere and the remainder recirculated.

An example calculation for the energy and economic operating costs of this system will be carried out. A single operating condition will be considered to illustrate the principles involved. The loads will be taken as representing occupancy on an average winter day at 20°F. For this condition, cooling is required in each zone. For simplicity, the four peripheral zones will be treated as a single zone. The fresh air requirement will be the recommended lower value of 5 cfm per person. The occupant density is uniformly distributed over each zone at 100 ft^2 per person. The system characteristics are given in Table 5.2.1.

5.2.2. Constant Air Volume System

The first system to be considered is termed the constant air volume system. This means that the air volume flow rate (cfm) will be constant over the day and the same, in terms of room air changes per hour, in each zone. The circulating air will

TABLE 5.2.1
Space Conditioning System Characteristics

Zone	Fresh Air (cfm)	Cooling (Btu/hr) Load
Core	500	92,000
Peripheral	1,000	98,000
Total	1,500	190,000

$$T_R = 75°F$$
$$T_{pl} = 55°F$$
$$T_A = 20°F$$

flow at a rate corresponding to three air changes per hour. This is 5000 and 10,000 cfm for the core and peripheral zones, respectively.

There are three energy flows to be evaluated. These are the central cooling requirement and the two reheat heating requirements. The cooling requirement is that required to cool air from the mixed temperature to the plenum temperature. The mixed temperature is determined from an energy balance on the mixing valve.

$$(\dot{m}h)_{rec} + (\dot{m}h)_A - (\dot{m}h)_{mix} = 0 \qquad (5.2.1)$$

For the conditions of 1500 cfm of fresh air and a recirculation flow of 13,500 cfm, the mixed temperature becomes 69.5°F.

The energy required to cool the air from 69.5 to 55°F is 235,000 Btu/hr. In a building, there would also be dehumidification, and this would increase the cooling load by approximately 30 percent to 300,000 Btu/hr.

The amount of energy required to reheat the air is determined by an energy balance on each zone.

$$\dot{m}_R h_{pl} + q_{rh} + q_R - (\dot{m}h)_R = 0 \qquad (5.2.2)$$

or, using the equations of state for an ideal gas

$$q_{rh} = \dot{m}_R c_p (T_R - T_{pl}) - q_R \qquad (5.2.3)$$

The flow rate through both zones is more than sufficient to meet the room load. The reheat requirements are 16,000 and 118,000 Btu/hr for the core and periphery, respectively.

The costs of providing the cooling and heating energy flows can be determined. For a vapor compression chiller operating at a COP of 2.5 with electrical costs of 4¢/kWh, the cost to provide cooling is $5/10^6$ Btu. The heating will be assumed to be supplied from an oil-fired boiler with a combustion efficiency of 0.75 with a delivered heating cost of $9/10^6$ Btu. The cost to provide cooling is $1.50/hr. The cost to provide reheat for the core zone is $0.14/hr, and that for the peripheral

SPACE CONDITIONING SYSTEMS

TABLE 5.2.2
Summary of System Operating Costs

System	Cooling Cost	Heating Cost	Total Cost
Constant air flow	$1.50/hr	$1.20/hr	$2.70/hr
Variable air flow	$0.65/hr	0	$0.65/hr
VAV with economizer	0	0	0

Cooling costs = $5.00/10^6$ Btu
Heating costs = $9.00/10^6$ Btu

zones is $1.06/hr. The total space conditioning costs are $2.70/hr. These values are summarized in Table 5.2.2.

It is obviously wasteful to first cool and then heat the air. The cooling is needed to remove water vapor and cannot be eliminated. The heating flow, however, is required only because the combination of air flow rate and plenum temperature provides a cooling potential greater than that needed in the zone. The plenum temperature must be low enough to remove moisture, and is restricted. The air flow rate, however, may be changed, and this would reduce the cooling potential. This idea will be explored in the next section.

5.2.3. Variable Air Volume System

In the variable air volume system, the air flow rate is set to exactly match the load. This means that the energy required for reheating is zero. The required flow rate is determined from Eq. (5.2.3) with the reheat energy set equal to zero. The room flow rate is then given by

$$\dot{m}_R = \frac{q_R}{c_p(T_R - T_{pl})} \tag{5.2.4}$$

The flow rates become 4260 and 4530 cfm for the core and peripheral zones, respectively. The reheat energy is now zero and the cooling energy is reduced since the flows are less. There is a further benefit to this approach. The total circulating flow rate is reduced to about 60 percent of that of the constant flow system. This reduces the fan power required to about 25 percent since the pressure drops are all lowered.

The cooling requirements are again those to cool the air after mixing. If the same fresh air requirements are kept, the recirculation flow rate drops and the mixed temperature lowers to 65.6°F. This reduces the load on the cooling coil to about 130,000 Btu/hr, or about 40 percent of the previous level. The operating costs are only those to provide cooling, and amount to $0.65/hr. The cost is also shown in Table 5.2.2.

The variable air flow rate system is so much more efficient in energy use than the constant flow rate system that it may seem surprising that constant flow systems were ever installed. The constant volume system was developed during the period in which energy costs were cheap. It simply and effectively controls the zone temperatures to meet changing loads. It is less costly to install and requires simpler equipment. However, there are probably no new installations of this system. This is a good example of the evolution of a system in response to increasing energy prices.

5.2.4. Economizer Circle

For the system in operation as shown in Fig. 5.2.1, outside air at 20°F is mixed with recirculation air at 75°F, and then the mixture is cooled to 55°F. Since the outdoor temperature is lower than the desired plenum temperature, it is possible to alter the ratio of recirculation air to outside air and obtain a mixed temperature of 55°F. This would mean that no cooling of the air supplying the zones was needed. A space conditioning system that has the controls to vary the air flows in this manner utilizes what is termed the "economizer cycle."

For the variable air volume (VAV) system, the amount of outside air that must be introduced is determined from the energy balance on the mixing valve, Eq. (5.2.1). The mass flow rate of the mixed flow is 8790 cfm to meet the zone loads, and this must equal the sum of the outside air and recirculation flows. Equation (5.2.1) can then be reformulated as

$$\dot{m}_A = \dot{m}_{mix} \frac{T_R - T_{mix}}{T_R - T_A} \qquad (5.2.5)$$

With values of T_R of 75°F, T_{mix} of 55°F, and T_A of 20°F, the outside air flow rate becomes 3200 cfm. This is about twice the minimum fresh air requirement, but the cooling costs are now reduced to zero, as shown in Table 5.2.2.

The principle of the economizer cycle can be extended to allow outside air for conditioning whenever the temperature is below room temperature. There are, however, some limits to this approach. With the ambient temperature close to room temperature, a large flow of air is required. The velocities may then be too large for comfort in the zone. Under some ambient conditions, the outside humidity may be too high to maintain room humidity in the comfort region even through the temperature is low. It is common to use an enthalpy controller to detect when the outside air enthalpy is low enough to meet both the latent and sensible loads, and to avoid this problem. Overall, this feature is a logical approach to providing cooling.

5.3. HEAT RECOVERY SYSTEMS

5.3.1. Waste-Heat Heat Exchanger

There are situations in which a heat exchanger could be used to transfer heat between the discharge and incoming air flows. For example, hospitals generally require 100 percent outside air for conditioning to avoid contamination. Kitchens, bathrooms, and many processes require large exhaust flows. In these examples, room air is discharged and ambient air, either hot or cold, is brought in for processing. A heat exchanger would either preheat or precool the incoming air and reduce the loads on the air conditioning system.

Waste-heat exchangers have become quite popular, and have been extended to include enthalpy exchange (total heat) as well as sensible energy. The methods for analyzing and optimizing these exchangers are covered in Chapter 10.

There are also many situations in which such an exchanger would seem feasible, but, in reality, may not be. The system of Fig. 5.2.1 discharges warm room air. However, recapturing this energy may eliminate the potential for cooling with the economizer cycle. However, if the ambient was warmer than room temperature, it would be advantageous to cool the incoming fresh air. The exchanger is also useful in heating very cold air to minimize condensation in the mixing chamber. The main applications, though, are those in which large amounts of conditioned air must be discharged.

5.3.2. Storage Systems

The load profile of Fig. 5.1.1 shows that both heating and cooling may be required during a day. If the heating and cooling energies at different times could be used to partially offset each other, the total costs could be reduced. This could be accomplished by using a storage tank. During operation of the air conditioner, the energy removed could be stored in the form of hot water. This would then be available for heating at night.

The use of this idea is restricted to conditions in which both heating and cooling are required daily. During summer, heating is not required, while during winter the economizer cycle can be used to meet the cooling loads. It is only during spring and fall seasons that storage is feasible.

A further constraint on storage is that storage systems are expensive, and they save heating energy which is relatively low in cost. Thus the first costs must be considered in detail. An example of a combined heat recovery–space conditioning system is given in Section 7.8.

5.4. ANNUAL ENERGY REQUIREMENTS

The annual heating and cooling requirements for the example building have also been estimated. The three systems of Section 5.2 were incorporated. These are systems with constant air volume flow rate, variable air volume flow rate, and variable air volume flow rate with economizer cycle.

The results are given in Table 5.4.1. Two heating energy quantities are calculated. The first is the gross heating requirement, which is the actual heating that must be supplied to the building to maintain room temperature at the thermostat setting. The net heating accounts for the fact that some of the heating energy may be offset by heat rejection from the cooling system. This heat rejection equals the

TABLE 5.4.1
Annual Energy Requirements[a]

	Constant Air Volume			Variable Air Volume			Variable Air Volume with Economizer		
Month	Cool	Htg	Net	Cool	Htg	Net	Cool	Htg	Net
Jan.	58	113	31	34	75	28	0	75	75
Feb.	53	100	25	31	65	22	0	65	65
Mar.	59	96	14	34	59	11	0	59	59
Apr.	64	78	0	38	41	0	0	41	41
May	81	66	0	47	29	0	20	29	0
June	85	36	0	50	0	0	50	0	0
Jul.	94	37	0	55	0	0	55	0	0
Aug.	85	37	0	50	0	0	50	0	0
Sept.	85	60	0	50	24	0	25	24	0
Oct.	80	73	0	47	36	0	0	36	36
Nov.	63	80	0	37	44	0	0	44	44
Dec.	65	108	17	38	71	17	0	71	71
Annual	872	884	87	511	444	78	200	444	391

[a]Energy quantities are in 10^6 Btu.

cooling supplied plus compressor work. The amount of heat recovery was calculated based on a cooling system COP of 2.5 and subtracted from the heating load during months when both heating and cooling were required.

In practice, all of the condenser heating may not be available to offset heating loads. The cooling energy is available during occupied periods and daytime, while heating is required at night and on weekends. Thus, a storage system must be provided. There are, inevitably, thermal losses from storage, and the cooling system performance may suffer as the storage tank temperature rises. The net heating term probably represents the minimum that may be required.

There are several interesting results indicated by Table 5.4.1. First, the annual cooling requirements of the constant volume system are reduced to 60 percent using a variable air volume system, and to 23 percent using the economizer cycle. The gross heating requirements are also reduced to about 50 percent by elimination of the reheat energy. Clearly, a constant air flow system is not as effective in energy use as the others. Further, the economizer saves significantly on cooling energy, although its advantage is less in more southerly locations where summers are longer and temperatures and humidities are higher.

The net heating energy column indicates some interesting trade-offs. If the heat rejection can be used to offset heating loads, heating energy can be reduced substantially for each system. The net for the constant air volume system is almost equal to that of the variable air volume systems, and about one-fourth that of the economizer system. The economizer cycle saves cooling energy but at the expense of increased heating costs. Depending on the price of the two energy forms and how much air conditioning energy can be reclaimed, an economizer cycle may or may not save money.

TABLE 5.4.2
Annual Operating Costs

	Constant Air Volume	Variable Air Volume	Variable Air Volume with Economizer
Without heat recovery:			
Heating	7,956	3,996	3,996
Cooling	4,360	2,555	1,000
Total	$12,316	$6,551	$4,996
With heat recovery:			
Net heating	783	702	3,519
Cooling	4,360	2,555	1,000
Total	$5,143	$3,257	$4,519

Cooling $5/10^6$ Btu
Heating $9/10^6$ Btu

The annual operating costs for each system are given in Table 5.4.2. Without heat recovery, the constant air volume system costs twice as much as the other two. The economizer cycle saves $1500/yr, which would definitely pay for the added controls. It is a cost-effective item in this situation.

If the heat recovery system which provides the net amount of heating is installed, the economizer cycle is no longer economical. It is more cost effective to use the cooling system to provide both heating and cooling. This conclusion is dependent on fuel costs. If, for example, coal were available for heating, the economizer cycle would be the most economical one.

This section has attempted to point out some of the considerations in providing space conditioning in multizone buildings. The situation is complicated when both heating and cooling are required. There are trade-offs between the different options that depend on fuel price and the local climate. Design point analyses are not sufficient for making a sound economic judgment, and annual performance must be evaluated.

5.5. ILLUMINATION

The energy and monetary costs of illuminating buildings can be very high. As shown in the previous examples, the thermal energy generated by electricity is a major portion of the heat gain of a commercial building. This energy is also paid for twice, first for the direct electrical use and second for the air conditioning energy. Thus measures that reduce lighting energy and allow the same functions to be carried out are valuable.

Overall, illumation energy requirements are a significant fraction of the total U.S. energy use. Illumination accounts for 2 percent of the total consumption and

15 percent of the electrical consumption. More efficient lighting techniques can alleviate the national energy problem.

In this section the basic ideas underlying illumination energy requirements will be discussed. Various light sources and the fixtures required to produce illumination on a surface will be considered. The monetary and energy costs for a typical commercial building will be illustrated. The emphasis will be on the fundamental relations; the *Illumination Engineering Society Handbook* has a wealth of information on the subject.

5.5.1. Basic Concepts in Illumination

The measure of illumination is the footcandle. This is the illumination on a surface one square foot in area on which there is a uniformly distributed flux of one lumen. One lumen equals the luminous flux through a unit solid angle from a point source of one candela. One candela is the luminous intensity from a blackbody radiator with a projected area of 1.67×10^{-6} m^2 and at the temperature of molten platinum (2042 K).

These basic definitions refer to illumination energy as opposed to thermal energy. They can be converted to provide an indication of the energy requirements to produce light.

The efficacy of a light source is given by the total illumination, in lumens, divided by the electrical power input. The theoretical maximum conversion of electricity to light is 683 l/W. Actual lamps provide 1 to 20 percent of the theoretical illumination. Illumination produced by heating methods such as incandescent bulbs and gas lamps is not too efficient on this basis. Relatively little of the thermal energy emitted is in the visible range. Very high temperatures such as found in arc discharge or fluorescence methods are needed for high efficacy.

The light source produces illumination measured in lumens. The illumination measured at the work surface is measured in footcandles (l/ft^2). The total illumination that must be produced by a lamp to illuminate a given area is determined by

$$\text{lamp lumens} = \frac{(\text{footcandles})(\text{area})}{\text{coefficient of utilization}} \quad (5.5.1)$$

The coefficient of utilization accounts for the conversion of light at the lamp surface to that at the work surface. It is a function of the complex interactions between the lamp, the lighting fixture (luminaire), the room, and the work surface. For typical office rooms where the walls and ceilings are reflective (painted in light colors), the value of the coefficient of utilization is about 0.6 to 0.7 for both incandescent and fluorescent lamp fixtures. For dark colored walls and ceilings, the coefficient is in the range of 0.4 to 0.5. There is, in addition, a degradation in utilization due to dirt on the room surface and luminaire and deterioration of the lamp which further reduces output. A value of 0.5 is probably a reasonable value to use for estimating illumination energy use.

TABLE 5.5.1
Recommended Illumination Levels

Area	IES (fc)	GSA (fc)
Office—Drafting areas	200	—
General offices	70-150	50
Conference Rooms	30	30
Hallways	20	10

The energy required to illuminate a given area per unit of floor area is given in terms of the lamp efficacy as

$$\frac{\text{electrical energy}}{\text{square feet}} = \frac{\text{footcandles}}{(\text{lumens/watt})(\text{coefficient of utilization})} \quad (5.5.2)$$

The recommended level of illumination depends on the tasks to be performed. The Illumination Engineering Society (IES) has prepared recommendations based on "worst case" situations. These levels have become maximum values for design. The Government Services Administration (GSA) and ASHRAE both recommend lower values in order to conserve electricity and reduce cooling loads. A brief summary of levels is given in Table 5.5.1. There is a large spread in recommended level depending on the tasks to be performed.

5.5.2. Lighting Sources

The electricity requirement depends strongly on the lamp used to provide illumination. Incandescent bulbs produce light by heating the tungsten filament to a temperature at which significant thermal emission occurs in the visible range. The filament temperature is about 3000 K, with about 10 percent of the energy in the visible range. These lamps are commonly used in homes and are the least efficacious in terms of lumens per watt. The lamp life is also an important factor in considering life cycle costs, and the rated life of most bulbs is 750-1000 hours. Longer life bulbs are available, but the greater life is achieved by operating the filament at a lower temperature. This decreases the illumination level per unit of electrical energy input.

Fluorescent lamps produce light by activating the fluorescent powder coating on the inside of the tube. This is accomplished by an electrical discharge through the mercury vapor that fills the tube. Since light is generated by nonthermal means, the temperature is lower and the efficacy is higher than that for incandescent bulbs. The output per unit length of the bulk is limited, and so bulbs must be long to provide sufficient illumination. Bulb life is reduced by turning the bulb on and off, but with normal usage the life is on the order of 7500 hours. The power requirement for the ballast to turn the lamp on adds about 20 percent to the total power.

TABLE 5.5.2
Performance and Cost Characteristics of Lamps

	Efficacy (lm/W)	Life (hr)
Incandescent		
25 W	9	1,000
100 W	13	1,000
500 W	20	1,000
Fluorescent (40 W)		
Cool white	70–80	7,500
Daylight	60–70	7,500
Mercury vapor	40–80	24,000
Metal halide	70	15,000
Sodium vapor		
High pressure	100	20,000
Low pressure	170	18,000

Mercury vapor and sodium vapor lamps are often used in outdoor situations. These are highly efficacious lamps, and have a relatively long life. The color produced is not suitable, though, for most indoor uses.

A brief summary of the performance is in Table 5.5.2. It is seen that for a given illumination level, fluorescent lamps consume about 25 percent as much electricity and last 7.5 times as long as incandescent lamps. It is for this reason that commercial buildings use fluorescent lamps almost entirely.

There are reasons why fluorescent lamps are not more accepted in homes. Technically, it has been difficult to construct fluorescent bulbs to be compatible with home lamp fixtures, although recently compact units have become available. Psychologically, fluorescent lamps are associated with work, and homes are retreats from work. The current energy situation may alter this view.

5.5.3. Example of Lighting Costs

The illumination of the building of Section 5.1 will be considered as an example of the energy and economic cost of lighting. The entire 30,000 ft^2 will be illuminated with fluorescent lights at the IES recommended level of 100 fc. Luminaires with a coefficient of utilization of 0.5 will be employed. The energy requirement is computed from Eq. (5.5.2) as

$$\frac{\text{electrical energy}}{\text{ft}^2} = \frac{(100 \text{ fc})}{(60 \text{ lm/W})(0.5)} = 3.3 \text{ W/ft}^2$$

This corresponds to the heat generation by lights given in Table 5.1.1. The total cost of lighting for the occupied period (2080 hr/yr) at \$0.04/kWh is \$8320.

If the lighting levels were reduced from the IES recommendation to that from the GSA (50 fc), the heat generation would be 1.6 W/ft^2. This would significantly reduce the air conditioning load as well as the lighting electrical bill.

Incandescent lamps could also have been used. The electricity required to provide 50 fc would be 7.7 W/ft^2, and the direct electrical bill would be $19,200. This large heat generation would significantly increase the air conditioning requirements as well. It is uneconomic to consider incandescent lighting in a commercial building.

There are methods available to further reduce lighting energy. Task or spot lighting in which each area is illuminated to the level needed for the task is better than uniform illumination throughout. Light switches for each small area rather than a single switch per floor allows lights to be on only when needed. Time clocks can be used to turn lights on and off at the start and end of the working day, while occupancy detectors turn lights on and off during the day. Controls to reduce illumination levels at night during cleanup are effective. These measures have reduced illumination levels from about 3 W/ft^2 to less than 2 W/ft^2, but further energy conservation is still possible.

5.6. COMFORT

The goal of the air conditioning system in a building is to provide a comfortable environment for the occupants. The definition of comfort is "that condition of mind which expresses satisfaction with the thermal environment." This definition shows that comfort is not really a physical state, but a psychological one. This is emphasized by the statement about thermal comfort that "a person who thinks he is uncomfortable is just as uncomfortable as if he really were uncomfortable."

Since it is the perception of comfort that is important, it is difficult for the engineer to design a thermal environment that is comfortable for all occupants. Comfort criteria are often determined by occupant votes, and even at the most comfortable condition, there are a significant number of people who are dissatisfied. The space conditioning system must be designed to satisfy most of the people most of the time.

In recent years the need to provide comfort has been tempered with the desire to conserve energy. The temperature criterion for thermal comfort has risen from the 65 to 70°F range in 1900 to the 75 to 78°F range in 1960. These increased levels reflect lighter weight clothing and changing living patterns, diets, and comfort expectations over the years. The recent government mandates for thermostat settings of 68°F in winter and 78°F in summer are in conflict with these "natural" criteria, and compromises are required.

The modified goal of the air conditioning system has become to achieve comfort and low energy use at the same time. In this section, the physiological criteria and physical constraints on comfort will be discussed. The requirements for achieving comfort in different environments will be determined. This section will be limited to the basic ideas; more material is available in the ASHRAE guides and *Thermal Comfort* by P. O. Fanger (Danish Technical Press, Copenhagen, Denmark, 1970).

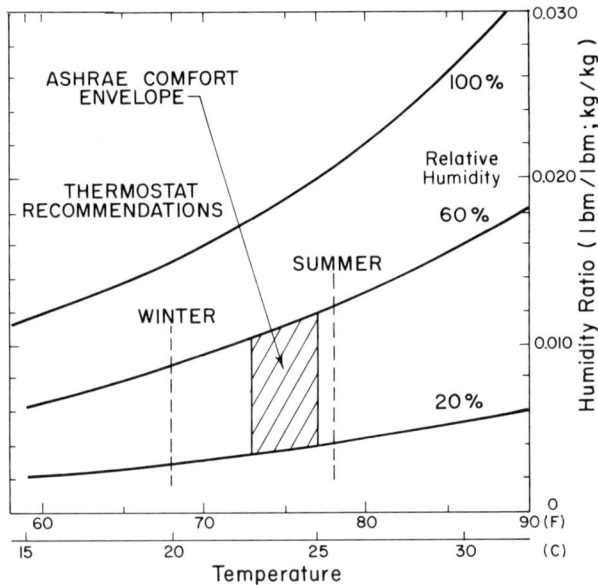

Fig. 5.6.1. ASHRAE comfort standard.

5.6.1. ASHRAE Comfort Standards

The comfort standards developed and recommended by ASHRAE have evolved over the years. They are based on extensive research on the thermal comfort of a wide range of subjects. The research has been to determine the temperature and humidity ranges that most people find comfortable. The subjects have been sedentary, and lightly clothed, and there has been a low air velocity in the test room.

The comfort conditions are shown on a psychrometric chart in Fig. 5.6.1. The temperature range is 73-77°F (22.8-25°C) and 20-60 percent relative humidity. Also shown are the thermostat recommendations for winter and summer. These are outside the comfort envelope. If comfort is to be achieved, the conditions of the occupants relative to those for the determination of comfort conditions must be changed.

The comfort criteria were determined for sedentary subjects. This corresponds to metabolic rates of 350 to 520 Btu/hr (100 to 120 W), which are comparable to the metabolic levels for office workers. The air velocities were low, 40-60 ft/min, which is in the range of air velocities used for space conditioning systems. The subjects were "lightly clothed." The level of clothing is the one variable that can be altered to achieve comfort in environments outside the comfort zone.

5.6.2. Physiological Relations

The thermal receptors in the human body are located in the skin and in the brain. The skin receptors are sensitive to skin temperature, and thus the thermal environment surrounding a person manifests itself through the skin temperature. The receptors located in the brain sense the core temperature. Core temperature changes occur due to changes in activity level such as increased metabolism which raises the body's temperature. The human thermal control system produces responses such as sweating or shivering in proportion to the excursions of these temperatures from their normal values.

Under normal comfort conditions for sedentary people, the main physiological control is vasomotor activity. The blood flow to the skin is either increased if the person feels warm or reduced if the person feels cool. This allows modulation of body temperatures and maintenance of comfort. If a person becomes cold, shivering initiates, which increases the metabolic heat production through muscle activity. If a person is hot, a signal is sent to the sweat glands to secrete sweat, and evaporative cooling on the skin occurs.

The physiological response to an increase in activity level is as follows. The metabolic rate increases which raises the core temperature. This rise in temperature is the signal that thermal energy must be transferred to the environment, and sweating initiates. The evaporation of sweat transfers heat. The skin temperature is lowered by evaporative cooling, and this maintains the feeling of comfort.

These physiological responses lead to the fact that under comfort conditions, skin temperature can be correlated with metabolic rate. This relation was determined by Fanger, and is shown in Fig. 5.6.2. This plot shows that sedentary people are comfortable if their skin temperature is in the range of 92-93°F (33-34°C). At higher activity levels, people are comfortable at lower skin temperatures.

A brief table of metabolic rates for different activity levels is given in Table 5.6.1. This shows that for normal office building operations, some occupants may work at metabolic levels of up to 1000-1600 Btu/hr. These people will be comfortable at skin temperatures in the range of 80-86°F, which is considerably lower than that for people with a lower activity level. The occupants can only alter their dress (clothing levels) to maintain comfort in this situation.

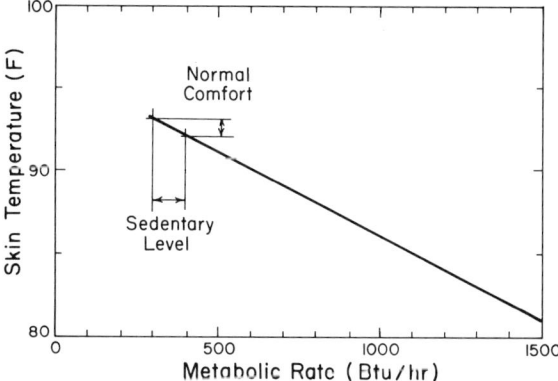

Fig. 5.6.2. Relation between skin temperature and metabolic rate for comfort.

TABLE 5.6.1
Metabolic Rates for Different
Activity Level

Activity Level	Metabolic Rate (Btu/hr)
Seated	350
Office work	420–510
Light work	780
Medium work	1040
Heavy work	1600

5.6.3. Heat Transfer Through Clothing

Clothing serves as a resistance to heat and water vapor transfer from the skin to the environment. The metabolic energy generated within the body is transferred, in part, through the clothing. There is then an interaction between the clothing and activity levels which allows comfort in a given environment. As would be expected, clothing levels should decrease as metabolic rates increase.

The thermal resistance of clothing is often expressed in units of "clo" where 1 clo equals 0.88 hr ft^2 °F/Btu. A brief table of clothing levels is given in Table 5.6.2. These show that thermal resistances vary by a factor of 3 or 4 for typical office dress. It is also seen that women are generally less warmly dressed than men. These considerations complicate the maintenance of comfort conditions in buildings.

The combined effect of activity and clothing level on the room temperature

TABLE 5.6.2
Thermal Resistance of Clothing

Dress	R Value	
	(clo)	(hr ft^2 °F/Btu)
Women		
Cool dress	0.20	0.18
Warm dress	0.50	0.44
Pantsuit	0.60	0.53
Dress, overcoat	1.10	0.97
Men		
Long pants, short sleeve shirt	0.45	0.40
Business suit	1.0	0.88
Suit, overcoat	1.5	1.30

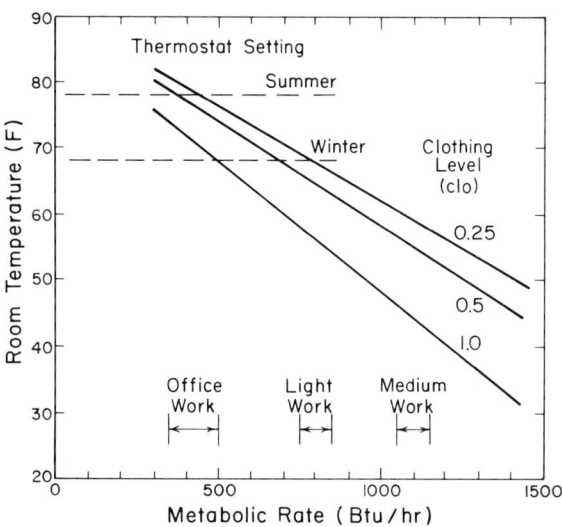

Fig. 5.6.3. Room temperature for comfort as a function of activity and clothing level.

necessary to maintain comfort is shown in Fig. 5.6.3. Also shown are the recommended thermostat settings for summer and winter. These relations are for low air velocities (less than 40 fpm) and 60 percent relative humidity in the room.

There are several interesting complications that can be shown from this figure. First, in an office with people dressed at different clothing levels, some people will probably be uncomfortable. There is not one thermostat setting that will satisfy all occupants. The stereotypical example of the well-dressed business executive and the lightly dressed secretary indicates that he will be too hot in summer while she will be too cold in winter. The actual thermostat setting may depend on office politics. Further, in a office with people doing different tasks, the dress must also be varied to achieve comfort. For example, people doing light manual labor would be comfortable at lower thermostat settings than typists. It is also an axiom that it is the uncomfortable person who complains about the room temperature.

The recommended thermostat settings undoubtedly save building heating and cooling energy. They are acceptable, though, only if people dress appropriately. During summer, office workers must dress lightly (0.25-0.5 clo), while in winter, heavier levels (1.0 clo) are in order. The occupants are an essential ingredient in the energy program for a building.

5.7. SUMMARY

Various aspects of the heating and cooling of multizone buildings have been considered. The loads differ from those in residences in that both heating and cooling may be required simultaneously and cooling is usually the bigger load. A significant proportion of the cooling load is due to electricity dissipation due to lighting. Techniques such as high-efficiency fluorescent lamps, task lighting, and the lowest feasible levels are all useful in reducing gains. Window treatments such as films and coatings play an important role in lowering solar gains.

The variable air volume system provides space heating and cooling at minimal energy costs. Refinements such as the economizer cycle and heat recovery heat exchangers enhance its operation in some situations. Storage units are helpful in using cooling system heat rejection to offset heating loads at other periods. The implementation of these ideas depends on the economics.

The ultimate goal of the space conditioning system is to provide a comfortable environment for the occupants at minimal energy cost. Low energy use requires relatively high thermostat settings in summer and low settings in winter. Building occupants must dress appropriately to match their activity with the interior environment. The current thermostat settings are near the practical limits of dress and comfort.

SUGGESTED READING

ASHRAE Handbook of Fundamentals, American Society of Heating, Refrigeration, and Air Conditioning Engineers, Atlanta, GA, 1977, 1981, Chapter 24, 25.

J. Ottenstein, Application of Solar Energy to Multizone Buildings, MS Thesis, Mechanical Engineering, University of Wisconsin–Madison, 1979.

P. O. Fanger, *Thermal Comfort*, Danish Technical Press, Copenhagen, Denmark, 1970.

F. C. McQuinston and J. D. Parker, *Heating, Ventilating, and Air Conditioning*, John Wiley & Sons, New York, 1982.

Illumination Engineering Society Lighting Handbook, Illumination Engineering Society, New York, 1981.

PROBLEMS

5.1. A commercial building is constructed with the characteristics given in Table 5.1.1. Assume that monthly average weather conditions are sufficient for estimating annual loads, and determine the annual heating and cooling requirements for Chicago, Illinois. Evaluate the following systems from an energy and economic viewpoint.

   (a) Constant air volume flow rate.
   (b) Variable air volume flow rate.
   (c) Variable air volume flow rate with economizer cycle.

5.2. For Problem 5.1, change the location to Washington, DC.

5.3. For Problem 5.1, change the location to Los Angeles, California.

5.4. For Problems 5.1-5.3, replace the uniform illumination system with task lighting. This reduces the lighting energy use by 40 percent at an increased first cost of $18,000.

5.5. Evaluate the cost effectiveness of heavier clothing to allow a reduced thermostat setting.

■ 6 ■

COMBUSTION FURNACES

The previous three chapters have been concerned with the determination of the heat losses and gains from buildings. The next two chapters will be concerned with the devices commonly used to meet these heating and cooling loads. Combustion furnaces will be studied in this chapter, and heat pumps and air conditioners will be the subject of Chapter 7.

The objective of these chapters is to first analyze the system to determine the present level of performance. Based on this analysis, the technical options for improvement will be considered, and the resulting improvements in performance evaluated. Lastly, the economics associated with the options will be briefly considered.

Virtually all homes and commercial buildings in the United States have some type of heating system. The vast majority provide heat through the combustion of fuel. The mix of fuels used for residential heating is given in Table 6.1. Natural gas and oil are the fuels predominantly used and account for about 80 percent of the total. Commercial uses have been in about the same proportion, although the recent price and availability uncertainty has created some shifts. Some larger operations

TABLE 6.1
Fuel Mix for Residential Heating (1974)

Fuel	Percent Use
Natural gas	55.7
Fuel oil	23.8
Electricity	11.9
LP gas	5.9
Coal	1.0
Wood	0.9
Other or none	0.8

have returned to coal. Electricity is a significant source only in the southeastern and northwestern regions where hydroelectric power is relatively plentiful and heating demands are not high. Recently, though, electric heating has taken a larger share of the new installations in areas where fuel oil was commonly used. For houses built since 1974, electric heating was installed in 36 percent of new homes, while natural gas, oil, and LP gas furnaces were installed in 63 percent of the homes. In spite of these shifts, it appears that combustion of fuel will be the main heat source for many dwellings in the future.

6.1. FURNACE OPERATION

The seasonal efficiency of a combustion furnace is the ratio of the heat delivered to the building and the energy value of the fuel entering the furnace. The maximum value of seasonal efficiency is unity. Seasonal efficiencies have been reported to be as low as 35-55 percent for natural gas and oil units, which means that up to two-thirds of the energy is not used as heat. In contrast, the measured steady-state efficiencies are usually about 75 percent. Large units such as factory boilers achieve higher performance and seasonal efficiencies may be in the range of 80-85 percent.

In this study of combustion furnaces, the limits on steady-state efficiency will be considered first. The nature of the additional losses associated with transient operation will then be discussed. A natural gas home furnace will be used as an example, but the same principles would apply to a furnace burning any other fuel and for any other application.

A typical residential furnace is shown schematically in Fig. 6.1.1. The flows of air and fuel and the relevant temperatures are also shown. In operation, fuel and air are mixed in the combustion chamber and the fuel is ignited. The source of ignition is usually either a pilot light for a natural gas furnace or an electric spark igniter for an oil furnace. Heat is transferred from the high temperature combustion gases to the circulating fluid, which is either air or water, in the house heat exchanger. The exhaust products are cooled in the process, but remain hot and buoyant enough to rise up the chimney and discharge to the atmosphere.

The flow of hot air up the chimney induces a flow of room air into the combustion chamber. A furnace based on this design is called a natural draft furnace or an atmospheric combustion furnace, and is the type commonly used in most residential applications. In mobile home furnaces, most commercial applications, and advanced residential designs, a blower is used to introduce the air flow into the furnace. The draft hood is a safety device to accommodate chimney downdrafts due to wind gusts. The draft hood allows a downdraft to go out into the room, and prevents a reverse flow of air in the furnace which could force the combustible mixture into the room. The draft hood is normally open to the surrounding room as shown. Room air exfiltrates up the chimney due to the higher temperature and lower density of room air compared to the colder ambient.

There are many potentially conflicting criteria in the design and operation of the furnace to achieve a safe, reliable, energy efficient device. Some of these are listed in Table 6.1.1. There are conflicts between the need for high temperatures to pro-

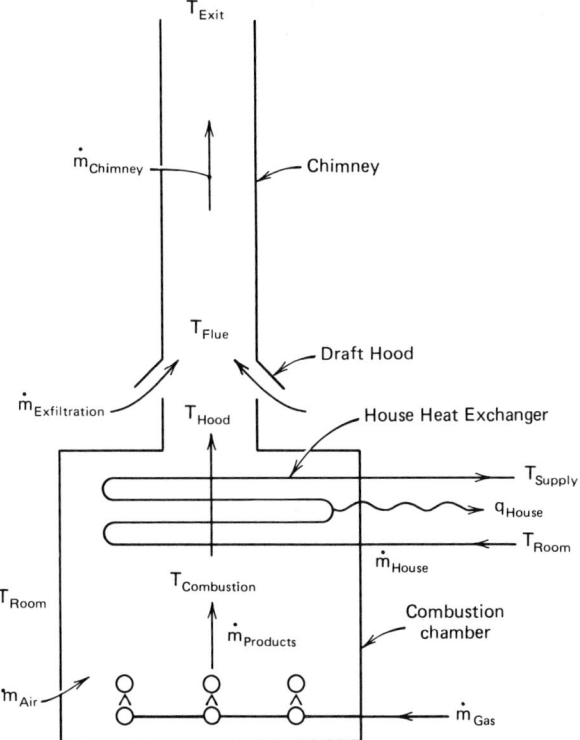

Fig. 6.1.1. Schematic of residential furnace in operation.

TABLE 6.1.1
Criteria for Furnace Operation

Combustion chamber	a. Sufficient excess air for complete combustion and no CO formation.
	b. Minimal excess air to reduce the flow of heated air up the chimney.
Heat exchanger	a. Low discharge temperature to maximize house heating.
	b. High discharge temperature to prevent condensation of water vapor.
Draft hood	a. Large opening to allow a downdraft to enter room rather than combustion chamber.
	b. Small opening to reduce exfiltration air flow.
Chimney	a. High exit temperatures to prevent condensation.
	b. High flue and exit temperatures to provide sufficient furnace draft for complete combustion.
	c. Low temperatures to reduce losses.

vide sufficient natural draft and to prevent condensation, and the need for low discharge temperatures to reduce thermal losses. There is also a conflict between the need to have high air flow rates for safety, and low air flow rates to reduce thermal losses. The furnace must operate satisfactorily over the course of the heating season, during which the ambient conditions change significantly. As a result of the many compromises necessary for operation, the thermal performance of the traditional natural draft furnace has suffered.

In order to see the interplay between the various components of a natural draft furnace, the physical mechanisms involved in furnace operation will be analyzed first. The chemical reactions and energy release in the heat exchanger and the flow relationships for the chimney will be studied to determine the energy flows. An overall energy balance for the furnace will then be formulated to determine the major contributors to low efficiency, and to determine those changes that would result in improved performance.

6.2. COMBUSTION REACTION

6.2.1. Combustion Equation

The chemical reaction of natural gas, oil, or coal with air produces water, carbon dioxide, and carbon monoxide as products. For safe furnace operation, the combustion process is controlled to eliminate carbon monoxide formation. This is accomplished by introducing more air than is needed into the combustion chamber and adjusting the air flow around the fuel nozzles. This also ensures that the combustion process is complete and that the entire heating value of the fuel is released in the combustion chamber.

The reaction of natural gas fuel with air is given by the chemical reaction relation. The components of a typical natural gas are given in Table 6.2.1. With this fuel, the reaction is given by

$$[0.89\ CH_4 + 0.08\ C_2H_6 + 0.03\ C_3H_8] + (1+E)(2.21)(O + 3.76\ N_2) \longrightarrow$$
$$1.14\ CO_2 + 2.14\ H_2O + 2.21\ EO_2 + 8.31\ (1+E)\ N_2 \qquad (6.2.1)$$

where E is the fraction of excess air supplied on a mole basis.

For satisfactory furnace operation, the amount of excess air should be about 50 percent. In practice, the flue gas is chemically analyzed to determine the percent by volume of CO_2 in the products. The percent CO_2 is then commonly used as a measure of operation rather than percent excess air. The relation between excess air and the percent CO_2 in the dry products is given in Fig. 6.2.1. Also shown for comparison is the relationship for a typical fuel oil. For furnace operation with 50 percent excess air, the corresponding CO_2 concentrations should be about 8 percent for natural gas and 10 percent for fuel oil. In terms of energy use, this means that 50 percent more air than is needed for combustion is heated and ultimately discharged to the ambient.

TABLE 6.2.1
Composition of a Typical Natural Gas[a]

Component	Percent by Volume	Percent by Mass
Methane (CH_4)	89	79.3
Ethane (C_2H_6)	8	13.4
Propane (C_3H_8)	3	7.3
	100	100

[a]Molecular weight = 18.0 lb_m/lb_{mole}.

6.2.2. Combustion Process

In order to analyze the thermal operation of the furnace, the products flow rate and the combustion temperature must be known. The combustion process will be assumed initially to be chemically complete and adiabatic. This will specify the maximum possible combustion temperature.

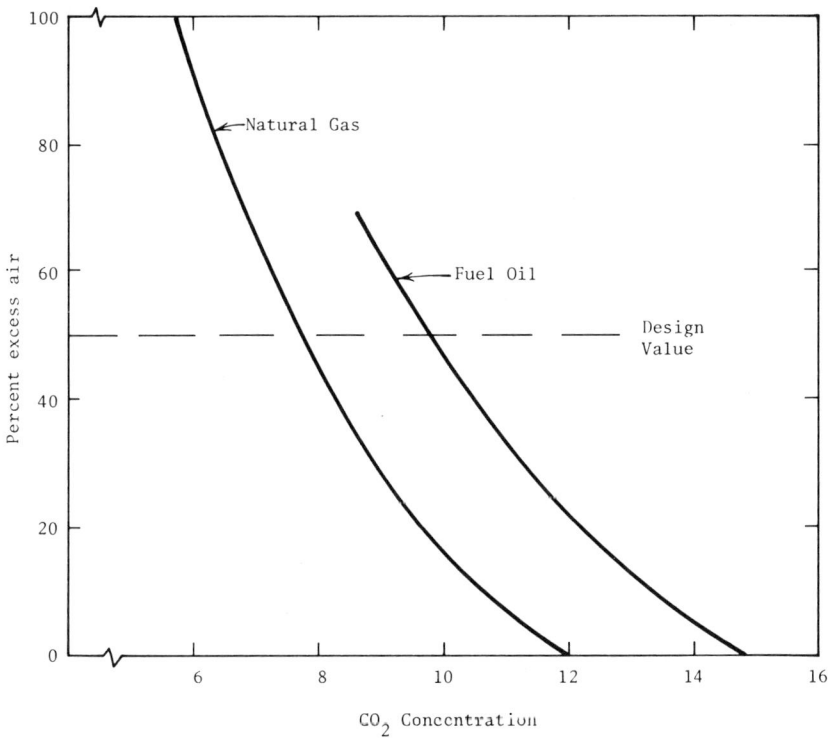

Fig. 6.2.1. Percent excess air as a function of CO_2 concentration.

COMBUSTION FURNACES

The conservation of mass principle for the combustion process is

$$\dot{m}_g + \dot{m}_a = \dot{m}_p \qquad (6.2.2)$$

where g, a, and p signify gas, air, and products, respectively. The conservation of energy principle applied to the same process is

$$(\dot{m}h)_g + (\dot{m}h)_a = (\dot{m}h)_p \qquad (6.2.3)$$

Equation (6.2.3) can be rearranged to introduce the enthalpy of the reaction h_{rp} at its reference temperature T_o. The rearrangement can be visualized by referring to Fig. 6.2.2. The actual combustion process occurs from the entering state at T_R to the combusted state at T_c. Thermodynamically, the reaction can be represented as a cooling (or heating) of the reactants from T_R to the reference temperature T_o, isothermal combustion at T_o, and heating of the products to temperature T_c. Equation 6.2.3 becomes

$$[(\dot{m}h)_g + (\dot{m}h)_a]_{T_R} - [(\dot{m}h)_g + (\dot{m}h)_a]_{T_o} + [(\dot{m}h)_g + (\dot{m}h)_a - (\dot{m}h)_p]_{T_o}$$
$$= [(\dot{m}h)_p]_{T_c} - [(\dot{m}h)_p]_{T_o} \qquad (6.2.4)$$

The third term in brackets is the product of the mass flow rate of fuel times the enthalpy change of the reaction.

$$[(\dot{m}h)_g + (\dot{m}h)_a - (\dot{m}h)_p]_{T_o} = \dot{m}_g h_{rp} \qquad (6.2.5)$$

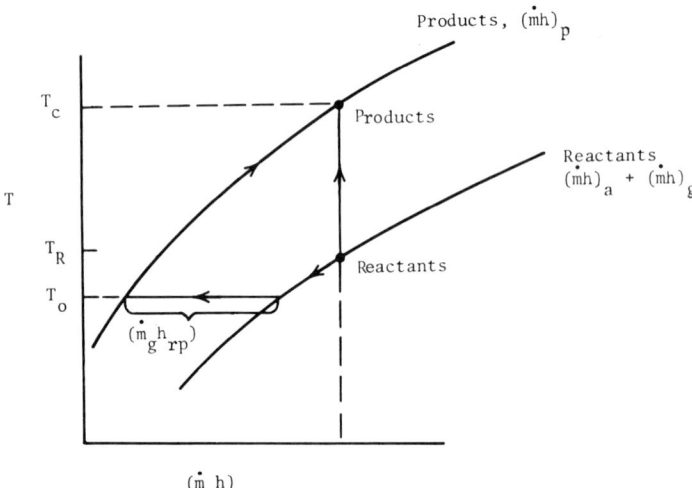

Fig. 6.2.2. Schematic of combustion process for temperature and enthalpy evaluation.

Equation (6.2.4) can be used to evaluate the combustion temperature under adiabatic conditions. For simplicity, the incoming, room temperature will be assumed equal to the reference temperature (77°F or 25°C) since these temperatures are fairly close for residential furnaces using room air for combustion. The complete equation could be employed to determine the effects of alternatives such as the use of outdoor air for combustion. For this simplified situation Eq. (6.2.4) can be solved for the combustion chamber temperature T_c. The products are assumed to behave as an ideal gas with a constant specific heat. Equation (6.2.4) becomes

$$T_c = T_o + \frac{\dot{m}_g h_{rp}}{(\dot{m}c_p)_p}$$

or, in terms of the air and gas flow rates

$$T_c = T_o + \frac{h_{rp}}{c_p(1 + \dot{m}_a/\dot{m}_g)} \quad (6.2.6)$$

The heating value to be used depends on whether the water vapor formed by combustion and present in the products is condensed. The lower heating value (LHV) assumes the water in the products is a vapor, while the higher heating value (HHV) assumes that the water is a liquid. In practice, condensation occurs in the chimney, or preferably, in the outside air, and the energy release is not available for heating the dwelling. The lower heating value is appropriate for residential furnaces. The properties of natural gas and fuel oil are given in Table 6.2.2.

The theoretical flame temperature as a function of percent excess air can be determined from Eq. (6.2.6). The ratio of the air flow rate to fuel flow rate is determined from the chemical reaction relation Eq. (6.2.1). For purposes of estimating combustion temperatures, the average specific heat of the products can be taken as 0.26 Btu/lb$_m$ °F.

For 50 percent excess air with natural gas as the fuel, the air fuel ratio is 25 to 1, and the theoretical flame temperature is about 3000°F. The actual tempera-

TABLE 6.2.2
Properties of Natural Gas and Fuel Oil

	Natural Gas	Fuel Oil
Density (lb$_m$/ft^3)	0.042	53
Lower heating value (Btu/lb$_m$)	21,500	18,500
Higher heating value (Btu/lb$_m$)	23,900	19,700

ture is less because of heat transfer by radiation from the hot gases to the surrounding furnace enclosure. However, this temperature is representative of the combustion temperature, and will be used as a base for evaluating the furnace heat exchanger.

6.3. HEAT EXCHANGER PROCESS

The room air heat exchanger is usually a parallel flow exchanger with both combustion products and room air or hot water rising vertically. This exchanger will be evaluated to see whether it is a major contributor to low furnace efficiency. The heat loss through the furnace walls is negligible and an energy balance on the exchanger yields the house heat flow as

$$q_{house} = (\dot{m}c_p)_p (T_c - T_h) = (\dot{m}c_p)_s (T_s - T_R) \qquad (6.3.1)$$

where h, R, and s denote hood, room, and supply, respectively.

The heat exchanger effectiveness ϵ is defined as the ratio of the actual heat transfer and the maximum possible.

$$\epsilon = \frac{q_{actual}}{q_{maximum\ possible}} \qquad (6.3.2)$$

The maximum heat transfer occurs when the fluid flow with the minimum flow rate-specific heat product is heated from one inlet temperature to the other. That is,

$$q_{maximum\ possible} = (\dot{m}c_p)_{min} (T_c - T_R) \qquad (6.3.3)$$

For a furnace, the combustion products have the minimum flow rate-specific heat product. This can be seen in that the temperature drop of the combustion products is much greater than the temperature rise of the room air in the heat exchanger. The actual heat transfer is that delivered to the house. Using Eq. (6.3.1), the exchanger effectiveness can be written as

$$\epsilon = \frac{T_c - T_h}{T_c - T_R} \qquad (6.3.4)$$

The highest value of effectiveness is unity and would be reached if the products were cooled to the room temperature. However, the house heat exchanger is usually a parallel flow exchanger to allow the heated air or water in the heat exchanger to rise and reduce the power required by the circulating fan or pump. The parallel flow nature restricts the lowest value of the hood temperature to be the supply temperature. In order to determine whether the heat exchange process is a major source of furnace inefficiency, the heat exchanger effectiveness will be estimated for a combustion temperature of 3000°F, a room temperature of 70°F, and a typical house supply temperature of 120°F. The effectiveness is then

$$\epsilon = \frac{3000 - 120}{3000 - 70} = 0.98$$

This shows that if a low hood temperature is obtained, 98 percent of the maximum possible heat transfer occurs.

The heat exchange process is, potentially, a very efficient one. However, hood temperatures in practice are on the order of 300-500°F, with resulting effectivenesses of 93-87 percent. The limit on the hood temperature is not due to the available heat transfer surface available. Additional area could be added, and it would probably be economically feasible to add heat transfer area to bring the hood temperature to within about 20°F of the supply temperature.

The chemical reaction equation shows that water is produced during combustion, and this is present in the flow as a vapor. The heat exchange is limited in that cooling the products below the dew point will cause condensation. The carbon dioxide formed will go into solution and corrosion could result.

The dew point of the products can be determined from the mole fraction of the water vapor. From the chemical reaction equation for combustion of natural gas with 50 percent excess air, the mole fraction of water vapor is 0.127. This corresponds to an absolute humidity of 0.079 lb_m/lb_m. The products' humidity is further increased since the room air used for combustion is not perfectly dry. At room conditions of 70°F and 50 percent relative humidity, room air would contain 0.008 lb_m/lb_m. The total humidity of the products would be 0.087 lb_m/lb_m. The partial pressure of the water vapor is 1.8 psia with a corresponding dew point of 123°F. If the products are cooled below 123°F, then condensation will occur. To prevent such condensation, the temperature leaving the exchanger must be greater than the dew point.

A plot of the dew point of the products as a function of percent excess air is shown in Fig. 6.3.1 for a typical room temperature and the limits of relative humidity. This shows that for an operating condition of 50 percent excess air, the products' temperature must be above about 130°F to prevent condensation. As will be shown, the lowest temperatures occur at the chimney exit, and the products must be above the dew point at this location. By increasing the percent excess air, the products could be cooled further. However, this would increase the flow of room air to the outside and increase the house heat loss.

Furnaces using fuel oil and coal are restricted to considerably higher flue gas temperatures than those burning natural gas. The presence of sulfur in fuel oil means that SO_2 is present in the products and reacts with water to form sulfurous acid. The hood temperature must be kept above the dew point of this mixture, which is in the range of 350-450°F.

If the products leaving the heat exchanger were directly vented to the outside, the temperature of the gas in the hood could be reduced to the dew point temperature. However, additional cooling of the products in the chimney occurs and in-

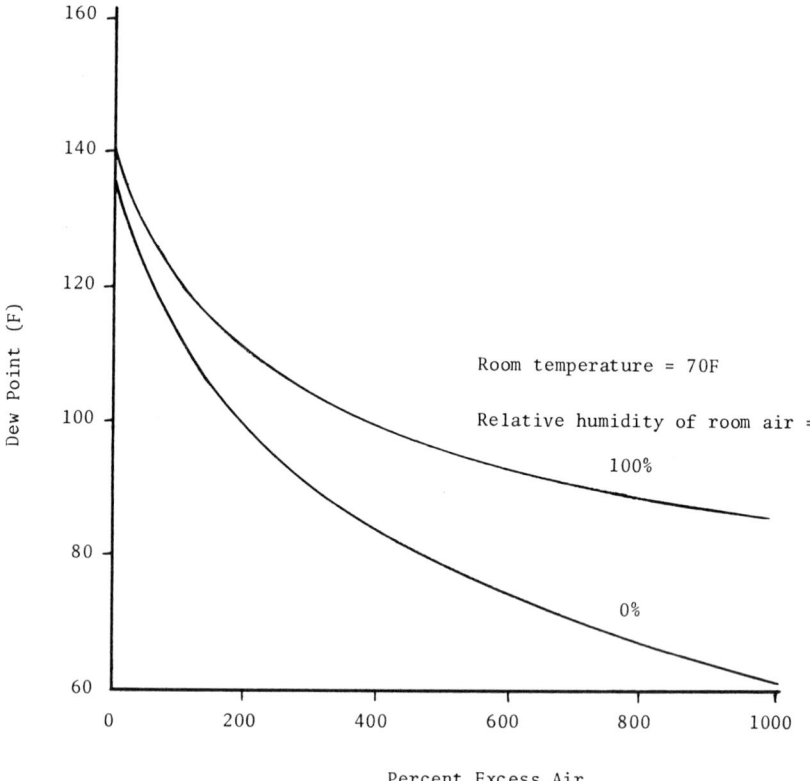

Fig. 6.3.1. Chimney gas dew point as a function of percent excess air.

creases the required hood temperature. There is also the need for sufficiently high temperatures in the chimney space to provide a natural draft. These two mechanisms will be considered in the next section.

6.4. CHIMNEY PROCESSES

Most residential furnaces and many commercial furnaces discharge the products of combustion through a natural draft chimney. The principle of operation is that the exhaust products are at a higher temperature than the ambient air, and thus are bouyant and naturally flow upward. This flow, or natural draft, induces the flow of air needed for combustion into the combustion chamber. In addition, air is induced through the draft hood. This mixes with the products and lowers the gas temperatures. The mean temperature of the flue gas must be high enough to provide sufficient air flow for combustion. The exhaust products are also cooled by heat transfer to the chimney wall. Both the flow and heat transfer processes will be formulated to determine the constraints on natural draft and dew point.

The flow rate up the chimney is the sum of the products and exfiltration flow rates:

$$\dot{m}_{ch} = \dot{m}_p + \dot{m}_{ex} \qquad (6.4.1)$$

The flue temperature is the result of mixing of the hot hood gases and cooler exfiltration flow. Its value is determined from an energy balance on the hood process:

$$(\dot{m}h)_h + (\dot{m}h)_{ex} - (\dot{m}h)_{ch} = 0 \qquad (6.4.2)$$

Assuming that all of the flows can be treated as ideal gases with constant and equal specific heats allows the flue temperature to be determined as

$$T_f = \frac{\dot{m}_p T_h + \dot{m}_{ex} T_R}{\dot{m}_{ch}} \qquad (6.4.3)$$

The chimney draft flow occurs due to a force unbalance between the weight of the air in the chimney and the pressure forces imposed by the ambient air. The situation is shown schematically on the left side of Fig. 6.4.1. The net force acting on the products inside the chimney is the difference between the pressure forces

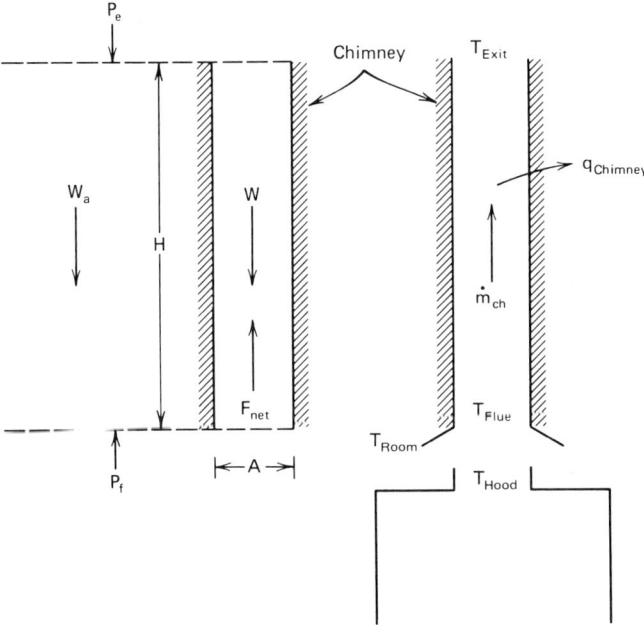

Fig. 6.4.1. Forces acting on air in a chimney.

and the weight of the air, or

$$F_{net} = (P_f - P_e)A - W \tag{6.4.4}$$

The weight of the air inside the chimney is given by

$$V = \rho_{ch} A H g \tag{6.4.5}$$

where ρ_{ch} is the average density of products in the chimney. The pressure difference is the same as that in the atmospheric air, and is given by a force balance on the ambient air as

$$(P_f - P_e)A = W_a = \rho_a A H g \tag{6.4.6}$$

where ρ_g is the ambient air density. The net force then becomes

$$F_{net} = (\rho_a - \rho_{ch}) A H g \tag{6.4.7}$$

The net force per unit area is the pressure difference creating the flow. The densities can be expressed using the equation of state for gases. The products have properties close to that of air and the absolute pressure is close to atmospheric pressure. The pressure difference creating flow can then be written as

$$\Delta P = \frac{F_{net}}{A} = \left(\frac{1}{T_a} - \frac{1}{T_{ch}}\right) \frac{P}{R} Hg \tag{6.4.8}$$

For small temperature differences between the chimney gas and outside air, this can be rewritten as

$$\Delta P = \frac{(T_{ch} - T_a)}{T} \rho H g \tag{6.4.9}$$

where ρ and T are average values of density and temperature, respectively. The chimney temperature is the average of the flue and exit temperatures:

$$T_{ch} = \frac{T_f + T_e}{2} \tag{6.4.10}$$

The upward flow of air through the chimney is given by the flow relation for orifices and nozzles:

$$\dot{m}_{ch} = C_d A \sqrt{2\rho \Delta P} \tag{6.4.11}$$

where C_d is an overall discharge coefficient for the chimney. The discharge coefficient accounts for friction along the chimney wall and the various restrictions at the chimney entrance and exit. The expression for the pressure difference from Eq. (6.4.9) can be introduced into Eq. (6.4.11) and the relation becomes

$$\dot{m}_{ch} = C_d A \rho \sqrt{\frac{2(T_{ch} - T_a) H g}{T}} \qquad (6.4.12)$$

This equation can be used either to size a chimney for a given furnace under given operating conditions, or to compute the chimney flow under operating conditions.

The ASHRAE guides indicate a value for C_d of about 0.45. Some limited tests in Wisconsin indicate a value of about 0.2. The actual value in any location depends heavily on chimney, vent pipe, draft hood, and ducting construction and is, therefore, very uncertain. An example calculation using Eq. (6.4.12) will be given at the end of this section.

The heat transfer occurring from exhaust products flowing through a chimney is shown schematically on the right side of Fig. 6.4.1. Heat is transferred from the hot products to and through the colder chimney wall to the house or ambient. An energy balance on the chimney yields

$$q_{ch} = (\dot{m} c_p)_{ch} (T_f - T_e) \qquad (6.4.13)$$

The heat flow can also be expressed by an overall conductance–surface area product. For a chimney passing through a heated space, the temperature difference is that between the chimney temperature and the room. For a chimney exposed to the ambient, the temperature difference would be that between chimney and ambient. Assuming the latter condition, the chimney heat flow is given by

$$q_{ch} = (UA)_{ch} (T_{ch} - T_a) \qquad (6.4.14)$$

A value of 1 Btu/hr ft^2 °F has been recommended for U_{ch}, but there is considerable uncertainty over the accuracy of the term. There is also uncertainty over whether a steady-state relation is valid. The furnace cycles and the chimney cools off. Equation (6.4.14) probably predicts the greatest value of heat loss and will be used in conservative estimates of chimney size.

There are two constraints on chimney design. First, the induced flow must provide sufficient air for combustion. Second, the exit temperature must be greater than the dew point. The common practice has been to provide a large margin of safety. This means a large air flow rate and a high exit temperature, both of which significantly degrade furnace performance.

The interaction between the flow, heat transfer, and temperature relations is complex. There are six coupled equations which are repeated below.

Mass balance on hood:

$$\dot{m}_{ch} = \dot{m}_p + \dot{m}_{ex} \qquad (6.4.1)$$

Energy balance on hood:

$$T_f = \frac{\dot{m}_p T_h + \dot{m}_{ex} T_R}{\dot{m}_{ch}} \qquad (6.4.3)$$

Energy balance on chimney:

$$q_{ch} = (\dot{m}c_p)_{ch} (T_f - T_e) \qquad (6.4.13)$$

Mechanism equation for chimney mass flow:

$$\dot{m}_{ch} = C_d A \rho \sqrt{\frac{2(T_{ch} - T_a) H g}{T}} \qquad (6.4.12)$$

Mechanism equation for chimney heat transfer:

$$q_{ch} = (UA)_{ch} (T_{ch} - T_a) \qquad (6.4.14)$$

Definition of average chimney temperature:

$$T_{ch} = \frac{T_f + T_c}{2} \qquad (6.4.10)$$

For a given furnace–chimney combination in operation, these relations can be used to determine the chimney and exfiltration flow rates, the chimney heat transfer, and the flue, exit, and chimney temperatures. There is in addition a measure of the loss associated with the chimney. This is the energy carried by the exfiltration air flow. It represents the additional fuel required to heat outdoor air up to room temperature, and is computed as

$$\dot{E}_{ex} = (\dot{m}c_p)_{ex} (T_R - T_a) \qquad (6.4.15)$$

An example will be carried out to illustrate the influence of the various parameters. The furnace will be taken to be a typical residential unit designed to supply about 70,000 Btu/hr to a house. For a steady-state efficiency of about 80 percent, this corresponds to a heat release in the combustion chamber of about 90,000 Btu/hr. The corresponding natural gas flow rate is 3.7 lb_m/hr. The air flow rate for 50 percent excess air is 94.1 lb_m/hr. The heat exchanger is designed so that the hood temperature is 400°F. The remaining parameters of the system are given in Table 6.4.1.

The solution to the set of equations is obtained by a trial-and-success technique. The procedure involves a guess for $\dot{m}_{ex}$ which then yields the chimney flow rate from Eq. (6.4.1) and the flue temperature from Eq. (6.4.3). An iteration is required

CHIMNEY PROCESSES

TABLE 6.4.1
Parameters for Chimney Example

Gas flow rate	$\dot{m}_g = 3.7\ \text{lb}_m/\text{hr}$
Air flow rate	$\dot{m}_a = 94.1\ \text{lb}_m/\text{hr}$
Products flow rate	$\dot{m}_p = 97.8\ \text{lb}_m/\text{hr}$
Room temperature	$T_R = 70°\text{F}$
Ambient temperature	$T_a = 20°\text{F}$
Hood temperature	$T_h = 400°\text{F}$
Chimney	
Diameter	= 6 in.
Height	= 20 ft
C_d	= 0.2
Flow area	= 0.2 ft^2
Surface area	= 31 ft^2

to determine the exit temperature and chimney heat flow from Eqs. (6.4.10), (6.4.13), and (6.4.14). The initial guess for $\dot{m}_{ex}$ can be checked using Eq. (6.4.12), and a better estimate then made.

The results are shown schematically in Fig. 6.4.2. The chimney allows enough air to be induced for combustion and also produces a large exfiltration flow rate. The amount of excess air at the chimney exit is 230 percent. The dew point from Fig. 6.3.1 is between 95 and 110°F. The exit temperature of 128°F is above this value. This shows that the chimney is properly sized to meet the natural draft and dew point criteria. The actual heat flux to the house will be calculated in Section 6.5, and is 71,200 Btu/hr.

The influence of the various design and atmospheric variables on furnace operation can be determined. The influence of ambient temperature is shown in Fig. 6.4.3. The chimney mass flow rate is not significantly affected by ambient temperature. The exfiltration loss, however, is a strong function of ambient temperature. At a 0°F ambient, the loss equals about 3 percent of the heat delivered by the furnace.

For all of the ambient temperatures shown on Fig. 6.4.3, the exit temperature is in the range of 120-140°F. The total percent excess air is 200-240 percent, and the corresponding dew point is below 110°F. This indicates that the furnace hood temperature could be reduced 10-20°F without condensation occurring. This would increase the heat delivered to the house and reduce exfiltration losses. However, this might not provide enough margin of safety considering the uncertainty in the calculations.

The effect of hood temperature is shown in Fig. 6.4.4. The chimney flow rate and the exfiltration loss both increase as hood temperature increases. At 600°F, the loss is about 3 percent of the furnace output. However, this is not the entire effect of hood temperature in that the delivered heat flow is about 8 percent lower than at 400°F. Correspondingly, at a 200°F hood temperature, the exfiltration losses are less and the heat delivered to the house is greater. The net effect is that as hood temperature is increased from 200 to 600°F, the heat delivered to the house

142   **COMBUSTION FURNACES**

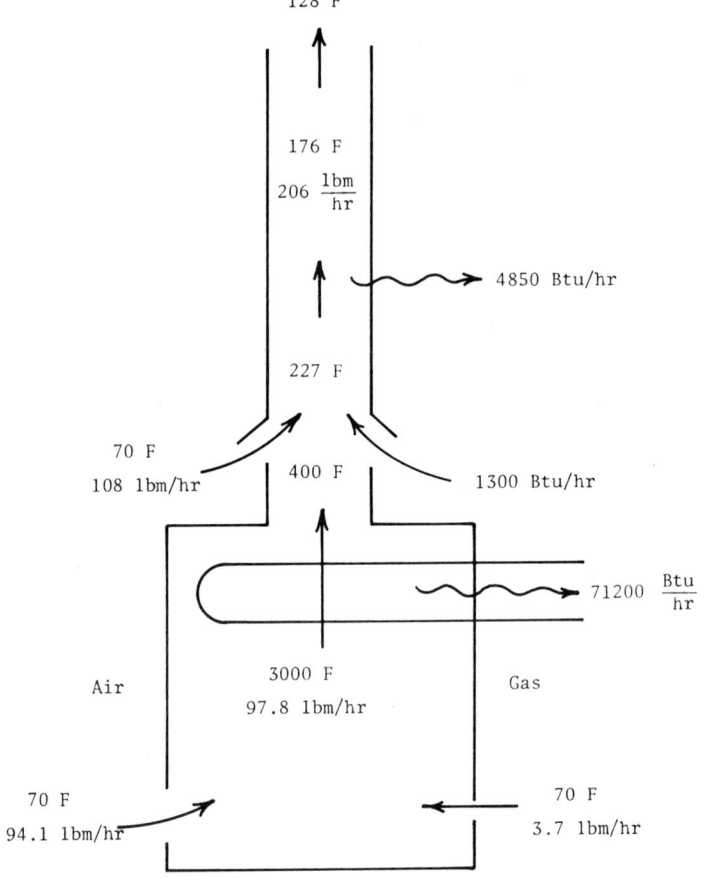

Fig. 6.4.2. Mass and energy flows for a furnace in operation.

is reduced about 18 percent. Clearly, the hood temperature should be as low as feasible to reduce these losses.

The constraint on the minimum hood temperature is the allowable exit dew point. For the example here, the exit temperature would equal the dew point for a hood temperature of about 300°F. Reducing the hood temperature would reduce exfiltration losses and increase the furnace heat transfer for a net improvement of 4 percent.

The exfiltration flow and heat loss through the chimney when the furnace is off are also shown in Fig. 6.4.4. The chimney flow rate is 50 percent of that when the furnace is on, but the exfiltration loss almost equals the on value. When the furnace is off, all of the chimney flow is an exfiltration loss, rather than just the flow induced through the draft hood. The flow rate when the furnace is off corresponds to an infiltration rate of about 0.2 air changes per hour for a house. This is one reason why homes heated with combustion furnaces have higher heat losses than those

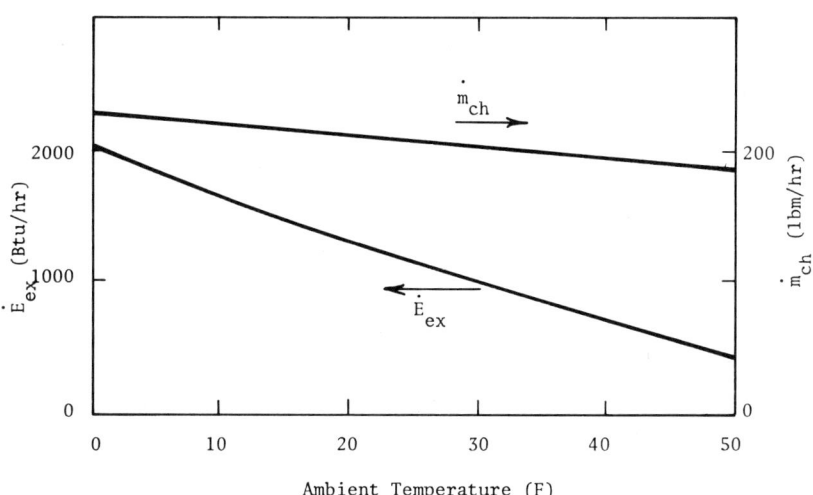

Fig. 6.4.3. Exfiltration loss and chimney flow rate as functions of ambient temperature.

heated by other noncombusting systems. This loss when the furnace is off is very important since most furnaces are off during most of the heating season. The exfiltration may be reduced through installation of dampers which would close off the chimney above the draft hood.

The chimney losses would be reduced by either a smaller-sized chimney or a restriction in the chimney above the draft hood. The performance for a reduced

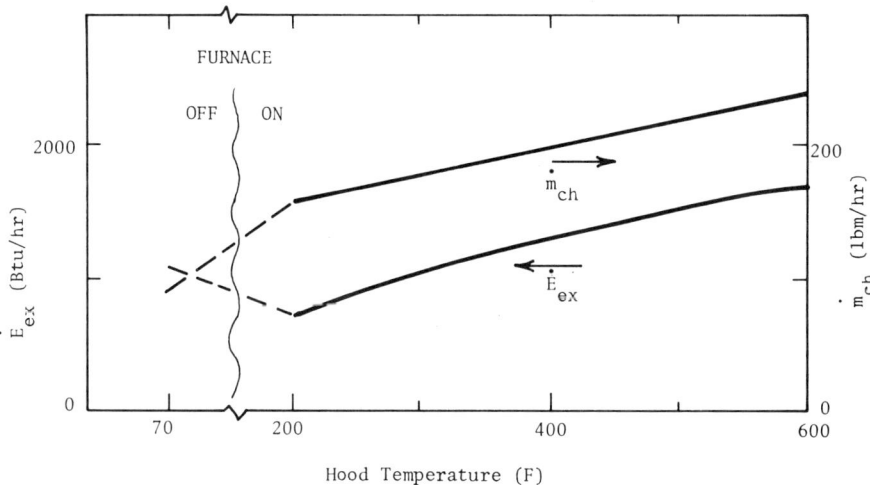

Fig. 6.4.4. Exfiltration loss and chimney flow rate as functions of hood temperature.

TABLE 6.4.2
Effect of Chimney Size on Performance

	Present	Improved
	6-in. diameter	4-in. diameter
	or $C_d = 0.2$	or $C_d = 0.1$
	or $H = 20$ ft	or $H = 5$ ft
Furnace on		
Exfiltration flow	108 lb_m/hr	21 lb_m/hr
Exfiltration loss	1300 Btu/hr	256 Btu/hr
Exit temperature	128°F	114°F
Furnace off		
Exfiltration flow	89 lb_m/hr	40 lb_m/hr
Exfiltration loss	1070 Btu/hr	480 Btu/hr

chimney size is given in Table 6.4.2. The reduction in size corresponds to a 4-inch-diameter chimney or a 5-ft-high chimney or to a permanent restriction with an orifice coefficient of 0.1. The exfiltration loss is reduced to about 20 percent when the furnace is on. However, the loss is about 50 percent of that for the larger chimney when the furnace is off. A flue damper would be a preferable way to further reduce these losses when the furnace is off.

In summary, a heated air flow is induced up the chimney both during operation and when the furnace is off. This flow exhausts room air into the outdoors and is replaced by cold air from the ambient. The temperature of the gas leaving the exchanger must be high enough to avoid condensation of the products during the flow up the chimney. The need for high temperatures reduces the heat transfer to the house. The chimney–hood–flue combination is simple but wasteful of energy. There are techniques available to reduce the losses associated with this system and improve furnace efficiency significantly. These techniques and their impact on seasonal efficiency will be discussed in the next section.

6.5. FURNACE EFFICIENCY DURING STEADY-STATE OPERATION

In the previous sections, the physical mechanisms and constraints for the combustion, heat exchanger, and chimney processes have been discussed. The furnace operation as a whole will now be studied to further identify those factors that contribute to low efficiency and to evaluate ways to improve performance. The performance under steady-state conditions will be studied first, and that under the cycling of actual operation next.

The steady-state efficiency is defined as the heat flow delivered by the furnace into the house divided by the fuel energy supplied. The energy supplied is the higher heating value since, potentially, all of this energy could be used as heat. The

FURNACE EFFICIENCY DURING STEADY-STATE OPERATION 145

efficiency becomes

$$\eta_{ss} = \frac{q_{house}}{\dot{m}_g(\text{HHV})} \quad (6.5.1)$$

The house heat flow can be evaluated from an energy balance on the furnace. The heat loss from the furnace jacket is negligible compared to the other heat flows. The energy balance becomes

$$(\dot{m}h)_g + (\dot{m}h)_a - (\dot{m}h)_p - q_{house} = 0 \quad (6.5.2)$$

It will be assumed that room air is used for combustion at a temperature close to the fuel reference temperature. The enthalpy difference between products and reactants is the fuel lower heating value since hood and flue temperatures are high and the water formed by combustion is a vapor. The house heat flow becomes

$$q_{house} = \dot{m}_g(\text{LHV}) - \dot{m}_p c_p(T_h - T_R) \quad (6.5.3)$$

The house heat flow can also be written in terms of the higher heating value as

$$q_{house} = \dot{m}_g(\text{HHV}) - \dot{m}_w(\Delta h_w) - \dot{m}_p c_p(T_h - T_R) \quad (6.5.4)$$

where $\dot{m}_w$ is the mass flow rate of water produced by combustion and Δh_w is the enthalpy change between vapor at T_f and liquid at T_R. The efficiency is then

$$\eta_{ss} = 1 - \frac{\dot{m}_w(\Delta h_w)}{\dot{m}_g(\text{HHV})} - \frac{\dot{m}_p c_p(T_h - T_R)}{\dot{m}_g(\text{HHV})} \quad (6.5.5)$$

The sum of the last two terms in Eq. (6.5.5) are termed the stack loss. The first term represents the loss due to the presence of water as a vapor instead of as a liquid in the products, while the second represents the sensible energy in the high temperature exhaust products.

In order to illustrate the relative size of the terms and the furnace efficiency, the previous furnace example will be considered. The parameters are as given in Table 6.4.1 and Fig. 6.4.2. Using the combustion reaction equation, there are calculated to be 7.9 lb_m/hr of water produced from burning 3.7 lb_m/hr of fuel. The difference in enthalpy of the water between vapor at 227°F and liquid at 70°F (T_o) is 1118 Btu/lb_m. The resulting furnace efficiency is then

$$\eta_{ss} = 1 - \frac{(7.9)(1118)}{(3.7)(23,900)} - \frac{(97.8)(0.26)(400 - 70)}{(3.7)(23,900)}$$

$$= 1 - 0.100 - 0.095$$

$$\eta_{ss} = 0.805$$

The actual heat delivered under steady-state operating conditions is calculated using Eq. (6.5.1) to be q_{house} = (0.805)(3.7)(23,900) = 71,200 Btu/hr.

This example shows that steady-state efficiencies on the order of 80 percent are possible with current design and careful control of combustion air and hood temperature. Certification requires that a 75 percent efficiency be achieved under steady-state conditions. This could be met in this example with a flue temperature as high as 600°F.

The loss due to water leaving the furnace as a vapor instead of a liquid in the products amounts to about 10 percent. Schemes have been proposed for recovering this energy and improving furnace efficiency. If, for the present example, the products could be cooled to 120°F, the furnace efficiency would be increased to 98 percent. Intermittent combustion furnaces (Section 6.7) condense the water vapor and recover the enthalpy of condensation. The seasonal efficiency is on the order of 95 percent.

It is probably not feasible to condense the vapor in oil or coal burning furnaces. The presence of sulfur in the fuel produces SO_2 in the products which reacts with water vapor to form sulfurous acid. This is very corrosive to the metal portions of the furnace. Oil and coal furnaces have inherently less potential for improvement due to the sulfur in the fuel.

The exfiltration flow rate does not contribute directly to furnace inefficiency. As shown by Eq. (6.4.2), the hood is an adiabatic device. The furnace can be thought of as discharging either products at the hood temperature or products plus exfiltration at the flue temperature. However, the exfiltration loss increases the house heating load since this exfiltration flow must be heated from outdoor to room temperature. As shown in Fig. 6.4.2, this loss is 1300 Btu/hr, or about 1.5 percent of the energy supplied. To eliminate exfiltration, the furnace could be converted to a forced draft system with a blower that forces combustion air only through the system.

The use of outdoor air for combustion has often been suggested as a way to increase furnace efficiency. This would increase the energy required to heat combustion air, but reduce the load on the house because heated room air would not be used for combustion. Since combustion air must be heated either directly in the combustion chamber or indirectly by way of the heat exchanger, the net effect on steady-state furnace efficiency is the same. The only advantage is the reduction in the exfiltration loss. This may be offset, though, by the presence of cold ambient air in the furnace room and the non use of jacket heat loss to keep the basement warm.

6.6. SEASONAL FURNACE EFFICIENCY

Steady-state efficiencies on the order of 80 percent are possible for combustion furnaces. However, furnaces do not operate continuously over a heating season. During spring and fall, the furnace output is larger than the heating load. A furnace for a particular application is often sized on the basis of the design load which is larger than the average load. Finally, a factor of safety is usually included to be sure that the furnace is sufficiently large. As a result, the heating output of the furnace is always considerably larger than the load and the furnace operates cyclically in an on-off fashion. Over the course of a heating season the furnace is on about 10 percent of the time. The cyclic operation produces low seasonal efficiencies.

The seasonal efficiency is defined as the total house heating required divided by the total fuel energy supplied over the heating season. It can be evaluated as

$$\eta_{\text{seas}} = \frac{Q_{\text{yr}}}{m_g(\text{HHV})} \qquad (6.6.1)$$

where Q_{yr} and m_g are the integrated values over the heating season of q_{house} and $\dot{m}_g$, respectively.

The typical sequence of events for a gas-fired residential furnace will be described to illustrate the mechanisms of loss. When the room temperature drops below the thermostat setting the main gas valve is opened. The gas is ignited, usually by a pilot light, and the furnace heats up. When the heat exchanger temperature exceeds between 125 and 140°F, the circulating fan turns on and warm air is circulated to the house. Combustion continues until the room temperature exceeds the thermostat setting at which time the gas valve closes. When the heat exchanger temperature drops below about 100°F, the circulating fan turns off.

This sequence of events is shown in Fig. 6.6.1 as a plot of temperature as functions of time. The only time that heat is delivered to the room is during the period when the circulating fan is on. The energy stored in the furnace and heat exchanger before the fan is on and after it is off is not delivered to the house. The area under the curve labeled "heat exchanger" during the fan-off periods reflects this loss since the energy stored in the exchanger does not get into the house. The relatively high hood and heat exchanger temperatures increase the driving force for natural convection through the combustion chamber and draft hood to the ambient during the time the furnace is heating up and cooling down.

There are, therefore, two losses associated directly with cyclic operation. The first is the nonutilization of thermal energy stored in the heat exchanger which may amount to 10 percent over the course of the season. The second is the exfiltration through the furnace chimney during the time the furnace is off which may be 10-20 percent.

The loss associated with the stored thermal energy may be significantly reduced by lowering the temperatures at which the circulating fan turns on and off. The present levels were choosen to give a desired sensation of comfort by the circulating air. The fan could be turned on when the fuel is ignited, and turned off when the exchanger is about 5°F above room temperature. This would significantly increase the amount of stored energy that is transferred to the house and reduce the exfiltration losses.

The exfiltration loss during the time the furnace is off can be eliminated by an atmospheric air source for both combustion and draft hood air. Either an exhaust damper installed in the flue which would be controlled to close when the main gas

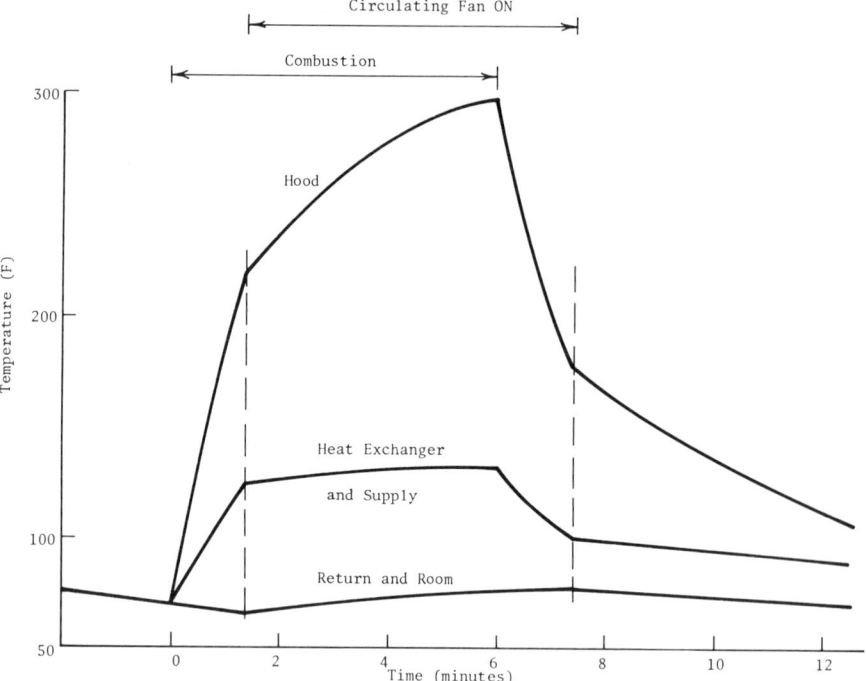

Fig. 6.6.1. Sequence of events during furnace operation.

valve is closed or a fixed restriction in the flue would reduce the loss. A sealed combustion furnace, which has a blower to force combustion air into the combustion chamber and does not have a draft hood, would also eliminate the cyclic exfiltration loss. Sealed combustion furnaces are currently used in mobile home operations where chimneys are not installed and are only recently being developed for other residential applications.

Another source of inefficiency is the pilot light. It is a continuous flame with a gas consumption of about 1 ft^3/hr, corresponding to about 8×10^6 Btu/yr. This is an appreciable energy usage, and may amount to between 5 and 15 percent of the total consumption for space heating. Essentially all of the heat released through combustion of the pilot light fuel is wasted in that very little adds to the heating of the house. Pilot lights burn with on the order of 10,000 percent excess air when the furnace is off. The corresponding temperature rise of the air in the combustion chamber is only about 2°F. The fan for the circulating fluid is off unless the main burner is on and so this heated air discharges through the chimney without transferring any heat to the house. When the furnace is on, the energy in the pilot light is recovered in the furnace. Since a furnace is on only about 10 percent of the year, 90 percent of the pilot light energy does not go into heating.

The installed capacity of a furnace also affects the efficiency. A furnace with a large heat output relative to the house load will cycle much more than a smaller furnace. The large furnace may never reach a steady state since it can heat up the house very quickly. As a result, the losses associated with cyclic operation are increased.

It has been estimated that a furnace sized to meet twice the design load has a seasonal efficiency 5 to 10 percent less than one sized to meet the design load. Since it is common practice to install a furnace larger than needed, this penalty is probably paid by a majority of the residences. It is possible to replace the existing fuel orifices by smaller ones to more correctly size the furnace to the load and reduce cycling.

Tests have been carried out by the Institute of Gas and Utilities Technology to evaluate the impact of different furnace modifications over a heating season. The results of these tests are summarized in Table 6.6.1. All of the modifications yielded an average improvement, although there were large differences between individual houses of up to 100 percent. The variations point out the difficulty of evaluating modifications from on-site tests. The greatest benefits were obtained from the most modified furnaces. Resizing the gas orifices or replacing the pilot light with electronic ignition produces an incremental improvement of about 6 percent. Fixed vent restrictors or automatic dampers produce about 3 percent improvement. Outdoor air for combustion has a marginal improvement of 1 percent. There is some interaction between the various improvements.

The results of these seasonal tests demonstrate that the relations developed earlier can be used to evaluate alternative techniques. However, there is a wide variation between different homes, furnaces, and chimneys, and prediction of individual savings is impossible. The ideas presented here should prove useful as guides to improved performance.

To demonstrate the role of these various factors on the overall performance of a furnace, an example calculation of the seasonal efficiency will be made. The house of Section 3.6 will be assumed to have the furnace of Section 6.4 installed in it. It will also be assumed that the performance during February is typical of the performance during the year.

The house load for February is 8.9×10^6 Btu (Table 3.6.3). This load does not include the heating of the combustion and exfiltration air flows from ambient to

TABLE 6.6.1
Improvement in Furnace Efficiency Through Modifications

Modifications	Average Increase in Furnace Efficiency (Percent)
Derate furnace, flue damper or restrictor, and electronic ignition	12
Derate furnace, flue damper or restriction	6
Automatic damper and electronic ignition	5
Automatic damper	3
Outdoor air supply	1

room temperature, and so these must be added to the load. The air flows depend on whether the furnace is on or off, and so the number of hours of operation must be determined. The number of hours that the furnace is on is estimated from the ratio of the house heating load and the furnace heating rate as

$$N_{on} = \frac{Q_m}{q_{house}} \qquad (6.6.2)$$

For the example, this becomes

$$N_{on} = \frac{8.9 \times 10^6 \text{ Btu/mo}}{71,200 \text{ Btu/hr}} = 125 \text{ hr/mo}$$

The furnace is off then 547 hours in February.
The actual load on the furnace is then

$$Q_{fur} = Q_m + [N_{on}(\dot{m}_A + \dot{m}_{ex}) + N_{off}\dot{m}_{ex}] C_p(T_R - T_A) \qquad (6.6.3)$$

Using the values for this example, the furnace load is

$$Q_{fur} = 8.9 \times 10^6 + [125(94.1 + 108) + 547(89)] (0.24)(70 - 20)$$

$$Q_{fur} = 9.8 \times 10^6 \text{ Btu/mo}$$

The combustion and exfiltration flows are seen to increase the house load by 10 percent.

The furnace meets this load while operating at a steady-state efficiency of 80.5 percent. The mass of natural gas required is given by

$$m_g = \frac{Q_{fur}}{\eta_{ss}\text{HHV}} \qquad (6.6.4)$$

For this example, the mass of natural gas is

$$m_g = \frac{9.8 \times 10^6}{(0.805)(23,900)} = 509 \text{ lb}_m$$

There is also natural gas consumed during the off period due to the pilot light which is

$$m_g = N_{off}\dot{m}_{pilot} \qquad (6.6.5)$$

which for this case is

$$m_g = (547 \text{ hr})(1 \text{ ft}^3/\text{hr})(0.042 \text{ lb}_m/\text{ft}^3) = 23 \text{ lb}_m$$

The total gas consumed during February is 532 lb$_m$, which is used to meet a house heating load of 8.9 × 10^6 Btu. The seasonal efficiency is then

$$\eta_{seas} = \frac{Q_m}{m_g \text{HHV}} \quad (6.6.6)$$

which becomes for this example

$$\eta_{seas} = \frac{8.9 \times 10^6 \text{ Btu/mo}}{(532 \text{ lb}_m)(23900 \text{ Btu/lb}_m)} = 0.700$$

The combustion and exfiltration air flows plus pilot light gas flow reduce the efficiency from a steady-state value of 80.5 percent to a seasonal value of 70.0 percent. There would be a further drop of 3-6 percent if the effect of the pilot light flow during summer was included.

The transient effect of cycling on the loss of energy stored in the furnace can be estimated. The furnace on-time is estimated to be 15 minutes, which means that there are 500 cycles during February. The furnace mass is estimated to be 200 lb$_m$ with a specific heat of 0.11 Btu/lb$_m$ °F and is heated from 70 to 200°F. It is assumed that this energy is used to heat the air which leaves the chimney by exfiltration. This energy to heat the furnace mass is

$$E = mc \, \Delta T (\text{no. of cycles}) \quad (6.6.7)$$

For this example:

$$E = (200 \text{ lb}_m)(0.11 \text{ Btu/lb}_m \text{ °F})(200 - 70)\text{°F} \,(500 \text{ cycles})$$

$$E = 1.43 \times 10^6 \text{ Btu/mo}$$

There is an additional 74 lb$_m$ of natural gas required to heat the metal in the furnace, and this use reduces the seasonal efficiency to 61.4 percent. These calculations indicate the magnitude of the effects that degrade furnace performance, and point the way to future improvements.

6.7. INTERMITTENT COMBUSTION FURNACES

The intermittent combustion furnace is an entirely new concept in furnace design that promises to produce significant energy savings. In this type of furnace, the steady combustion of a gas flame is replaced by intermittent combustion similar to that in a pulse jet engine or internal combustion engine. The furnace is shown schematically in Fig. 6.7.1.

In operation, the gas-air mixture is introduced into the combustion chamber and then ignited by a spark plug. The combustion process forces the hot gases through the heat exchanger. In addition, a partial vacuum is created which allows another

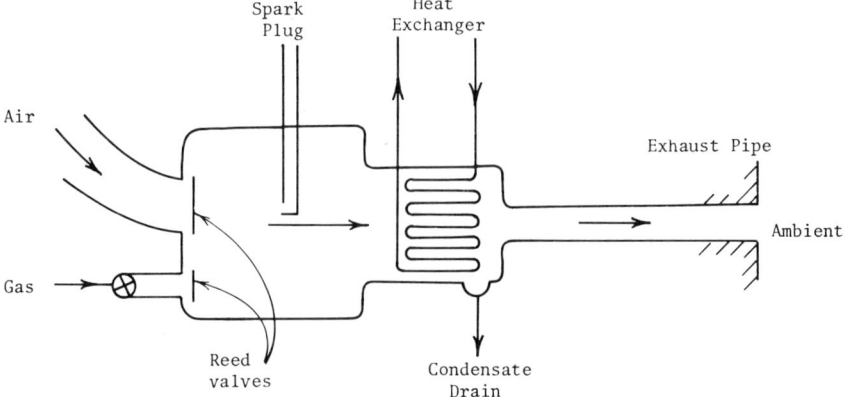

Fig. 6.7.1. Schematic of intermittent combustion furnace.

charge of air and gas to be drawn in through floating reed valves. Combustion occurs at a rate of about 65 cps.

There are several features of this furnace which reduce energy consumption. The turbulence created by the combustion process increases the heat transfer coefficient in the heat exchanger, and increases its effectiveness. This allows the combustion products to be cooled to the room supply temperature. In this process, the water vapor is condensed and the energy of condensation is transferred to the circulating air. This means that essentially all of the higher heating value of the fuel is available for heating.

The combustion process allows outside air to be drawn in for combustion. The process also forces the exhaust products out through a horizontal small diameter (1.5-inch) exhaust pipe. This pipe can be made of plastic because of the low discharge temperatures, and is also of a relatively small diameter. As a result, all of the penalties associated with a natural draft chimney are eliminated. There is no exfiltration loss during the time the furnace is off, and no excess air induced through a draft hood during combustion. There are no storage losses associated with heating and cooling of the structure since temperatures are all low. Condensation is drained to the sanitary sewer and so corrosion is not a problem.

The preliminary field tests on these units indicate that seasonal efficiencies are in the range of 95 percent. The problems encountered in its development are mainly concerned with the noise levels. In operation, the furnace hums at a level comparable to conventional blower levels. The costs of these devices are estimated to initially be about 50 percent higher than conventional units, or to cost in the range of $1500–$2000 installed. There does not appear to be any reasons why these units would not be widely accepted for new and replacement markets.

6.8. SUMMARY OF NATURAL DRAFT FURNACE PERFORMANCE

The seasonal efficiency of a conventional natural draft furnace depends on both its steady and cyclic operation. The contribution of each of these factors to the loss is

SUMMARY OF NATURAL DRAFT FURNACE PERFORMANCE

TABLE 6.8.1
Furnace Performance

Heating value of the fuel	100%
Steady-state losses	
Sensible stack loss	−15%
Latent stack loss	−10%
Steady-state efficiency	75%
Cyclic losses	
Pilot light	−5 to −10%
Stored energy	−5 to −10%
Exfiltration	−5 to −10%
Seasonal efficiency	60 to 45%

given in Table 6.8.1 for a representative atmospheric combustion furnace with a nominal 75 percent steady-state efficiency. The values are for a furnace sized correctly for the house design load. The steady-state and cyclic components each contribute about equally to the drop in seasonal efficiency from 100 percent.

A summary of modifications and improvements are given in Table 6.8.2. The improvements are probably representative of northern climates. They indicate the effect of any one improvement, and since some effects are related, the potential improvement might not be the sum of the individual values.

The estimated life cycle savings and first costs for each of the improvements are also given in Table 6.8.2. The savings are estimated for a furnace that consumes 100×10^6 Btu/yr for a total fuel cost of \$500/yr. The economic factor P_2 for a 20-year life is 22.17. A 15 percent federal tax credit would probably be allowed for each change and reduces first cost. The life cycle savings of all improvements except

TABLE 6.8.2
Improvements to Furnace Performance

Modification	Improvement in Efficiency	First Costs (\$) New	First Costs (\$) Retrofit	Life Cycle Fuel Savings (\$)
Electric ignition	6%	30	200	660
Pilot light off during nonheating season	3%	0	0	330
Outside air for combustion and exfiltration	1%	400	400	50
Circulating fan thermostat adjustment	7%	0	0	750
Flue damper	3%	50	200	330
Fixed restriction	3%	25	200	330
Sealed combustion furnace	10–15%	1500 to 2000		1000
Intermittent combustion furnace	40%	1500 to 2000		2800

the use of outside air for combustion are greater than the first cost and this indicates that these measures are cost effective. They should be considered both in retrofit and new installations.

In an existing furnace installation, the most effective techniques are to turn off the pilot light in the nonheating season and to adjust the circulating fan thermostat. The combined effect is to raise seasonal furnace efficiency about 10 percent. The life cycle savings are on the order of $1000, and achieved at no expense. Either a fixed restriction or a damper could be installed in the flue which would increase efficiency another 3 percent at an expense of $200 less any tax credits.

Recently, furnace manufacturers have responded to the need for improved furnaces, and a wide variety of furnaces are available. Many natural draft furnaces are now equipped with electric ignition and flue dampers. Their cost is marginally higher ($100-$200) than conventional ones and the seasonal efficiency is increased to 65-75 percent.

Residential units are also available with electric ignition and a forced draft fan. In these, the heat exchanger is designed to allow condensation of the water vapor and to recover the heat of vaporization. The fan then forces the combustion products out of the building through a small duct which replaces the chimney. The exfiltration losses both when the furnaces is on and off are eliminated. The efficiency is very close to that of the intermittent combustion furnace, and in the range of 85-90 percent. The cost is $300-$500 higher than conventional units.

The seasonal efficiencies of these various types of new residential furnaces for use with natural gas are given in Table 6.8.3. New furnaces have significantly better performance than older, conventional units and should definitely be considered for both new and retrofit situations.

This section has focused on residential furnaces. Furnaces conventionally used in commercial or industrial applications are commonly of higher quality with better seasonal performance. Electric igniters have been standard for some time. Combustion air is often introduced from outside the building with fans, and there are minimal exfiltration losses. Combustion air is often preheated by the exhaust gases to reduce stack losses. Finally, the combustion air is carefully metered so that minimal excess air is introduced. As a result conventional furnaces of this type have seasonal efficiencies of 75-85 percent. Improvements such as those being implemented in residential furnaces are also feasible for these units.

TABLE 6.8.3
Seasonal Efficiency of New Residential Furnaces

Type	Seasonal Efficiency
Conventional	50-60%
Conventional with flue damper, electric ignition	65-75%
Forced draft with condensing exchanger, electric ignition	85-90%
Intermittent	90-95%

PROBLEMS 155

6.9. SUMMARY

The devices for heating a building are quite inefficient in their use of energy. The evolution of the furnace has been toward a simple, reliable device with minimal regard to efficiency. The change in energy price and availability has motivated many changes, and significant improvements are feasible.

SUGGESTED READING

ASHRAE Handbook, Equipment Volume, American Society of Heating, Refrigeration, and Air Conditioning Engineers, Atlanta, GA, 1979, Chapters 25, 26.

G. J. Van Wylen and R. E. Sonntag, *Fundamentals of Classical Thermodynamics*, John Wiley & Sons, New York, 1976.

E. C. Hise and A. S. Holman, *Heat Balance and Efficiency Measurements of Central, Forced Air, and Residential Gas Furnaces*, ORNL-NSF-EP-88, Oak Ridge Natl. Lab., October 1975.

R. A. Macriss, T. S. Zauacki, M. T. Kovo, and P. A. Ketels, *Analysis and Correlations of Seasonal Performance Data from the Gas Industries Space Heating Efficiency Program*, Inst. of Gas Tech., Chicago, IL, May 1980.

J. Chi, *DEPAF—A Computer Model for Design and Performance Analysis of Furnaces*, ASME Paper 77-HT-11, ASME WAM, 1977.

PROBLEMS

6.1. A house has a natural gas furnace installed in it sized to meet the design heating load. The furnace is a standard natural draft furnace with a pilot light. The chimney diameter is 5 inches, the combustion occurs with 50 percent excess air, and the hood temperature is 400°F.

Using February as a representative winter condition, estimate the furnace seasonal efficiency. Determine the effects due to exfiltration when the furnace is on and off and due to the pilot light. Neglect transient effects due to energy storage. Assume the location is Madison, Wisconsin, and the house characteristics are as given in Tables 3.6.1 through 3.6.3.

6.2. The following are options that are available for improving the efficiency of a furnace. Determine the most cost-effective option for the furnace in Problem 6.1.

(a) *Electrically Operated Flue Damper*—This reduces exfiltration flow during off time to 0.1 of the unrestricted value and costs $200 installed.

(b) *Thermally Activated Flue Damper*—This reduces exfiltration flow during off time to 0.2 of the unrestricted value and costs $120 installed.

(c) *Fixed Restriction*—This reduces all exfiltration to 0.5 of the unrestricted value and costs $60 installed.

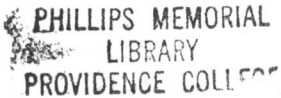

(d) *Outside Air for Combustion*—This requires a $500 modification of the furnace room.

6.3. For Problems 6.1 and 6.2, change the location to Chicago, Illinois.

6.4. For Problems 6.1 and 6.2, change the location to Washington, DC.

6.5. For Problems 6.1 and 6.2, change the location to Los Angeles, California.

■ 7 ■

HEAT PUMPS

A heat pump is a refrigeration system whose purpose is to remove heat from one area and supply it to another. In a conventional refrigerator, cooling is the only desired effect, while in a heat pump, either heating or cooling may be the desired effect. In most residential and commercial heat pump applications, heat is taken from the cool outside air and "pumped" into the room for heating or heat is removed from the room and "pumped" to the warmer outside air for cooling. In the cooling mode the heat pump operates in the same manner as a conventional air conditioner. There has been recent interest in heat pumps which use either groundwater or ice for either a source or sink, and there are many areas which could use these to advantage. At present, though, the vast majority of units are air-to-air heat pumps.

In the 1950s, many heat pumps were installed in residences as the primary heating source. However, within a few years there were so many failures that heat pumps dropped in popularity. The recent concern over energy prices and the diminishing availability of natural gas and oil has renewed interest in heat pump systems.

The early installations suffered from failure of the components. Many manufacturers converted air conditioners into heat pumps without redesigning the various parts. Many of the components, and in particular the compressor, could not withstand the greater loads imposed during the heating season and failed. In addition, many installations were sized in the same way as conventional furnace systems, which produced rapid cycling and accelerated failure.

A comprehensive study by the Alabama Power Company was conducted on 11,000 heat pumps serviced by the utility that had been installed between 1968 and 1976. The yearly failure rates for compressors were 5 to 7 percent which shows that compressor life is about 20 years. Compressor failures are the most costly repair item, but are not the only component to fail. Fan blades and fan motors had a yearly failure rate of 18 percent, refrigerant leakage problems occurred at a rate of 10 percent, and defrost circuits failed at a 3 percent rate. These high failure rates and the resulting high costs have made heat pumps a questionable alternative to the furnace. On the basis of this early experience, manufacturers have improved the construction of heat pumps for heating only. The procedure for sizing heat pumps has evolved, and satisfactory performance can be achieved.

Heat pumps have been successfully employed in commercial buildings for many years. These units are used in transferring energy from cooled to heated areas. In a typical large building, the interior zones are often cooled throughout the year while heating of the zones on the building perimeter occurs during the winter. Heat generated by lights, people, and machinery may then be pumped from the interior to heat the perimeter. Since the temperatures of the fluids used to heat and cool are fairly close, the heat pump can effect this transfer efficiently and at a lower cost than for separate heating and cooling systems.

In this chapter the basic thermodynamics of a heat pump will be considered. The heat pump will be compared to alternative methods of heating. The design of a residential air-to-air heat pump system to meet a given load in a given climate will be considered, and the overall seasonal performance will be determined. Various alternatives will then be considered. The economic and energy feasibility of heat pumps will be assessed for residential and commercial applications.

7.1. THERMODYNAMICS OF OPERATION

The components of a heat pump system that provide both refrigeration and heating are shown schematically in Fig. 7.1.1. The refrigerant in the vapor state is pumped by a reciprocating compressor to a relatively high pressure and temperature. The refrigerant condenses to a liquid as it passes through the condenser heat exchanger coil and, in the heating mode, the heat of condensation is transferred to the room air circulating over the condenser coil. The warm liquid refrigerant flows through tubing to the expansion valve. The large pressure drop across the valve causes the

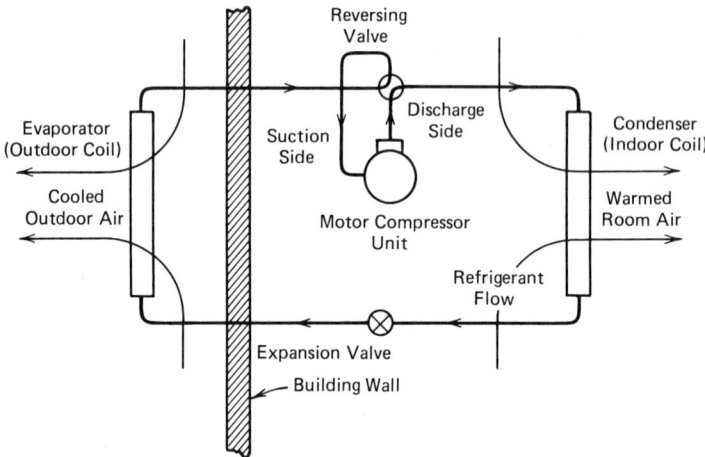

Fig. 7.1.1. Schematic of air-to-air heat pump.

fluid temperature to drop. This cold refrigerant then enters the evaporator heat exchanger coil located outdoors. Heat is transferred from the cold ambient air to the even colder refrigerant as it passes through the evaporator. The liquid in the evaporator vaporizes and flows back to the suction side of the compressor to repeat the cycle.

The heat pump can be used to cool the room air by reversing the functions of the evaporator and condenser coils. Rotation of the reversing valve would send hot refrigerant from the compressor discharge to the outdoor coil first (now the condenser) so that heat would be rejected outdoors. After passing through the expansion valve, the cold refrigerant would pass through the indoor coil (now the evaporator) cooling the room air.

The air-to-air heat pump described above is the most common type of heat pump for residential applications. Other types are water-to-air units in which water from a lake or well is used as the heat source or sink. Water-to-water units are also available and widely used in commercial installations. In these, the water rather than the refrigerant is switched between heat exchangers to convert the unit from heating to air conditioning.

The heat supplied to the room by a heat pump comes from two sources, the outside air and the electrical work supplied to the compressor fans. For the heat pump, the energy absorbed in the evaporator is transferred from the condenser by the input of compressor work. The energy balance on the heat pump itself is

$$q_{ev} + \dot{W}_{comp} = q_{cond} \tag{7.1.1}$$

The heat pump coefficient of performance is the ratio of the desired effect (heating or cooling) to the power input. This is given for heating by

$$\text{COP}_{hp} = \frac{q_{cond}}{\dot{W}_{comp}} \tag{7.1.2}$$

The COP depends strongly on the outdoor and indoor temperatures. As the difference between the two increases, more power is required to pump the heat from the lower to the higher temperature, with a resulting decrease in COP.

The system coefficient of performance in a heating situation is not the same as the heat pump coefficient of performance. The total electrical power input is greater than the compressor power because the motor efficiency is less than unity and power is required for the heat exchanger fans. In addition, the heating effect is greater than the condenser heat transfer since some of the electrical losses occur in the heated space. For many installations, the compressor is installed in the house so that the motor losses are also recovered for heating. The electrical energy for the condenser fan is also a heat input. For heating, the system coefficient of performance becomes

$$\text{COP}_{htg} = \frac{q_{htg}}{\dot{W}_{elec}} \tag{7.1.3}$$

The system coefficient of performance for heating is usually in the range of 2-4. This means that two to four times as much heat is delivered as electricity supplied.

For cooling, the coefficient of performance is defined as

$$\text{COP}_{cl} = \frac{q_{ev}}{\dot{W}_{comp}} \qquad (7.1.4)$$

For the same source and sink temperatures, the COP for air conditioning is less than that for heating by about 1, and is usually in the range of 1-3. The system COP for air conditioning is defined analogously to Eq. (7.1.3).

The seasonal COP, denoted $\overline{\text{COP}}_{htg}$, is the total seasonal heating supplied divided by the sum of the seasonal auxiliary energy and the electricity used in running the compressor and fans, and is the range of 1.6-2.5. One reason that the seasonal COP is less than the system COP is due to the formation of frost on the outdoor coil at low outside temperatures. To remove this frost, the heat pump is operated briefly in the cooling mode and the heating COP is then negative. Another reason for a lower seasonal COP is that auxiliary energy is delivered at a COP of unity. Another reason is that the heat pump cycles on and off during the heating season. This reduces the COP by 5-10 percent below the steady state value.

A simple comparison can be made between the heat pump and conventional furnace to determine the economics of operation in both energy and dollar terms. To meet a given annual house heating load Q_{yr}, the total electrical work required for a heat pump is

$$E_{elec} = \frac{Q_{yr}}{\overline{\text{COP}}_{htg}} \qquad (7.1.5)$$

Electricity is produced at a power plant from resource energy at a thermal efficiency of η_{pp}. The amount of resource energy required by the heat pump to meet the house load is

$$E_{res,hp} = \frac{Q_{yr}}{\eta_{pp}\overline{\text{COP}}_{htg}} \qquad (7.1.6)$$

The resource energy required for a furnace to meet the same house load is given in terms of furnace efficiency as

$$E_{res,fur} = \frac{Q_{yr}}{\eta_{fur}} \qquad (7.1.7)$$

For a heat pump to use less resource energy than a furnace, Eqs. (7.1.6) and (7.1.7) show that the $\overline{\text{COP}}_{htg}$ must be greater than the following:

$$\overline{\text{COP}}_{htg} > \frac{\eta_{fur}}{\eta_{pp}} \qquad (7.1.8)$$

A typical power plant efficiency is 0.35. The seasonal efficiencies of new residential furnaces are 0.75-0.95 while those for commercial units are about 0.85. In order to save energy compared to a furnace, the seasonal COP must be greater than about 2.3-2.7. This is a relatively high value for northern climates, but if it can be achieved, the heat pump will save resource energy.

A similar analysis can be made to see if the heat pump can compete economically with a furnace even though the cost of electricity is higher than that of conventional fuel. The annual cost to operate the heap pump is

$$C_{hp} = C_{F,elec} E_{elec} \tag{7.1.9}$$

or, using Eq. (7.1.5)

$$C_{hp} = \frac{C_{F,elec} Q_{yr}}{\overline{COP}_{htg}} \tag{7.1.10}$$

The annual cost to operate the furnace is

$$C_{fur} \frac{C_{F,fuel} Q_{yr}}{\eta_{fur}} \tag{7.1.11}$$

For the annual cost of heat pump operation to be less than that of the conventional furnace, the $\overline{COP}_{htg}$ must be greater than the following:

$$\overline{COP}_{htg} > \frac{C_{F,elec} \eta_{fur}}{C_{F,fuel}} \tag{7.1.12}$$

Current electricity costs are about 5¢/kWh corresponding to about $15/10^6$ Btu. Natural gas costs are about $5/10^6$ Btu, and with a furnace efficiency of 0.75-0.95 the $\overline{COP}_{htg}$ must then be greater than about 2.3-2.9 in order for the heat pump to save money.

This simple comparison shows that a heat pump must have a $\overline{COP}_{htg}$ of greater than about 2.5 to save both money and energy. Depending on fuel cost, the heat pump may save resource energy, but cost the consumer money. Most home owners and businesses would chose to save money, not energy.

The economic picture is further complicated in that the first costs of the heat pump are considerably greater than that of the furnace. This increases the COP needed for the heat pump to be cost effective. A more complete economic evaluation will be performed following the calculation of seasonal heat pump performance.

7.2. HEAT PUMP PERFORMANCE

The heat pump coefficient of performance is important in determining the amount of electricity required for a given amount of heating. For a given unit, the amount

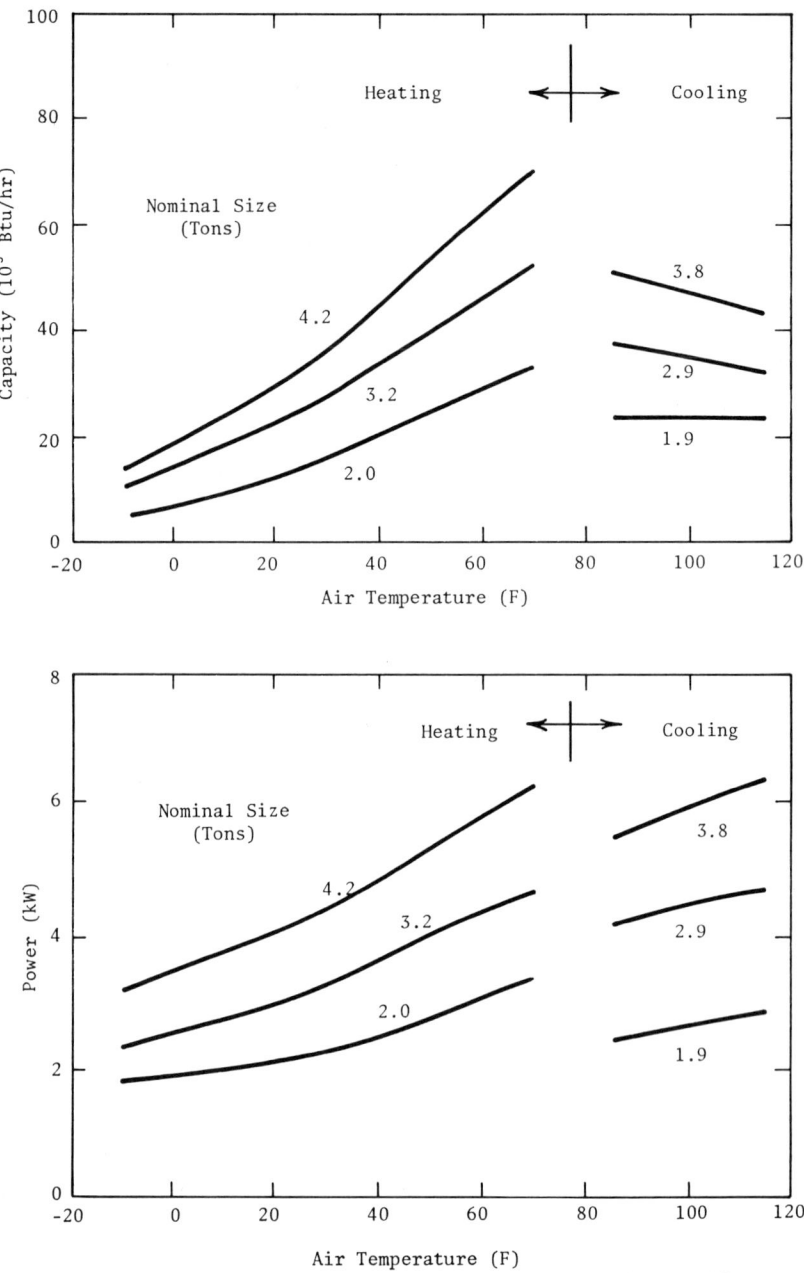

Fig. 7.2.1. Heat pump performance as a function of ambient temperature.

of heat that can be delivered is also important. Both COP and the heating capacity q_{htg} depend on the temperature of the source and decrease as the source temperature decreases. The cooling performance is also a function of the source temperature, and both cooling capacity and COP decrease as source temperature increases.

Typical steady-state performance curves for three heat pump sizes suitable for residential application are given in Fig. 7.2.1. The nominal heating capacity given on each curve is the American Refrigeration Institute (ARI) rating at 70°F indoor and 47°F outdoor air temperatures for heating, and 80°F indoor and 95°F outdoor temperatures for cooling. These are systems in which the compressor, expansion valve, and condenser (during heating) are located indoors, and the evaporator is outdoors. Locating the compressor indoors means that motor losses are heat gains into the house, which increases the COP during heating. However, these losses add to the heat gains during the cooling season and more air conditioning must be supplied.

The curves show that heating capacity increases significantly with air temperature. At 70°F, the units are capable of delivering about seven times as much heat as at 0°F. The curves reflect a decrease in capacity below 40°F due to the need to defrost the evaporator coils.

The curves for cooling are the total (latent plus sensible) capacity of the unit. The sensible capacity is very closely 0.54 of the total heat capacity. For residential applications, the heat pump latent capacity is more than sufficient to meet the moisture load usually encountered and the sensible heat capacity determines the in-

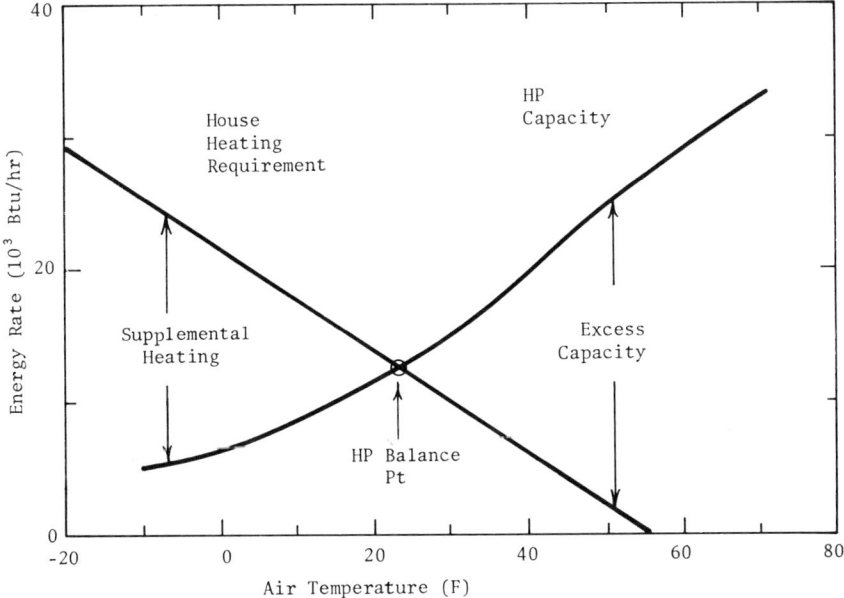

Fig. 7.2.2. Heat pump and house operating characteristics as functions of ambient temperature.

stalled size. The electrical power curves are the total power required for the compressor motor and the fans that circulate air over the evaporator and condenser.

A disadvantage of a heat pump is that it is not well matched to house heating requirements over a season. Figure 7.2.2 shows the heating requirements for a typical home as a function of outside air temperature in comparison with the capacity curve for the 2-ton heat pump. The intersection of the two curves is the heat pump balance point. At this outside air temperature the house demand and heat pump capacity are matched. At higher outside temperatures, the heat pump has greater capacity than required and needs to operate only part of the time. At temperatures below the balance point, the heat pump is unable to meet the load alone. Supplemental heating, usually in the form of electrical resistance heating, is required. A larger heat pump will shift the balance point to a lower temperature, but there may still be a need for supplemental heating. Because of the high initial cost of a large heat pump it is uneconomical to install a heat pump that can meet the largest load expected. As a result electric resistance heaters are commonly installed with the heat pump to provide supplemental heating.

7.3. CALCULATION OF SYSTEM PERFORMANCE

The dependency of heat pump capacity and house heating requirements on outside air temperature complicates the calculation of the performance of the heat pump system over the course of a heating season. A procedure termed the bin method that accounts for these variations has been developed. In this method the outdoor temperature range is divided up into 5°F increments, or "bins," and the number of hours during the heating season that the outdoor temperature is in each bin is determined. The heating demand for the house and the heat pump capacity and power requirements for the average temperature of each bin are determined.

For the bins in which the outdoor temperature is below the heat pump balance point, the heat pump is on continuously and supplemental heating is required. The rate of supplemental heating is the difference between the house heating requirement and the heat pump capacity. The annual energy requirements for the bin are the sum of the products of the rates and the number of hours.

For those bins above the heat pump balance point the heat pump capacity is larger than the house load and supplemental heating is not required. The heat pump operates cyclically and is on only a fraction of the number of hours in each bin. This fraction is the ratio of the house heating requirement to the heat pump capacity. The annual electrical requirement for this bin is the product of the heat pump power, the number of hours in the bin, and the fractional time the unit is on.

Annual totals of heat pump and supplemental heating and power consumption are obtained by summing up the bin values. From these, the seasonal performance of the system can be obtained. The method differs from the degree-day approach of Chapter 3. The bin method does not distinguish between different months or time periods, but uses ambient temperature as the sole determinant of house load. As a result, the method is not as accurate as a more detailed calculation would be.

TABLE 7.3.1
Worksheet for Heat Pump Performance Calculation

Heating season: $T_{ave} = $ _____ °F House: $\overline{UA_o} = $ _____ Btu/hr °F
$N_{total} = $ _____ hr $\overline{T}_{Bal} = $ _____ °F

①	②	③	④	⑤	⑥	⑦	⑧	⑨	⑩
Heating Season		House			Heat Pump				
T_A (°F)	ΔN (hr)	$\overline{T}_{Bal} - T_A$ (°F)	Demand (10^3 Btu/hr)	Capacity (10^3 Btu/hr)	Power (kW)	Time (fraction)	HP Heating (10^6 Btu)	Supplemental Heating (10^6 Btu)	Electricity (kWh)

Total

House load = _____ Btu Heat pump $\overline{COP}_{hp} = $ _____
Supplemental heating = _____ Btu System $\overline{COP}_{htg} = $ _____
Heat pump electricity = _____ kWh Total electricity = _____ kWh

A worksheet has been developed to aid the bin method calculations, and this is given in Table 7.3.1. The specific climate and house information required will be discussed in Sections 7.4 and 7.5. Columns 1 and 2 are the bin data for the average temperature and number of hours for each bin, respectively. Column 3 is the temperature difference between the average house balance temperature and the average bin temperature. The house heating demand is the product of the average conductance-area product and temperature difference and is entered in column 4.

Columns 5 and 6 are the capacity and power requirements from the heat pump performance curves. The heat pump balance point is found by comparing the house heating requirement with the capacity. For bins below the balance point, the time fraction entered in column 7 is unity. Above the balance point the fraction equals the ratio of column 4 to column 5.

The heat supplied by the heat pump (column 8) is the product of the number of hours (column 2), capacity (column 5), and time fraction (column 7). The supplemental heating (column 9) is zero above the balance point. Below the balance point it equals the difference between demand (column 4) and capacity (column 5) times the number of hours (column 2). The heat pump work (column 10) is the product of number of hours (column 2), the power (column 6), and the time fraction (column 7).

Columns 8, 9, and 10 are summed to yield the annual values of the heating and electrical requirements. The house load is the sum of columns 8 and 9. The average heat pump $\overline{COP}_{hp}$ is the ratio of heat pump heating (column 8) to electricity (column 10). The seasonal $\overline{COP}_{htg}$ is the ratio of house load (column 8 plus column 9) to energy supplied (column 9 plus column 10). An example calculation will be given in Section 7.6.

7.4. GENERALIZED WEATHER DISTRIBUTIONS FOR HEAT PUMP CALCULATIONS

The application of the traditional bin method requires the frequency distribution of ambient temperature over the course of the year. Such data are available for various locations. However, a large amount of information is required in order to assess heat pump performance in different locations. A generalized ambient temperature distribution has been obtained that accurately predicts heat pump performance. It was originally developed for use on a monthly basis, and has been computerized for that purpose. However, it is tedious to perform monthly calculations by hand. An annual method has been developed that is sufficiently accurate for performance estimates.

The generalized distribution is symmetric about the seasonal average temperature, and has a spread of 70°F. The fraction of the total number of hours in the heating season is given in Table 7.4.1. With this table, only the average temperature and heating season length need be known for each location.

The performance using this generalized temperature distribution in the bin method was compared to that using actual temperature distributions. A wide combination of heat pump and house sizes was tested in 14 U.S. locations varying from

TABLE 7.4.1
Bin Data for Generalized
Temperature Distribution[a]

$T - T_{ave}$	$\dfrac{\Delta N}{N_{total}}$
0	0.1378
±5	0.1224
±10	0.1021
±15	0.0816
±20	0.0612
±25	0.0408
±30	0.0204
±35	0.0026

[a] T is the average temperature of the bin. T_{ave} is the heating season average temperature. N_{total} is the total number of hours in the heating season. ΔN is the number of hours in each bin.

Bismark, North Dakota in the north to Lake Charles, Louisiana in the south. The agreement between the two methods was found to be within 3 percent using monthly distributions and 6 percent using the annual distribution. This is within the accuracy for which heat pump and house characteristics are known.

The heating season comprises those months in which residents would actually turn on their heating system. This roughly corresponds to those months for which the monthly average temperature is below the balance temperature. This is not a very precise definition but the results using the generalized distribution are not too sensitive to this division. For example, including a spring month such as March in the heating season increases the length of the season but raises the season average temperature. These two compensate and the results are about the same. The heating season information for the four locations of Appendix A and three additional locations are given in Table 7.4.2.

TABLE 7.4.2
Heating Season Data

Location	Months	T_{ave} (°F)	N_{total} (hr)
Madison, WI	Oct.–Apr.	32	5088
Chicago, IL	Nov.–Apr.	36	4344
Washington, DC	Nov.–Mar.	40	3624
Albuquerque, NM	Oct.–Apr.	45	5088
Memphis, TN	Nov.–Apr.	47	4344
Charleston, SC	Nov.–Mar.	50	3624
Los Angeles, CA	Dec.–Feb.	55	2160

7.5. HOUSE HEATING REQUIREMENTS

The detailed procedure for calculating house heating loads is described in Chapter 3. The loads are calculated for each month and for day and night periods. The loads from this approach need to be modified for use with the heat pump performance calculation procedure.

The monthly heating requirement is evaluated using Eq. (3.5.16):

$$Q_{\text{htg sys},m} = \sum_i \sum_{D_m} UA_{o,i}(T_{\text{Bal},i} - T_A) N_i \qquad (3.5.16)$$

The sum is performed over daytime, nighttime, and night setback periods. The monthly values are summed over the heating season to yield the annual load.

In the heat pump procedure there is no way to distinguish between day and nighttime periods. With the yearly temperature distribution, the temperatures are not categorized by month. This means that Eq. (3.5.16) must be modified to use average values of the conductance-area product and balance temperature. The average hourly rate of heat loss is determined from Eq. (3.5.16) by dividing by the number of hours in the month and using average values to yield

$$\bar{q}_{\text{htg sys}} = \overline{UA_o}(\bar{T}_{\text{Bal}} - T_A) \qquad (7.5.1)$$

The average value of UA should be the averages, weighted by the appropriate day, night, and night setback fractions of the entries in Table 3.5.1. The average heating load should be the annual load divided by the number of hours in the heating season. The annual heating load calculated using Eq. (7.5.1) should agree within about 5 percent of that given in Table 3.5.1.

As an example of this calculation, the house of Section 3.6 will be considered. The annual heating load for the 7-month heating season (Oct.-Apr.) is 45.3×10^6 Btu, which yields an average rate of 8900 Btu/hr. The average $\overline{UA_o}$ from Table 3.6.3 is 368 Btu/hr °F. The average air temperature for the heating season is 32°F. From Eq. (7.5.1), the average balance temperature becomes 56°F.

7.6. EXAMPLE CALCULATION

A example of the heat pump calculation procedure will be carried out using the example house of Section 3.6 located in Madison, Wisconsin. The heat pump will be the 2.0-ton unit of Fig. 7.2.1. The balance temperature and $\overline{UA_o}$ values for the heating season months as computed in Section 7.5, and are 56°F and 368 Btu/hr °F, respectively. These numbers are entered in Table 7.6.1.

The average temperature and number of hours for the heating season are taken from Table 7.4.2, and are 32°F and 5088 hours, respectively. The entries in column 1 are at 5°F increments up to the bin (57°F) that is greater in temperature than the house balance temperature (55°F). Since there is no heating required at this and

TABLE 7.6.1
Worksheet for Heat Pump Performance Calculation

Heating season: T_{ave} = __32__ °F House: $\overline{UA_o}$ = __368__ Btu/hr °F
N_{total} = __5088__ hr T_{Bal} = __56__ °F

①	②	③	④	⑤	⑥	⑦	⑧	⑨	⑩
Heating Season		House			Heat Pump				
T_A (°F)	N (hr)	$\overline{T}_{Bal} - T_A$ (°F)	Demand (10^3 Btu/hr)	Capacity (10^3 Btu/hr)	Power (kW)	Time (fraction)	HP Heating (10^6 Btu)	Supplemental Heating (10^6 Btu)	Electricity (kWh)
52	311	3	1.1	25	2.9	0.04	0.3	0	40
47	415	8	2.9	23	2.7	0.13	1.2	0	141
42	519	13	4.8	21	2.5	0.23	2.5	0	297
37	623	18	6.6	18	2.4	0.37	4.1	0	548
32	701	23	8.4	16	2.3	0.53	5.9	0	846
27	623	28	10.3	15	2.2	0.69	6.4	0	941
22	519	33	12.1	13	2.1	0.93	6.3	0	1014
17	415	38	14.0	11	2.1	1	4.6	1.2	871
12	311	43	15.8	10	2.0	1	3.1	1.8	622
7	213	48	17.6	8	2.0	1	1.7	2.0	426
2	104	53	19.5	7	1.9	1	0.7	1.3	198
−3	13	58	21.3	6	1.8	1	0.1	0.2	23
						Total	36.9	6.5	5967

House load = __43.4 × 10⁶__ Btu Heat pump $\overline{COP}_{hp}$ = __1.8__
Supplemental heating = __6.5 × 10⁶__ Btu System $\overline{COP}_{htg}$ = __1.6__
Heat pump electricity = __5967__ Total electricity = __7872__ kWh

greater temperatures, these bins are not entered. The number of hours in each bin are evaluated using Table 7.4.1. Column 3 is computed directly and the house demand, which is the product of $\overline{UA_o}$ and column 3, is entered.

The heat pump capacity and power are taken from Fig. 7.2.1 at the temperatures corresponding to the bin temperatures. The heat pump balance point is about 20°F. This is quite a bit lower than the heating season average temperature, which means that the heat pump will meet most of the heating requirements. The time fraction that the heat pump is on and the heat pump heating, supplemental heating, and electrical work terms are calculated as described in Section 7.3.

The seasonal totals are at the bottom of the columns in Table 7.6.1. The 2-ton heat pump is able to meet 85 percent of the heating load with the remaining 15 percent needed from a supplemental source. The supplemental heat is 32 percent of the purchased energy. The total purchased energy is 5967 kWh for the heat pump compressor and 1904 kWh for supplemental heating for a total of 7872 kWh. The system seasonal COP is 1.6. The criteria of Section 7.1 indicates that the heat pump will not save resource energy relative to a natural gas furnace and will cost more to operate. The heat pump COP is higher than the system COP but this is really not relevant.

These calculations are based on the steady state performance of the heat pump. In operation, cycling of the heat pump would occur which would reduce the COP. There is refrigerant migration when the system is off and losses due to compressor inefficiency during start up. The effect is most pronounced when the ambient is considerably above the balance point. There is no effect below the balance point since the heat pump is operating continuously.

The effect of cycling is characterized by a degradation coefficient defined as:

$$C_D = \frac{1 - \frac{COP_{cyc}}{COP_{ss}}}{1 - \frac{q_{house}}{q_{htg}}} \qquad (7.6.1)$$

Values of C_D range from 0.05 to 0.30. Manufacturers are now required to test heat pumps in a cyclic manner and report values of C_D, and an average value for many heat pumps is 0.25.

Equation (7.6.1) may be rearranged to evaluate the COP and power during cycling. The cycling COP is given by:

$$COP_{cyc} = COP_{ss} \left[1 - C_D \left(1 - \frac{q_{house}}{q_{htg}} \right) \right] \qquad (7.6.2)$$

The power during cycling is calculated from the capacity divided by the COP_{cyc}. This yields the power in terms of the steady state power as:

$$\dot{W}_{elec,cyc} = \frac{\dot{W}_{elec,ss}}{\left[1 - C_D \left(1 - \frac{q_{house}}{q_{htg}} \right) \right]} \qquad (7.6.3)$$

As an example of the effect of degradation due to cycling, the results presented in Table 7.6.1 will be re-evaluated. A value of C_D of 0.25 will be assumed. For the 42F bin, the COP_{ss} is 2.46. (column 5 divided by column 6).

The ratio of q_{house} to q_{htg} is the time fraction, column 7. The COP_{cyc} is then:

$$COP_{cyc} = 2.46(1 - 0.25(1 - 0.23)) = 1.98$$

There is a 20 percent drop in COP at this temperature due to cycling. The resulting electrical power required to produce this same heating effect increases from 2.5 kW to 3.1 kW.

The evaluation of the increase in electricity due to cycling has been made for the heating season example, Table 7.6.1. An additional 433 kWh are required to deliver the same heating effect. The average COP_{hp} drops from 1.8 to 1.7 and the system $\overline{COP}_{htg}$ drops from 1.6 to 1.5. The effect of cycling is to reduce the performance of the heat pump system about 5 percent. This further reduces the attractiveness of heat pump systems.

7.7. COMPARISON BETWEEN SYSTEMS

The energy and dollar economics of the heat pump system can be evaluated and compared to those for an electric furnace and a natural gas furnace. For the heat pump system, the case where supplementary heat will be supplied by either electricity or natural gas will be considered. The rates are 5¢/kWh and $3.50/$10^6$ Btu for electricity and natural gas, respectively. A power plant efficiency of 0.35 and a furnace efficiency of 0.70 will be assumed. The first costs are representative of current system prices, and are $1500 plus $500 per ton for the heat pump, $1000 for a natural gas furnace, and $500 for an electric one. The values for P_1 and P_2 are 22.17 and 1.0, respectively.

The systems are compared in Table 7.7.1. Also included is the system performance with the 3.2-ton heat pump, which has a seasonal COP of 2.2. The best system in terms of resource use is the 3.2-ton heat pump. The natural gas furnace is

TABLE 7.7.1
Comparison of Systems

System	Resource Energy Use (10^6 Btu)	End Use (10^6 Btu)	Annual Operating Cost ($)	First Cost ($)	LCC ($)
Electric resistance heat	124	43.4	637	500	14,600
2-ton heat pump	77	26.9	394	2500	11,200
2-ton heat pump with natural gas supplement	67	26.9	331	3500	10,800
3.2-ton heat pump	57	19.9	291	3100	9,500
Natural gas furnace	62	43.4	218	1000	5,800

second, and the 2-ton units follow. The electric resistance heating system uses the greatest resource energy by a factor of 2 over the other systems.

In terms of operating costs, the natural gas furnace is again the least expensive by far. The three heat pump systems are comparable. The electric resistance heating system is almost twice as expensive per year as any of the other systems. The life cycle costs are in the same relation as the annual operating costs. The first costs of the various options have a relatively small effect on the total costs.

For a heat pump to be economically viable in a northern climate such as Madison, Wisconsin, the heat pump coefficient of performance must be improved significantly. A seasonal COP of about 3 is needed in order for the heat pump to be competitive with the gas furnace in terms of annual cost. It is feasible to improve the units by increasing condenser and evaporator areas. Compressors in which variable capacity is achieved through staging or variable speed drives also improve performance. It is probably possible to increase the COP by about 50 percent at an added first cost of about $1000. However, these improved heat pumps would still not compete with improved gas furnaces economically.

The use of the heat pump in more favorable climates was explored by calculating its performance for the other three locations of Section 3.7.2. The house was taken to be the same insulated house as in Sections 3.7.2 and 7.6, and the 2-ton heat pump was assumed to be installed. The heat pump was compared to the natural gas furnace on both a resource and economic basis. The results are given in Table 7.7.2. The house loads are slightly different from those in Table 3.7.3 due to the use of the generalized temperature distributions.

The heat pump saves resource energy compared to natural gas furnaces only in the warmest climate of Los Angeles. There is very close to the same use in Washington, DC. Heat pumps with 50 percent higher COP would save resource energy in all locations. Currently, heat pumps do not save resource energy over natural gas furnaces, but are definitely valuable when the alternative is electric resistance heating.

Heat pumps are not cost effective in any location compared to natural gas furnaces. Even with improved COPs, this situation would hold. A heat pump with a 50 percent higher COP would save on annual fuel costs in the warmer locations (Washington, DC and Los Angeles). However, the first cost of the heat pump is much higher than that of the furnace and dominates the life cycle cost. For those

TABLE 7.7.2
Heat Pump Performance in Different Locations

Location	House Load (10^6 Btu)	Gas Furnace Resource (10^6 Btu)	LCC ($)	COP_{sys}	Heat Pump Resource (10^6 Btu)	LCC ($)
Madison, WI	44	62	5800	1.6	77	11,200
Chicago, IL	31	45	4500	1.8	50	8,200
Washington, DC	21	31	3400	1.9	33	6,200
Los Angeles, CA	4.5	6.4	1500	2.2	5.9	3,100

locations where heat pumps work best, the heating loads are small and the systems cannot save much money.

There are circumstances that may make heat pumps economically attractive. In many areas of the country air conditioning is necessary. The reverse cycle heat pump can be used to meet those air conditioning loads. An air conditioner does not have to be purchased and this reduces the total life cycle cost of the space heating and cooling system.

Finally, these comparisons have been made against natural gas at current (1980) prices. If natural gas prices increase drastically relative to electricity, the economics will change. The heat pump is definitely cost competitive with fuel oil and electricity, and should be considered in those areas in which only those fuels are available. The heat pump also has potential in many specialized circumstances.

7.8. WATER SOURCE HEAT PUMPS

In many areas of the country, ground- or surface water is plentiful. It may, potentially, serve as a heat source or sink for a heat pump. However, there are often laws and codes relating to the use of water for any purpose. It is sometimes illegal to return water used for a process to the water table. In urban areas, there are often restrictions on the use of city-supplied water, and prohibitions against using either storm sewers or sanitary sewers for disposing of process water. In many locations, lakes, streams, and rivers are under the jurisdiction of various natural resources boards, and these sources may not be used for heat pump heating. Thus, there are many impediments in the way of water source heat pumps.

If a source is available, though, the heat pump performance is significantly improved over that of an air-to-air unit. Well water is at a uniform and relatively high temperature throughout the winter season since the deep ground temperature is essentially equal to the annual average temperature. Its use would increase the seasonal COP over that of air units. A water source unit may also inherently have a higher COP since the better heat transfer in the evaporator reduces the temperature difference between the refrigerant and the source. The capacity is then constant over the heating season. This means that a unit sized only to meet the design heating load using a relatively warm source is required. As a result, the size of a water source heat pump is smaller than an air source one for the same application. A further benefit occurs in the cooling season, when the heat pump can reject heat to cold water rather than the warm ambient. Thus, there are strong incentives for the use of water as a heat source.

The thermal and economic performance of water source heat pumps can be estimated for residential applications. The capacity, power, and COP of a representative 2-ton unit is given in Table 7.8.1. Comparison of these performance figures with the curves of Fig. 7.2.1 show that the water source unit has considerably higher capacity and COP at a given source temperature.

The performance of water source units in residential applications has been determined for the locations given in Table 7.7.2. For each location, a 2-ton unit is large

TABLE 7.8.1
Performance of 2-Ton Water Source Heat Pump

Water Temperature (°F)	Capacity (10^3 Btu/hr)	Power (kW)	COP
32	20	2.4	2.4
40	25	2.7	2.7
45	28	2.9	2.8
50	31	3.1	2.9
55	34	3.3	3.0
60	37	3.5	3.1
65	40	3.7	3.2

enough to meet the house load without auxiliary heating. The thermal and economic performance of the systems is summarized in Table 7.8.2.

Comparison of these results for the water source units with those for the air-to-air units demonstrates the better performance possible if water is available. These heat pump systems are less expensive than air-to-air units but more expensive than natural gas for heating only. However, if air conditioning is a consideration, the water source unit would be more economical than a combination of gas furnace and conventional air conditioners.

The results in Table 7.8.2 do not include the cost of wells or other water transport systems. Deep wells may cost $2000 to $5000, while piping to and from lakes or streams and pumps may cost $500 and up. Comparing the costs in Table 7.7.2 with those in Table 7.8.2 shows that these additional costs must be less than $100 for the water source heat pump to be cost effective in Los Angeles, and less than $3500 for Madison, Wisconsin. Clearly, the costs to supply the water has a significant effect on the economics of these systems.

In summary, water source units have the potential to economically meet the heating and cooling loads of buildings. Water availability and the cost of providing it are significant limiting factors. Water source units are feasible only for certain areas of the country where these factors are favorable.

TABLE 7.8.2
Water Source Heat Pump Performance in Different Locations

Location	$\bar{T}_A$ (°F)	COP	P (kWh)	LCC ($)
Madison, WI	45	2.8	4600	7600
Chicago, IL	51	2.9	3110	6000
Washington, DC	56	3.0	2040	4800
Los Angeles, CA	61	3.1	420	3000

7.9. HEAT PUMP AND STORAGE SYSTEMS

The effects of the mismatch on either side of the balance point between the heat pump capacity and the house heating demand can be alleviated by a storage system. The heat pump would pump energy into storage during periods in which there is excess capacity, and this energy would be used for heating during periods in which capacity is insufficient. The storage unit could be either a water tank in a water system or a rock bed in an air system. The size of the storage can range from two days' capacity for a small system up to systems large enough to store energy collected in one season for use in another season.

In a system with a relatively small storage capacity, the heat pump would operate continuously during the winter. Heat in excess of that needed for the house would be delivered to storage for later use. The maximum improvement in performance would occur if all supplemental heating were eliminated completely.

In order to illustrate the improvement in system performance with a small storage system, the example of Section 7.6 will be considered. It will be assumed that the storage system is large enough to store the total supplemental heating requirement of 6.5×10^6 Btu. The heat pump could supply this energy to storage during those days when its capacity is greater than demand. This will be assumed to occur for those bins just above the balance point. This implies that during the winter there are enough days occurring in which the daily temperature exceeds the balance point to allow the storage to be charged, and that these days occur frequently enough so that the storage system is not depleted in-between. The size of the storage needed for this depends on weather sequences.

With this storage system, the heat pump would need to operate continuously during the bins with average temperatures of 22 and 27°F, and 60 percent of the time at 32°F. The heat pump would deliver the 6.5×10^6 Btu of heat to storage using 816 kWh of compressor work at a COP of 2.3. The system seasonal COP would rise from 1.6 to 1.9, and the yearly fuel cost would drop from \$394 with the electric auxiliary to \$339. Resource use would drop by 14 percent to 66×10^6 Btu. Clearly, this system is better from an energy standpoint.

From an economic standpoint, the savings of \$55/yr are achieved at the cost of a storage system. The amount of energy to be stored will be assumed to be enough for one day when the weather is at design conditions. This amounts to about 700,000 Btu. This energy could be stored in a 1500-gal water tank in which the water temperature change during storage was 50°F. Alternatively, a 800-ft^3 rock bed could be used in which the rock temperature change was 50°F. These large temperature changes in the stores would lower heat pump COP somewhat.

The costs of large water tanks are about \$1/gal, and those of rock bed storage systems are about \$1/ft^3. The added cost of storage alone is then in the range of \$1000–\$2000. There would be additional costs due to added ducting or piping and controls, and there might be a cost associated with the loss of living space for the storage.

The life cycle savings for the fuel saved due to storage are about \$1200. The first costs would probably exceed this by a factor of 2. Further, the storage size may be optimistically small because of weather sequences which either might cause the storage to become depleted or might not allow energy to be stored. Thermal storage systems are probably not economical for residential applications.

Commercial buildings are more suitable for combined heat pump-storage systems. The internal gains due to lighting, people, and equipment are appreciable and may be more than sufficient to heat the building in cold weather. Excess heat generated during occupied periods may then be stored for nighttime and early morning. A diurnal storage system could be cost effective. This will be considered in Section 7.10.

A seasonal storage system that has received attention recently is the ice-maker system. In this approach, a large water reservoir is used as the source. During winter, heat is extracted and the water is frozen. In summer, the ice is melted to provide air conditioning. For most locations, the yearly heating load is different from the cooling load, and so additional energy must either be added or subtracted to balance the system over the course of the year. The installed heat pump has sufficient capacity to meet the design load at the source temperature required and thus there is no auxiliary energy supply in such a system. This heat pump would be larger than that normally installed for an ambient air source system.

To illustrate the performance of such a system, the example of Table 7.6.1 will again be considered. The design heating load for the house is for an ambient temperature of -11°F, and is about 31,000 Btu/hr. Using the performance data in Table 7.8.1, it is seen that a 3-ton heat pump would be appropriate. The power requirement is 3.6 kW, and the heat pump and seasonal COP are the same and equal to 2.4.

The total house heating requirement of 43.4×10^6 Btu is met by the heat pump with the expenditure of 5298 kWh. The annual cost amounts to $265 for annual savings over the 2-ton air-air unit of $129. The resource energy use is 52×10^6 Btu. Compared to the other systems listed in Table 7.7.1, there are significant savings both in resource energy use and operating costs.

The life cycle cost of the ice maker heat pump system can be estimated. The life cycle fuel costs are $5900, and the heat pump first cost is about $3100 for a subtotal of about $9000. Even without considering the cost of storage the ice-maker system is not competitive with natural gas. In order for the ice-maker heat pump to compete economically with the other heat pump systems, the storage costs must be in the range of $1000-$3000. The volume of storage needed is 35,000 gal, which is about one-third the volume of a home swimming pool or two-thirds the volume of a full-sized basement. The cost to put in a swimming pool is on the order of 5-10¢/gal. At this price range, the ice-maker system is barely competitive with the other heat pump systems.

The stored ice can be used for air conditioning during the summer. The operating costs are low since only pumping of cold water is required. However, the amount of ice available is about four times larger than that needed, and during summer heat must be supplied to melt the ice. The concept is best suited for localities in which the heating and cooling loads are comparable.

7.10. COMMERCIAL APPLICATIONS

Heat pumps have been used widely in large buildings to distribute energy throughout the structure. Heat generated by lights, people, and machinery in interior zones must be removed for comfort. A heat pump is commonly used to transfer this heat to a perimeter zone where it may be used for heating.

An example of a heating and cooling system using a heat pump is the system of Section 5.2, which is shown here as Fig. 7.10.1. The heat pump evaporator is used to cool the air entering the interior space. The energy removed plus the compressor work is used to heat the perimeter zones. The heating and cooling loads are rarely of the correct proportion and do not occur at the same times so that the heat pump

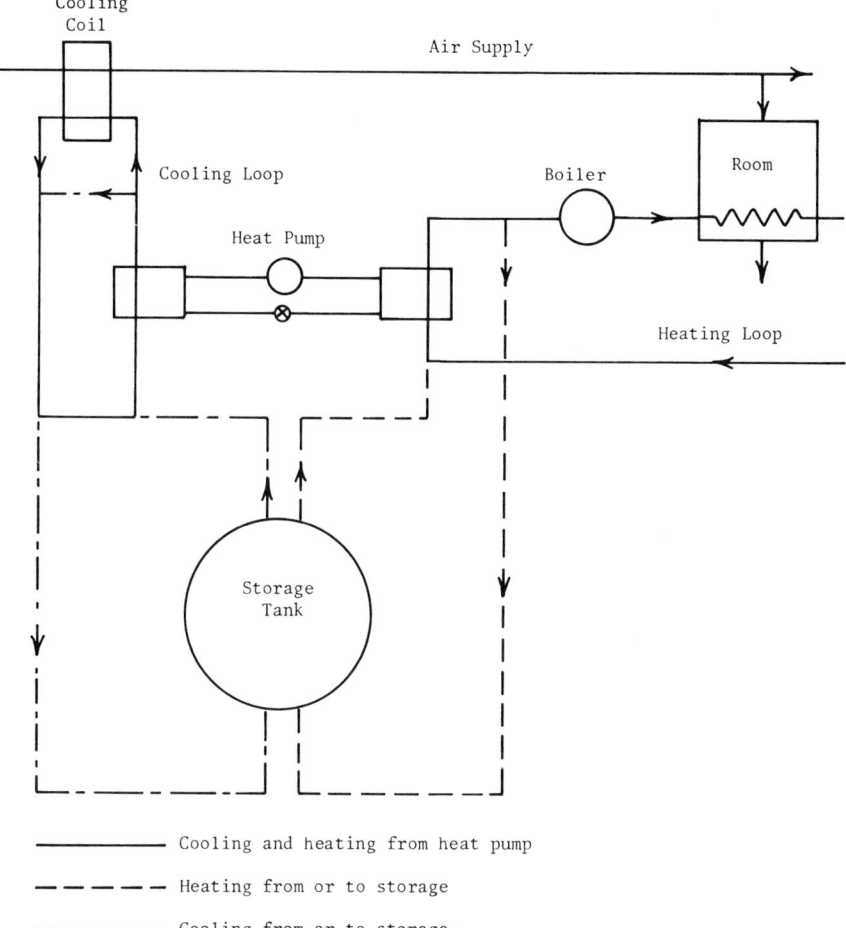

Fig. 7.10.1. Schematic of commercial building system using a heat pump.

can not accomplish both functions alone. A storage tank is needed to match supply and demand.

In winter, auxiliary heating with a conventional boiler would be required to supplement the condenser output. In spring and fall, the daytime heating loads are small, the air conditioning loads large, and heating is required at night. The energy obtained from air conditioning is then stored in the storage tank during the day and used for heating at night. In some systems in which the stored energy is not sufficient for direct heating, the heat pump is used to cool down the tank and pump heat to the building. In summer all zones would require cooling and the condenser heat flow would be transferred to the environment through a cooling tower (which is not shown). Some systems also take advantage of off-peak electric rates to cool down the storage tank in summer and use the cold water for cooling later on. These options are all indicated in Fig. 7.10.1. In this manner, the heat pump can be used to reduce heating costs throughout the heating season.

The operation of the system at the conditions chosen in Chapter 5 will be considered. During the 9-hour occupied period, the air conditioning load is 1.9×10^5 Btu/hr for a total cooling load of 1.7×10^6 Btu. The heating demand is 1.2×10^5 Btu/hr for the remaining 15 hours for a total of 1.8×10^6 Btu. For heating load and supply air temperatures of 120 and 55°F, respectively, the heating COP of the heat pump would be about 3.2.

During the occupied period, the heat rejected from the condenser would be stored in the tank. The amount of energy that could be stored is the sum of the cooling load and compressor work. In terms of the COP, this can be written as

$$Q_{\text{cond}} = \frac{Q_{ev} \text{COP}_{hp}}{\text{COP}_{hp} - 1} \qquad (7.10.1)$$

For the total cooling load of 1.7×10^6 Btu, a total of 2.6×10^6 Btu could be stored. This is more than the heating requirement of 1.8×10^6 Btu, and the heat pump would only have to operate for 6.2 hours to charge the tank. The system would then switch to the economizer cycle and the heat pump would not be used.

The operating costs of the system in this mode are the heat pump costs only. The total electrical work required for the 6.2 hours of operation is 160 kWh. At 4¢/kWh, this amounts to $6.40 per day. The alternative is to use the economizer cycle to meet all of the cooling load and to use the boiler for heating. The boiler would deliver the 1.8×10^6 Btu of heat using oil at $9/10^6$ Btu for a total cost of $16.20 per day.

The savings due to using the heat pump system are $9.80 per day. These savings must pay off the added costs of the storage tank and controls. A storage tank of about 5000 gal is needed, and the added costs are then on the order of $5000. The system would pay for itself in about 500 days of operation, which is about 4-6 heating seasons. The system would also use less resource energy than the oil furnace. Thus, the heat pump in this application is both more energy and cost efficient than the alternative.

It is probably obvious that the system for commercial buildings can be quite complex. Many decisions as to when to switch to a given mode are required. The savings potentials are high, and so detailed analyses considering all situations are needed. This example is given to illustrate the potential of a heat pump in a commercial application.

7.11. SUMMARY

The basic thermodynamic relations and performance characteristics of heat pumps have been described in this chapter. The heating or cooling capacity are strong functions of ambient temperature, and decrease as house heating or cooling demands increase. This mismatch means that supplemental heating is always required and that the unit may not be able to supply enough cooling under extreme conditions.

A method that uses generalized weather data has been developed for calculating the seasonal performance of heat pump systems. The information required is the average temperature of the heating season and its length. With the performance determined, the heat pump economics can be evaluated.

Compared to electric resistance heating, heat pumps save both resource energy and money. The same is generally true compared to oil furnaces. At current natural gas prices and with improved gas furnaces, heat pumps are less energy and cost effective. Improved heat pumps are technically possible and would improve their competitiveness. There are also considerations such as air conditioning needs and circumstances such as lake or other water sources which encourage the installation of heat pumps. Heat pumps have been widely used in the space conditioning systems for commercial structures, and are both energy and cost effective here.

SUGGESTED READING

W. F. Stoecker and J. W. Jones, *Refrigeration and Air Conditioning*, McGraw-Hill Book Co., New York, 1982.

Alabama Power Company Reports and personal communications.

J. V. Anderson, J. W. Mitchell, and W. A. Beckman, Performance Predictions of Air-to-Air Heat Pumps Using Generalized Weather Distributions, to be published in *ASHRAE Transactions*.

H. C. Fisher, Ice Maker Heat Pump: A New Tool for Energy Conservation, *Refrigeration Service and Contracting Magazine*, Jan.–Feb. 1977, p. 23.

R. D. Heap, *Heat Pumps*, John Wiley & Sons, New York, 1979.

PROBLEMS

7.1. A house with characteristics of Table 7.6.1 is heated with a heat pump. Compare the life cycle savings of the heat pump system against an all electric home and one with a natural gas furnace (70 percent seasonal efficiency). Fuel prices are $0.05/kWh for electricity and $5/10^6$ Btu for natural gas. Choose a house location from Table 7.4.2.

7.2. For Problem 7.1, vary the heat pump size to determine the optimum heat pump.

7.3. For Problem 7.1, compare the performance against one using seasonal storage and ice.

■ 8 ■

ELECTRICAL POWER PRODUCTION

Electricity production comprises about 25 percent of the total U.S. energy use. Of this, about 40 percent is used in the residential and commercial sector with the remaining 60 percent used by industry. Conventional power plant efficiencies are approximately 35 percent, which means that only about one-third of the resource energy ends up as usable energy. The remaining two-thirds is "wasted" as heat transfer to the environment. The industrial, residential, and commercial sectors also have large demands for heating. Methods that simultaneously use fuel for electricity and heat production would be effective.

In this chapter, the potential for producing electricity from different fuel sources will be explored first. This requires an application of the second law of thermodynamics to determine the thermodynamic availability of a fuel, which is a measure of its potential to produce electricity. The sources of inefficiency in conventional power plants will be studied to determine the limits on efficiency. The improvements through production of "waste heat" from a power producing plant, or of "waste electricity" from a heating plant, will be analyzed. The economic and energy savings of such cogeneration plants will be estimated, and some problems of load management discussed.

8.1. THERMODYNAMIC PRINCIPLES

An electrical generating plant converts fuel energy into useful work (electricity). As such, both the first and second laws of thermodynamics are applicable. The first law deals with the energy transformations and conservation, while the second law deals with availability (available energy) transformations and the irreversibility associated with the process.

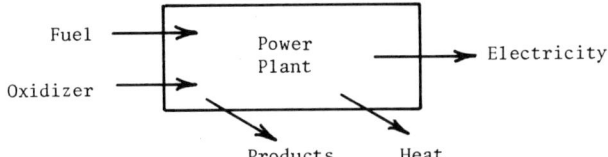

Fig. 8.1.1. Representation of a power plant.

A representation of a power plant is shown in Fig. 8.1.1. Fuel and oxidizer enter the plant, electricity is produced, products leave, and heat is rejected to the environment. The corresponding energy flows are shown in Fig. 8.1.2. The conservation of energy principle applied to a power plant for steady flows of energy is

$$\dot{E}_r - \dot{W}_{elec} - \dot{E}_p - q = 0 \tag{8.1.1}$$

The thermal efficiency is defined in terms of the amount of electricity produced per unit of fuel energy supplied as

$$\eta_{th} = \frac{\dot{W}_{elec}}{\dot{E}_r} \tag{8.1.2}$$

The thermal efficiency can approach unity if the products and rejected energies approach zero. This may be possible for some energy sources and types of power plants.

The second law of thermodynamics can also be applied to the power plant. The key concept of the second law is availability, or available energy, defined as

availability = the maximum useful work a system can do when it interacts only with its surroundings

It is important to realize that availability deals with *useful* work, not energy. It is a defined quantity reflecting man's desire to use machines to replace human labor.

Availability is not conserved in the same way energy is. The second law states that availability is always destroyed in any real process. This destruction is sometimes termed the irreversibility of the process. Although availability destruction or irreversibility has the units of energy, it represents the loss in the potential for pro-

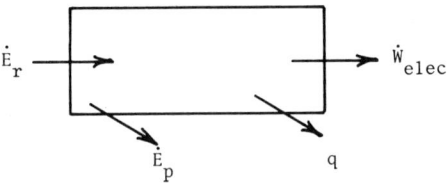

Fig. 8.1.2. Energy flows for a power plant.

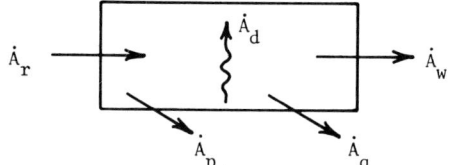

Fig. 8.1.3. Availability flows for a power plant.

ducing *useful* work. In general, then, the second law applied to a system undergoing steady-state processes is

$$\dot{A}_{in} - \dot{A}_{out} - \dot{A}_d = 0 \tag{8.1.3}$$

The availability flows for a power plant are shown schematically in Fig. 8.1.3. The second law applied to the power plant states that

$$\dot{A}_r - \dot{A}_p - \dot{A}_w - \dot{A}_q - \dot{A}_d = 0 \tag{8.1.4}$$

The maximum utilization of fuel will occur for the smallest values of $\dot{A}_p$, $\dot{A}_q$, and $\dot{A}_d$. The goal of the following discussion is to determine what contributes to these terms and how they can be made small in practice.

In order to make a second law analysis, the relations between availability and energy content for various forms must be established. For mechanical shaft work or electrical work, the relation is

$$\dot{A}_w = \dot{W} \tag{8.1.5}$$

All of the work done is useful work. In a steady-flow process, no work is done against the atmosphere.

The relation between the availability transfer due to heat and the heat flow itself is

$$\dot{A}_q = q\left(1 - \frac{T_o}{T_s}\right) \tag{8.1.6}$$

The term in parentheses is the Carnot efficiency for an ideal cycle operating between the source (T_s) and the environment (T_o). The maximum amount of useful work that a given amount of heat can produce is obtained through use of a Carnot engine.

The availability of the flow of a substance such as air is given by

$$\dot{A}_f = \dot{m}\left[(h - T_o s) + \frac{V^2}{2} + zg\right] \tag{8.1.7}$$

The first term in parentheses represents the work that could be done through expansion of the substance against the atmosphere and heat transfer to a Carnot engine. All of the enthalpy is not available for producing useful work as represented by the term $T_o s$. However, the kinetic and potential energy forms could be converted entirely into useful work.

These availability forms given are only for flows of heat, work, and mass. In power plant applications, various energy sources are employed as fuel to produce electricity. The availability of these sources in comparison to their energy content needs to be determined.

8.2. AVAILABILITY OF ENERGY SOURCES

In this section, the thermodynamic availability of energy sources will be evaluated. This will establish whether it is thermodynamically possible to completely convert the energy contained in the source to electricity. This allows consideration of whether the definition of efficiency [Eq. (8.1.2)] is reasonable in that the efficiency could, theoretically, approach unity.

There are two types of energy sources that will be considered. The first are those that are directly utilized for producing work such as wind and water sources. The second category are those that produce electricity indirectly through heat production such as petroleum, nuclear, and solar sources. These are the major sources at present and also those proposed for the near future.

For the directly convertible forms, the availability flow is equal to the energy flow. The availability in wind is the kinetic energy, or

$$\dot{A}_{\text{wind}} = \dot{m} \left(\frac{V^2}{2} \right) \quad (8.2.1)$$

The availability of water behind a dam is its potential energy or

$$\dot{A}_{\text{hydro}} = \dot{m}(zg) \quad (8.2.2)$$

For the second class, the availability value is not equal to the energy value. For solar energy used in a solar thermal power plant, the source is heat from the sun and given by

$$\dot{A}_{\text{solar}} = \dot{E}_{\text{solar}} \left(1 - \frac{T_o}{T_{\text{sun}}} \right) \quad (8.2.3)$$

Here, the energy is potentially available at the temperature of the sun through the use of an ideal collector. The sun temperature is about $10,000°R$, and, for an earth environment of $60°F$ ($520°R$) the availability flow is equal to about 0.95 of the energy flow. For photovoltaic or solar thermal power plants, essentially all of the energy flow could potentially be completely converted into electricity.

Nuclear fusion and fission processes produce heat which is potentially at a tem-

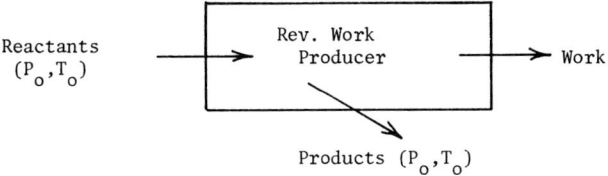

Fig. 8.2.1. Schematic of an ideal work producing device.

perature comparable to that of the sun. The availability is given by

$$\dot{A}_{nuc} = \dot{E}_{nuc} \left(1 - \frac{T_o}{T_{nuc}}\right) \quad (8.2.4)$$

The term in parentheses is about 0.95, and here the availability is essentially equal to the energy.

The available energy associated with a petroleum fuel is a little more difficult to determine. To evaluate it, an ideal, reversible, work producing device will be analyzed. The reactants (petroleum fuel and oxidizer) enter at the reference state (surroundings pressure and temperature) and the products leave at the reference state. Work is produced and transferred out. The system is shown in Fig. 8.2.1.

Since the device is ideal, the availability destroyed is zero. This is shown schematically in Fig. 8.2.2. The second law (Eq. 8.1.3) states that

$$\dot{A}_r - \dot{A}_w - \dot{A}_p = 0 \quad (8.2.5)$$

The availability flow due to work equals the work since this is also the maximum work that could be produced. Its value is given by

$$\dot{A}_w = \dot{W}_{max} = \dot{A}_r - \dot{A}_p \quad (8.2.6)$$

Neglecting the changes in kinetic and potential energy of the flows and using Eq. (8.1.7) allows the maximum work to be expressed as

$$\dot{W}_{max} = [\dot{m}(h - T_o s)]_r - [\dot{m}(h - T_o s)]_p$$

or

$$\dot{W}_{max} = [(\dot{m}h)_r - (\dot{m}h)_p] - T_o[(\dot{m}s)_r - (\dot{m}s)_p] \quad (8.2.7)$$

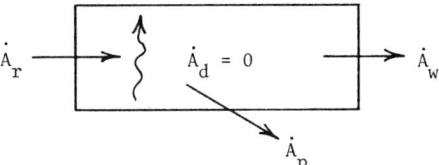

Fig. 8.2.2. Availability flows for ideal work producing device.

The first term in brackets is the enthalpy difference between reactants and products at T_o. This is the enthalpy of combustion of the fuel on a per unit mass of fuel basis (h_{rp}) at T_o and P_o multiplied by the fuel flow rate

$$[(\dot{m}h)_r - (\dot{m}h)_p] = \dot{m}_{fuel}h_{rp} \qquad (8.2.8)$$

If T_o is also the reference temperature at which h_{rp} has been measured, h_{rp} is the fuel higher heating value HHV. Equation (8.2.7) can be written using Eq. (8.2.8) as

$$\dot{W}_{max} = \dot{m}_{fuel}h_{rp} - T_o[(\dot{m}s)_r - (\dot{m}s)_p] \qquad (8.2.9)$$

The availability of combustion of the fuel can be defined analogous to the enthalpy of combustion. It is the maximum useful work produced in the combustion process per unit mass of fuel, and is defined as

$$a_{rp} = \frac{\dot{W}_{max}}{\dot{m}_{fuel}} \qquad (8.2.10)$$

Using Eq. (8.2.9), the availability of combustion is evaluated as

$$a_{rp} = h_{rp} - \frac{T_o[(\dot{m}s)_r - (\dot{m}s)_p]}{\dot{m}_{fuel}} \qquad (8.2.11)$$

The availability of combustion differs from the enthalpy of combustion by the entropy difference between reactants and products at T_o and P_o. This difference can be calculated using the chemical combustion relations and property values.

The availability of combustion for three different petroleum fuels is compared to the enthalpy of combustion in Table 8.2.1. This shows that the numerical value of the availability of a petroleum fuel is essentially equal to its energy value.

The development in this section shows that the availability of most fuels or sources currently used to produce electricity is essentially equal to the energy content. This means that all of the fuel sources have the potential to completely convert the energy content into useful work (electricity). For sources such as wind and water, conversion efficiencies are high. However, for petroleum, nuclear, and solar

TABLE 8.2.1
Availability of Petroleum Fuels

Fuel	h_{rp} (Btu/lb$_m$)	a_{rp} (Btu/lb$_m$)	a_{rp}/h_{rp}
Coal	14,087	14,118	1.002
Methane	21,502	21,069	0.980
Octane	19,256	19,647	1.020

AVAILABILITY DESTRUCTION DUE TO COMBUSTION

sources, the conversion efficiencies are relatively low. To determine why the efficiencies are so low, the conventional power plant will be analyzed from a second law point of view. To do this, the availability destroyed in a combustion process will need to be evaluated first.

8.3. AVAILABILITY DESTRUCTION DUE TO COMBUSTION

In this section the availability destruction due to combustion will be determined. An adiabatic and chemically complete combustion process will be analyzed. This is a limit of what occurs in a fuel-fired boiler or internal combustion engine. Combustion occurs first and then the high temperature products either transfer heat, as in a boiler, or expand and do work, as in an engine. A schematic of the system is given in Fig. 8.3.1. The reactants enter at T_o and P_o and the products leave at a high temperature (T_p) and atmospheric pressure.

The second law [Eq. (8.1.3)] applied to the system of Fig. 8.3.1 is

$$(\dot{m}a)_{rT_o} - (\dot{m}a)_{pT_p} - \dot{A}_d = 0 \qquad (8.3.1)$$

where the subscripted temperatures denote that the reactants and products are at T_o and T_p, respectively. The availability destroyed is

$$\dot{A}_d = (\dot{m}a)_{rT_o} - (\dot{m}a)_{pT_p} \qquad (8.3.2)$$

The difference in availability can be rewritten in terms of the availability difference between reactants and products at T_o plus the difference between products at T_o and products at T_p. Thus

$$\dot{A}_d = [(\dot{m}a)_r - (\dot{m}a)_p]_{T_o} + \dot{m}_p [a_{T_o} - a_{T_p}] \qquad (8.3.3)$$

The first term in brackets is the availability of combustion of the fuel times the fuel flow rate, $\dot{m}_{\text{fuel}} a_{rp}$. The second term is the difference in the availability of the products between T_o and T_p. In order to evaluate it, the products are assumed to be gaseous. Using ideal gas relations with an average specific heat $\bar{c}_p$ for the mixture and recognizing that the products are at atmospheric pressure allows the second

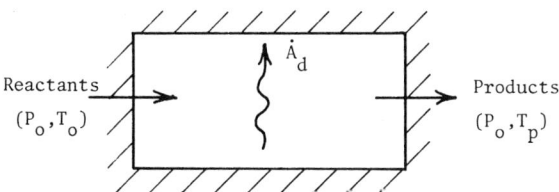

Fig. 8.3.1. Schematic of an adiabatic combustion chamber.

ELECTRICAL POWER PRODUCTION

term in brackets to be rewritten as

$$\dot{m}_p [a_{T_o} - a_{T_p}] = \dot{m}_p [(h - T_o s)_{T_o} - (h - T_o s)_{T_p}]$$

$$= -\dot{m}_p \bar{c}_p \left[(T_p - T_o) - T_o \ln \frac{T_p}{T_o} \right] \qquad (8.3.4)$$

Equation (8.3.3) then yields the availability destroyed per pound of fuel as

$$\frac{\dot{A}_d}{\dot{m}_{\text{fuel}}} = a_{rp} - \frac{\dot{m}_p \bar{c}_p}{\dot{m}_{\text{fuel}}} \left[(T_p - T_o) - T_o \ln \frac{T_p}{T_o} \right] \qquad (8.3.5)$$

Equation (8.3.5) states that the availability destroyed equals the fuel availability less the availability of the products. In a power cycle, it is desirable to minimize the availability destruction. This would occur with high $\dot{m}_p$ and high T_p. However it is not possible to increase $\dot{m}_p$ and T_p independently. For a given fuel flow rate, increasing $\dot{m}_p$ is achieved by increasing the amount of excess air, which lowers the products temperature T_p.

The relation between the availability destruction and the products temperature is shown in Fig. 8.3.2 for methane. The destruction decreases with increasing products temperature. The maximum products temperature is obtained for complete combustion at stoichiometric conditions. Under these conditions, there is still a destruction of about 25 percent of the fuel availability. For a typical combustion temperature of 3000°F, the destruction amounts to about one-third of the fuel availability.

The combustion process inherently destroys a significant amount of fuel availability. The potential for complete conversion of fuel energy to electricity is possible for noncombusting systems such as fuel cells, but the conventional cycles such as

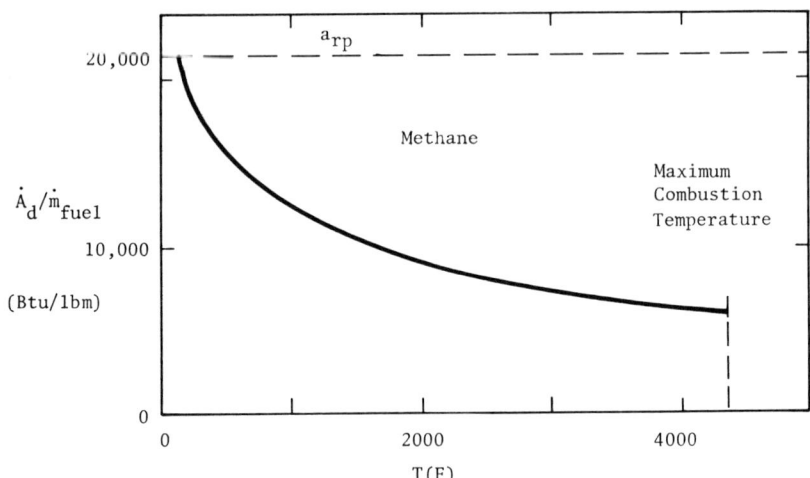

Fig. 8.3.2. Availability destroyed as a function of products temperature.

Rankine and Diesel cycles in which fuel is combusted have a limit considerably below unity. It is helpful to define the second law effectiveness of the combustion process as

$$\epsilon_c = \frac{\dot{A}_p}{\dot{A}_{\text{fuel}}}$$

or

$$\epsilon_c = 1 - \frac{\dot{A}_d}{\dot{m}_{\text{fuel}} a_{rp}} \tag{8.3.6}$$

The effectiveness of a boiler in transferring availability, in contrast to transferring heat, can also be defined. The availability transferred by heat transfer to the circulating fluid at T_f is

$$\dot{A}_q = q\left(1 - \frac{T_o}{T_f}\right) \tag{8.3.7}$$

The effectiveness of the boiler in transferring availability from fuel to the circulating fluid is defined as

$$\epsilon_b = \frac{\dot{A}_q}{\dot{m}_{\text{fuel}} a_{rp}} \tag{8.3.8}$$

In contrast, the boiler energy efficiency is defined as the heat transfer per unit of fuel energy

$$\eta_b = \frac{q}{\dot{m}_{\text{fuel}} h_{rp}} \tag{8.3.9}$$

Combining Eqs. (8.3.7) and (8.3.8) with (8.3.9), and utilizing the fact that a_{rp} and h_{rp} are essentially equal, allows Eq. (8.3.9) to be written as

$$\epsilon_b = \eta_b\left(1 - \frac{T_o}{T_f}\right) \tag{8.3.10}$$

The combustion process effectiveness, the boiler effectiveness, and the boiler efficiency are all shown in Fig. 8.3.3. The temperature on the ordinate is that of the products for ϵ_c, and the steam for ϵ_b. The combustion effectiveness increases as the products temperature increases, and the boiler effectiveness increases as the steam temperature rises. As these temperatures decrease the effectiveness approach zero. In contrast the boiler energy efficiency remains high regardless of temperature.

The significance of Fig. 8.3.3 can be shown by example. For 50 percent excess air, the products temperature would be about 3000°F, and the corresponding com-

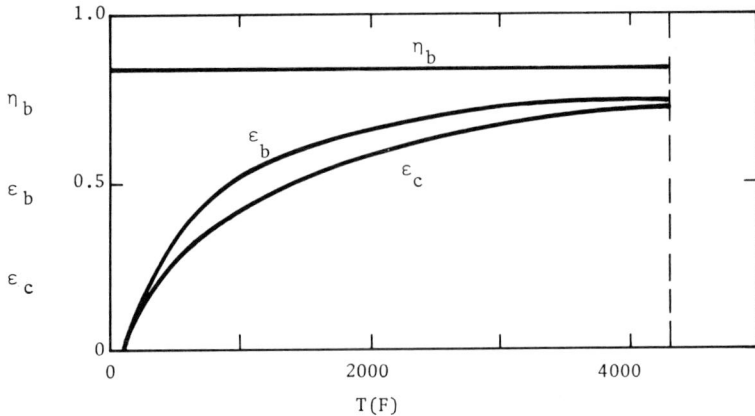

Fig. 8.3.3. Boiler efficiency, effectiveness, and combustion effectiveness as functions of temperature.

bustion effectiveness is 0.68. The circulating steam would leave the boiler at around 1200°F and the corresponding effectiveness is 0.57. This shows that 32 percent of the availability is destroyed in the combustion process itself, and only another 11 percent in transferring the heat from the high temperature products to the steam.

The energy and availability transfers in a typical power cycle are shown in Figs. 8.3.4 and 8.3.5. The values are the percent of the entering fuel energy and availability. These figures reveal that while energy efficiency of the components may be high, the availability effectiveness is not. Of the entering fuel energy, 85 percent of the energy is transferred in the boiler, while only 57 percent of the availability is. The electricity accounts for 35 percent of the energy and the availability. In the condenser 50 percent of the heat is transferred at zero availability.

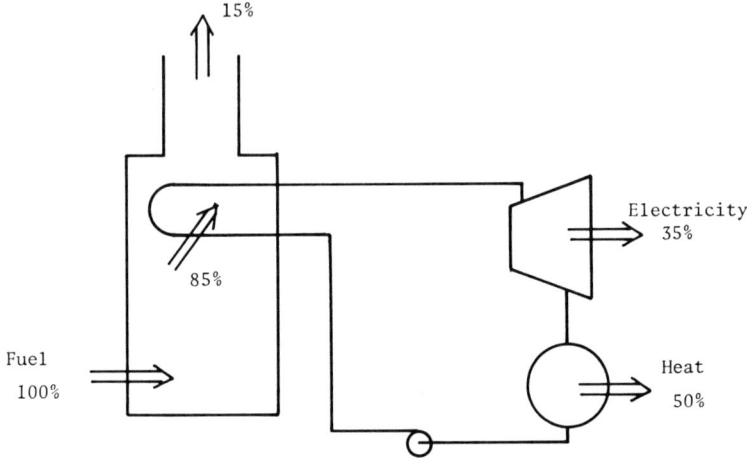

Fig. 8.3.4. Energy flows for a power plant.

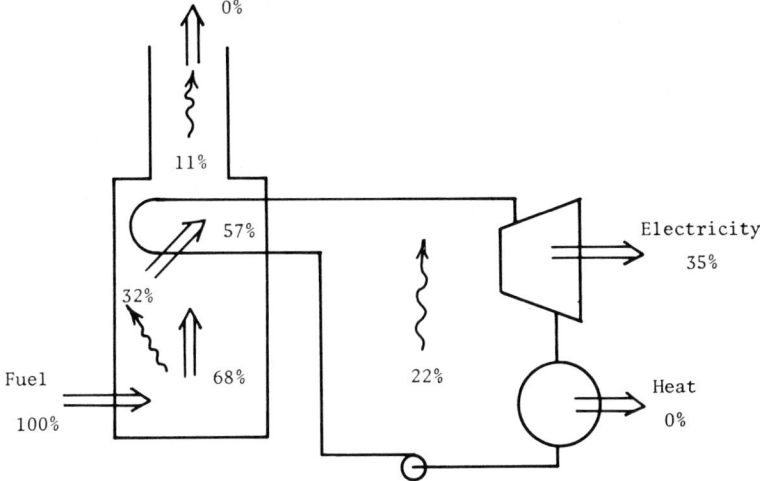

Fig. 8.3.5. Availability flows for a power plant.

The destruction of availability in the steam cycle itself amounts to 22 percent of the fuel availability. Of this, about one-half is due to the operation of the cycle as a Rankine cycle, where heat is added at varying temperatures, in contrast to a Carnot cycle where heat is added at a constant temperature. The destruction due to the condenser heat transfer is only 2 percent. Only 10 percent of the losses are due to turbine inefficiencies, heat losses, and pipe friction. Power plant development has reduced these losses to low values.

In summary, the combustion process is inherently irreversible and accounts for a major destruction of availability. For any power cycle using combustion of fuel the loss is on the order of 30 to 40 percent of that available in the fuel. Better engineering cannot avoid this loss for combusting systems. The alternative for power generation is to develop direct conversion devices such as fuel cells. Another approach is to use fuel energy for both electrical generation and heating. This idea will be studied in the next section.

8.4. COGENERATION OF ELECTRICITY AND HEAT

A conventional central power plant has a large availability destruction due to the combustion process. Although most of the fuel energy is transferred to the steam, only 50–60 percent of the availability is transferred of which 30–40 percent ends up as useful work. Thus, there is a significant amount of energy transferred that has little potential to produce useful work, but that could be used for heating.

In contrast, a heating plant destroys all of the availability of the fuel. The potential to produce electricity exists after combustion, but since the need is for heat, the potential is generally not utilized. There is in a heating plant, then, the possibility of generating electricity.

Power plants and heating plants do not separately utilize the full potential of fuel to produce electricity and heat. Combining the two functions in one plant will increase the overall utilization of fuel. Historically, this has been done in industries that require process steam. "Waste electricity" is produced either by high temperature steam prior to its use in a process, or by low temperature steam exhausting from a process. Some central power plants have produced steam in excess of that needed for electrical production, sold the steam for heating, and used the returning condensate for reheating. These practices are becoming more common, and recently the term "cogeneration" was coined to describe the production of both electricity and heat from the same facility.

In order to illustrate the ideas behind cogeneration for different purposes, a comparison of four different systems will be carried out. The goal will be to understand the constraints and performance of each. As such, a simplified cycle analysis will be included to contain the essence of the more complex systems. The energy use, availability destruction, and economics of each will be discussed.

The four systems that will be studied are:

1. A conventional system consisting of a separate power plant and a heating plant.
2. A single power plant that produces heat through increased condensing temperatures (termed waste heat utilization).
3. A single boiler from which power and heating are produced in parallel (termed district heating, and patterned after European practice).
4. A heating plant producing electricity as a by-product and purchasing and selling electricity to a power grid (termed waste electricity production).

In order to compare these systems under similar operating conditions, a specific load will be assumed. The load will be representative of either a city or an industry for which the yearly heating requirements are three times the yearly electrical requirements. (This is representative of Madison, Wisconsin.) The heating load is taken as 15×10^{12} Btu/hr, and the electrical load as 5×10^{12} Btu/yr (1.5×10^9 kWh/yr).

The steam cycle chosen for all systems is a simple Rankine cycle without reheat or feed water heaters. In practice, a more complex cycle with higher thermal efficiency would be employed but the simple cycle will allow an easier determination of performance. The trends and relative comparison between the systems will be the same for the more complex cycles.

The boiler pressure and temperature are taken to be 1250 psig and 950°F, respectively, and the condensing temperature is 75°F. Boiler and turbine efficiencies are 0.85 and 0.9, respectively. The fuel will be coal, with a heating value of 12,000 Btu/lb$_m$ at a cost of \$50/ton, corresponding to \$2.10/10^6 Btu.

Supplementary heating will be provided by natural gas furnaces with an efficiency of 0.7. While this might be low for large industrial furnaces, this is representative of industrial unit heaters and home furnaces. The price of natural gas is taken as \$3.50/10^6 Btu.

8.4.1. Conventional System

The conventional system, consisting of a conventional power plant and a conventional heating system, is shown schematically in Fig. 8.4.1. The power plant only produces an annual total electrical work (W_{elec}) equal to the specified load, while the heating system only produces a total heat flow (Q_{htg}) equal to the annual demand.

There are various quantities of importance in evaluating the system. The total energy flow reflects the fuel costs. The steam flow reflects the size of the system. Since all of the combined systems will operate at the same boiler outlet conditions, larger steam flows will require larger units and increased capital costs.

The total resource energy used is the sum of the coal and natural gas used. The power plant consumes 15.0×10^{12} Btu/yr and the furnace 21.4×10^{12} Btu/yr for a total of 36.4×10^{12} Btu. The annual fuel cost is $\$106 \times 10^6$.

The power plant cycle efficiency is 0.33. Of more relevance to combined cycle studies is an overall energy utilization efficiency defined as the total useful energy delivered divided by the total fuel energy consumed.

$$\eta_o = \frac{W_{elec} + Q_{htg}}{E_b + E_{fur}} \qquad (8.4.1)$$

The overall efficiency for this cycle is 0.55.

Also relevant is the overall effectiveness of the power plant. This is defined as the availability of the work and heat delivered divided by the fuel availability input.

$$\epsilon_o = \frac{A_{elec} + A_{htg}}{A_b + A_{fur}} \qquad (8.4.2)$$

The availability of the work is the energy value, while that for the heat depends on the delivery temperature T_d and the environmental temperature T_o. It will be assumed that heat is delivered at 120°F to the house and the surroundings are at

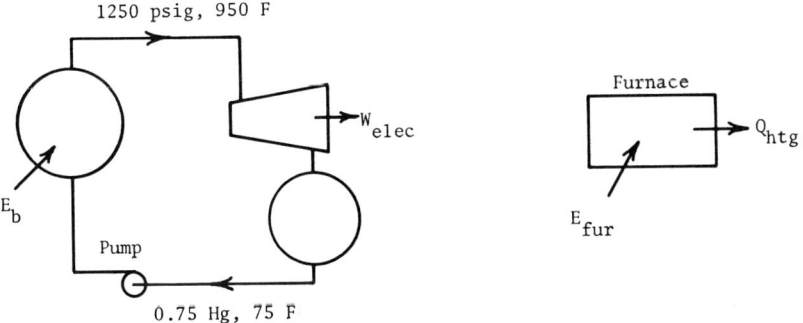

Fig. 8.4.1. Conventional power plant and heating system.

20°F. The fuel availability flows are their energy values, which allow Eq. (8.4.2) to be written as

$$\epsilon_o = \frac{W_{elec} + Q_{htg}(1 + T_o/T_d)}{E_b + E_{fur}} \tag{8.4.3}$$

The value for this cycle is 0.21. These values are also given in Table 8.4.1 for comparison with the other cycles that will be described in the following sections.

8.4.2. Waste Heat Utilization by Increased Condensing Temperature

In the conventional system, the condenser heat transfer is at too low a temperature (75°F) to be useful for space or process heating. By raising the condenser temperature, the heat transfer can be utilized. It will be assumed that a condensing temperature of 220°F is required. This accounts for heat exchanger or transmission losses in a space heating application or process heating in an industry.

The schematic of the system is shown in Fig. 8.4.2. An auxiliary furnace may still be required since the condenser temperature of 220°F may not produce a heat flow that exactly matches the load. The power cycle meets the electrical demand exactly and the sum of the annual condenser heat transfer and furnace heat transfer equals the annual heating load.

The relevant parameters for this cycle are given in Table 8.4.1. About 90 percent of the space heating requirement is met by the condenser. The total resource energy used is the sum of the coal and natural gas energy and equals 24.1×10^{12} Btu/yr. This is a reduction of 35 percent in the total energy required to meet the same load

TABLE 8.4.1
Comparison Between Power and Heating Systems

	Energy Use (10^{12} Btu/yr)					Cost (10^6/yr)
System	Coal	Natural Gas	Total	η_o	ϵ_o	
Conventional	15.0	21.4	36.4	0.55	0.21	106
Waste heat	21.5	2.6	24.1	0.83	0.31	54
District heat	32.7	–	32.7	0.61	0.23	69
Cogeneration	24.3	–	24.3	0.82	0.31	51

	Steam Flow (10^9 lb$_m$/yr)		20-yr Life Cycle Costs (10^9)		
System	Boiler	Turbine	Fuel	Capital	Total
Conventional	8.9	8.9	1.41	0.11	1.52
Waste heat	14.2	14.2	0.72	0.32	1.04
District heat	20.6	8.9	0.92	0.24	1.17
Cogeneration	16.1	16.1	0.69	0.32	1.01

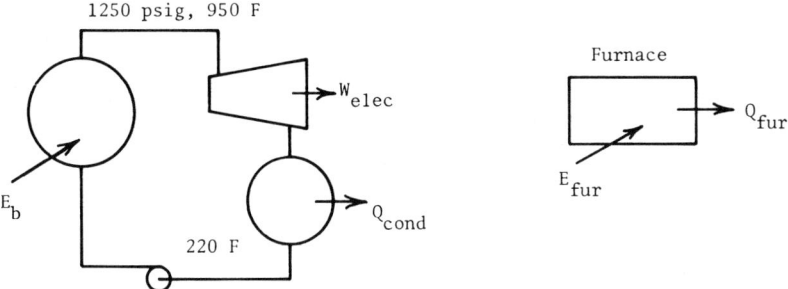

Fig. 8.4.2. Waste heat utilization system with auxiliary furnace.

as the conventional system. Further, coal, which is currently more readily available, supplies about 90 percent of the total energy. The utilization efficiency increases to 0.83 and the effectiveness increases to 0.31. The yearly cost drops to 54×10^6. The cost reduction is proportionately greater than the utilization efficiency increase due to a greater use of the cheaper fuel, coal. This increase in fuel utilization and reduction in fuel costs are in part offset by the increased size of the boiler and turbine. The total steam flow increases 60 percent, which increases the size of the power plant components. The corresponding capital cost increases will be discussed later.

The choice of the condensing temperature affects the performance of the system. Figure 8.4.3 shows the effect of condensing temperature on the condenser and furnace heat flows and the total energy requirements. As the condensing temperature increases, there is more energy available from the condenser and, consequently, less required from the furnace. At a temperature of about 260°F, the condenser meets the entire demand and the furnace is not needed. For higher condensing temperatures, excess energy would be available for heating but could not be used. This situation could also occur at a lower condenser temperature if the heating demand were lower. This illustrates a problem in that at off-design conditions either heat is not used and is transferred to the environment or the work output is reduced.

The total energy use drops slightly as condenser temperature increases. This is a result of replacing the furnace with the higher efficiency boiler. The total steam flow rate increases with condenser temperature, however, which increases capital costs.

8.4.3. District Heating

In a district heating system, a single boiler supplies both the power producing turbine and the heating system. Such a system is shown schematically in Fig. 8.4.4. The only real interaction between the power and heating systems is the use of hot condensate returning from the heating system in a feed water heater. The boiler output can be controlled to meet the entire power and heating demand.

The performance parameters for the district heating system are given in Table 8.4.1. The system uses only coal, and consumes about 10 percent less energy than

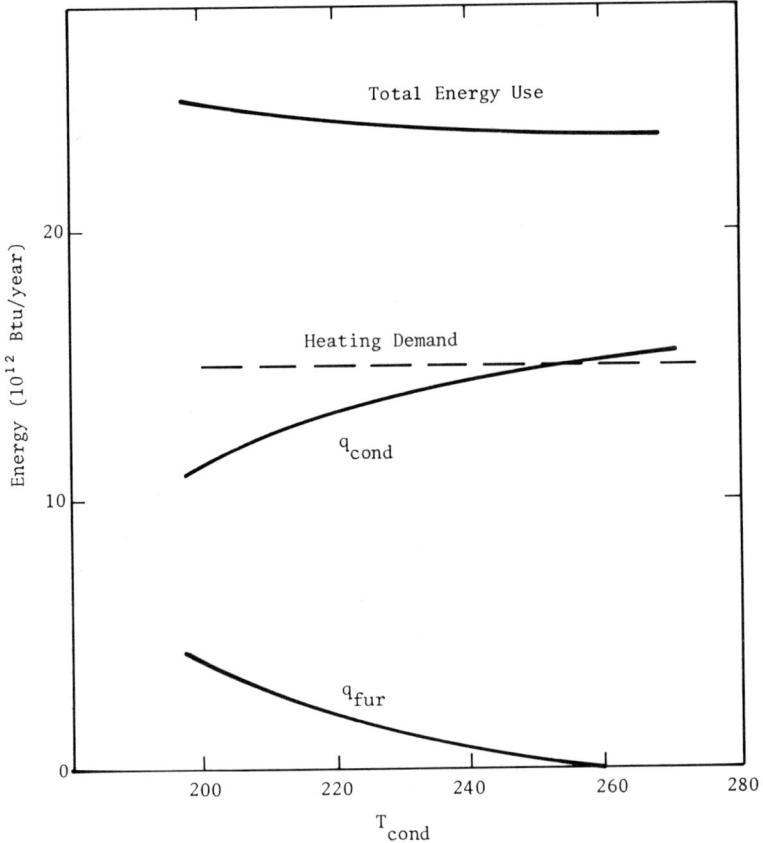

Fig. 8.4.3. Energy flows as a function of condensing temperature.

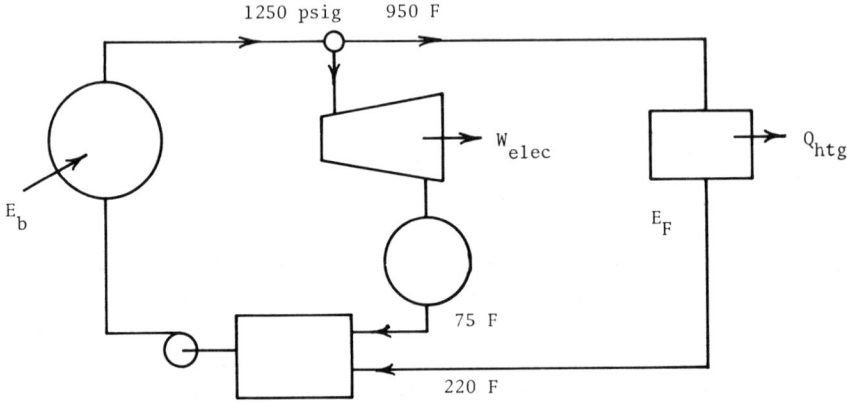

Fig. 8.4.4. District heating system.

COGENERATION OF ELECTRICITY AND HEAT

the conventional system. The cost reduction is 30 percent due to the use of coal as the sole fuel source. The turbine steam flow is the same as for the conventional system since the power plant is the same.

8.4.4. Cogeneration System

A cogeneration system in an industrial application would supply all of the plant heating needs. Steam for the heating system would first flow through the turbine and then to the condenser. The needed heat flow from the condenser would specify the steam flow, and the electrical power produced would result from that flow. If the power produced was insufficient to meet the demand, additional power would be purchased from the external utility grid. Presumably, excess power generated would be sold back to the grid, although, at present, this is the exception rather than the rule. The schematic for this system is given in Fig. 8.4.5.

The performance parameters of the cogeneration system are given in Table 8.4.1. For the particular demands chosen and the given cycle, excess power amounting to 0.65×10^{12} Btu/yr is produced. If this power can be sold back to the power company at \$0.01/kWh, a credit of $\$1.9 \times 10^6$ can be realized against the cost of providing heating which reduces fuel costs 2 percent.

The influence of condensing temperature is to change the amount of electricity that can be sold or must be purchased. This is shown in Fig. 8.4.6. For this load and cycle, the power and heating demands are matched at about 260°F. Also given is the total operating cost based on sales to the grid at \$0.01/kWh, and purchases at \$0.02/kWh. The cost of purchasing is greater than the revenues from sales.

8.4.5. Comparison Between Systems

The various systems are compared on both energy and economic basis in Table 8.4.1. The conventional system uses the most energy. The district heating system saves about 10 percent, while the waste heat and cogeneration systems save 35 per-

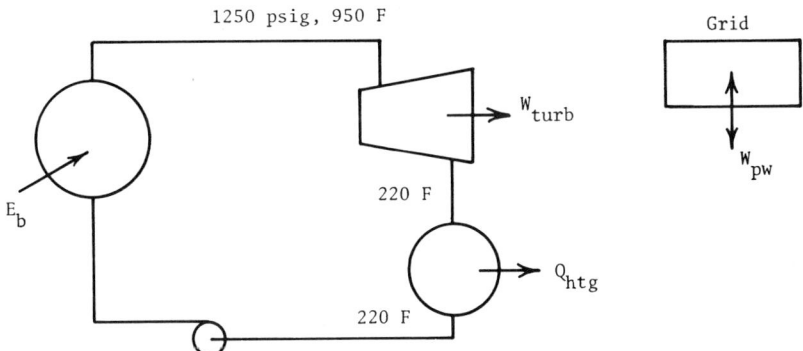

Fig. 8.4.5. Cogeneration system with auxiliary power from a grid.

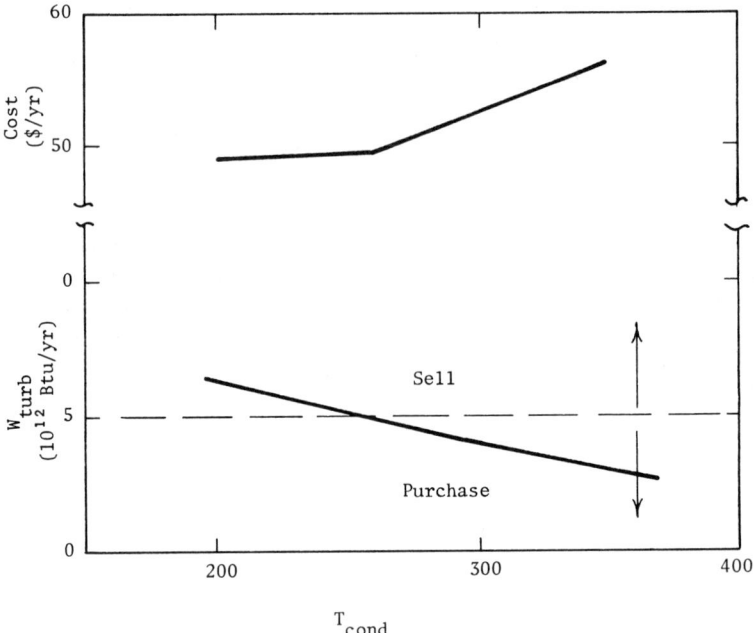

Fig. 8.4.6. Influence of condenser temperature on electricity production and total costs.

cent. The overall efficiencies given in Table 8.4.1 reflect the energy consumption. The more efficient systems are also more effective.

Annual operating costs reflect both energy use and fuel type. The substitution of coal for natural gas in the district heating system reduces cost 37 percent. This presumably allows natural gas to be used for other purposes. The waste heat system is more costly to operate than the cogeneration system due to the use of natural gas to meet the remainder of the heating demand. However, these two are comparable in energy use with the difference being whether heat or electricity is the main desired output. It is clear that combined systems reduce annual cost and energy use over conventional systems.

The energy utilization efficiency of the systems is relatively high compared to the effectiveness. The effectiveness of the combined systems is about 50 percent higher than that of the conventional system, which indicates improvement. However the value is still only one-third or the potential. Ways to increase the effectiveness of heat production will be explored in Chapter 9.

Annual operating costs are only a portion of the economic evaluation. Capital costs must also be included and these are much harder to estimate. The costs given in Table 8.4.2 will be assumed in order to estimate the overall costs. A 20-year life and 10 percent fuel inflation rate will be assumed. The corresponding P_1 and P_2 are 13.3 and 0.884, respectively. The installed capacity of the power plant is assumed to be 200,000 kW, while that of the heating system is 8×10^9 Btu/hr. These values reflect peak, not average, demands, and would be used to size the plants.

TABLE 8.4.2
Capital Costs for Combined Systems

Home furnace	$1000/(installed 10^6 Btu/hr)
Condensate heat exchanger and distribution system	$20,000/(installed 10^6 Btu/hr)
Power plant	$600/(installed kW) 75°F condensing temperature
	$900/(installed kW) at 220°F condensing temperature

The life cycle costs reveal that fuel is the major cost. The conventional system is the most costly, the district heating system second, and waste heat and cogeneration the least costly. The capital costs of the conventional system are the least due to the low cost of home furnaces. On a delivered energy basis, a home furnace costs about one-tenth that of an industrial boiler.

The cost comparisons are probably less realistic and relevant than the energy comparisons. For any given application all four alternatives would probably not be feasible. Conventional, district heating, and possibly waste heat systems are feasible for cities. Industrial applications would consider only cogeneration and a district heat type of system.

For industrial applications there are alternatives to the Rankine cycle steam system. Gas turbines with "waste heat" boilers in the exhaust stream provide both power and heat. The cost is relatively low and in the range of $250–$350 per installed kilowatt. For smaller applications and for those in which the heating load is high relative to the electricity load, Diesel engines are used. The relatively low thermal efficiency is not a severe detriment since the exhaust and cooling water energy is used for heating.

This section has attempted to describe the general characteristics of different heat and electricity producing systems. It is clear that the conventional method with separate power and heating systems is not as effective in utilizing energy as are combined systems. The best system depends, obviously, on the application. The perspective of a utility which must be able to meet all customer power demand is different from that of an industry which can purchase needed electrical power. The relative heating and electrical loads are important in the decision as to how to size the components. It is hoped that the methods of analysis presented are of value in arriving at a sound decision.

8.5. PRICE OF DELIVERED ELECTRICITY AND HEAT FROM COGENERATION PLANTS

For a utility or industrial plant that produces both heat and electricity from a single fuel source, the price to charge for each output needs to be determined. Two schemes are feasible. The outputs could be priced on the basis of either their energy content or their availability content. In the first scheme, the basis is that only the quantity of energy is important. The second scheme recognizes that electricity is inherently worth more because of its ability to do work. Pricing the output on the basis of its availability is more reasonable.

As an example of the difference in the two schemes, the outputs of the cogeneration plant of Section 8.4.4 will be priced. Only fuel costs will be considered. From Table 8.4.1, the annual fuel bill is 51×10^6. The plant produces 5×10^{12} Btu of electricity and 15×10^{12} Btu of heat. The average fuel cost per unit of energy output is then $2.55/10^6$ Btu. This means that both electricity and heat cost $2.55/10^6$ Btu, which corresponds to an electrical cost of 0.87¢/kWh. The pricing on the energy basis seems unrealistic since the electrical cost is much less than a utility would charge for its output. Further, the heating cost on this basis is 20 percent higher than the conventional fuel cost of $2.10/10^6$ Btu even though electricity is being produced.

The price of energy on the basis of its availability can also be determined. For heat delivered at 120°F and an outside air temperature of 20°F, the availability transfer due to the 15×10^{12} Btu of heat transfer is 2.6×10^{12} Btu. The total availability transferred by the system is 7.6×10^{12} Btu. The average cost per unit of availability is $6.70/10^6$ Btu. Since the energy and availability of electricity are equivalent, the cost of electricity is $6.70/10^6$ Btu or 2.3¢/kWh. The availability transfer due to heat flow is only 17 percent of the heat flow, and thus the cost of heating energy is $1.15/10^6$ Btu. These prices are much more in agreement with conventional energy costs. It appears that pricing the outputs on the basis of availability is a rational way to apportion costs.

8.6. LOAD MANAGEMENT

In Section 8.4 a design point analysis was conducted for the four alternatives. In any actual application, loads would vary over time. Residential and commercial heating loads change dramatically between summer and winter. Industrial loads vary over the course of the day, and depend on the process and the production schedule. In this section, some of the aspects of load management will be considered.

As a representative example of the load variations that may be expected, the electrical and natural gas consumption for Madison, Wisconsin will be used. These are primarily residential and commercial loads, and change over the course of the year due to seasonal heating demands. There are large heating loads in winter, and smaller, but significant air conditioning loads in summer. The average monthly energy use is given in Fig. 8.6.1.

The electrical demand is relatively constant with the minimum loads during spring and fall. In the winter months, the loads are 5-10 percent higher due to a shorter day length (more artificial lighting), less outdoor activity (more TV), and more cooking. In summer, the demand increases about 20 percent due to air conditioning. In contrast, the heating demand varies by about a factor of 5 over the course of the season. The low, nearly constant load in summer is due to water

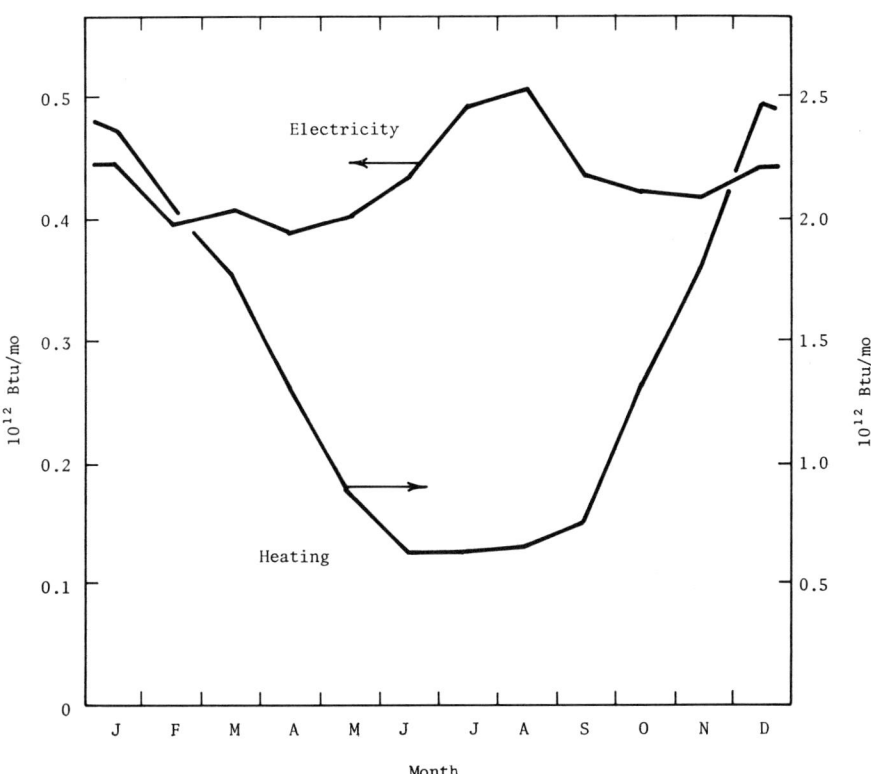

Fig. 8.6.1. Representative monthly heating and electrical demands.

heaters and gas lights. The remainder of the loads follows the changes in ambient temperature.

Conventional and district heat systems have little trouble meeting these time varying loads. The flow rates of steam and natural gas are adjusted, as needed, to meet the demand. There is no thermal interaction between the heating and power production for the conventional system, and a minimal one for the district heating system. For the other two systems, there is a significant interaction, and definite load management procedures are needed.

One method of control for both systems is to change condensing temperature. In Section 8.4.4 the influence of condenser temperature on the electrical power produced was evaluated. This showed that a small increase in condensing temperature produced a large decrease in turbine work. There is a corresponding increase in condenser heat flow. For these two cycles the influence of changing condenser temperature on the heat flow and power output are given in Fig. 8.6.2.

Although the increase in heat production equals in amount the decrease in electrical production as temperature increases, the relative changes are quite different. The percent change in work for a given temperature change is about three to five times the percent change in heat output. Cogeneration systems such as these are

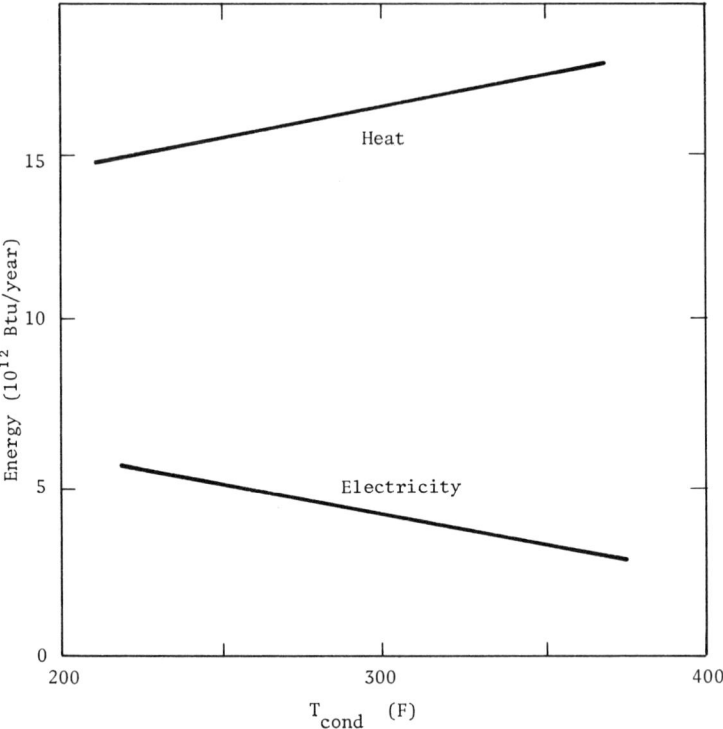

Fig. 8.6.2. Influence of condensing temperature on heat and electricity production.

readily able to meet changes in electrical demand at nearly constant heating demand. They are not able to meet changes in heating demand at nearly constant electrical demand, which is more nearly the case for industries and communities with heating loads dependent on season. Methods of control other than altering the condensing temperature are needed.

The annual performance of a waste heat system such as an electrical utility which is operated to always meet the electrical load is shown in Fig. 8.6.3. In winter, the demand is greater than the condenser heat flow and additional heating must be supplied. This would be done either by separate furnaces in each home or a steam flow bypassing the turbine as in the district heating system. In summer, more heat is produced than demanded and some must be rejected to the environment. There is some demand in summer which means that if all of the condensation occurs at the heating temperature ($220°F$), the power plant electrical production efficiency will suffer. Two condensers might be employed with one to reject heat at available cooling water temperatures and the other to supply the required heating. In any event, it is difficult for a utility to meet the combined demands of a city using one system.

The situation for an industrial heating system facing the same demands is shown

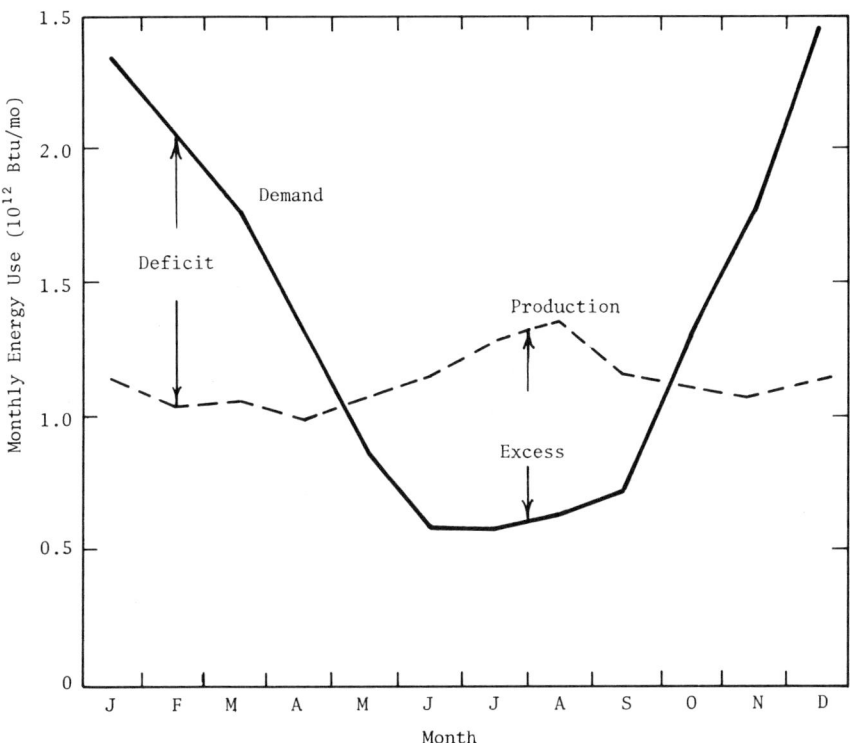

Fig. 8.6.3. Monthly heat flows for waste heat system.

in Fig. 8.6.4. The system is operated to always meet the heating demand. Excess electricity is produced in winter, but a deficit exists in summer. The industrial customer can take advantage of the presence of the utility network by selling the excess power and purchasing additional power as needed. If sales to the utility are not possible, the turbines can be bypassed and the steam generated solely for heating at no economic penalty. The industrial user, which has the utility network available to it, is in a good position to meet time-varying loads.

These same considerations apply on a daily basis as well. An industry may require heat to warm up equipment in the morning, electrical power for processes during midday, and hot water for cleanup in the evening. Residential space heating loads would be high at night, while electrical consumption would peak in morning and early evening. A cogeneration system could meet these loads if supplementary power was available.

A current method for industries to meet the varying loads involves load shedding. Certain loads are designated as having a lower priority than others and are shut down during periods in which they cannot be met without incurring additional costs. For example, a plant with a number of air compressors might never run them all at once and shut them off in sequence as the peak power consumption is ap-

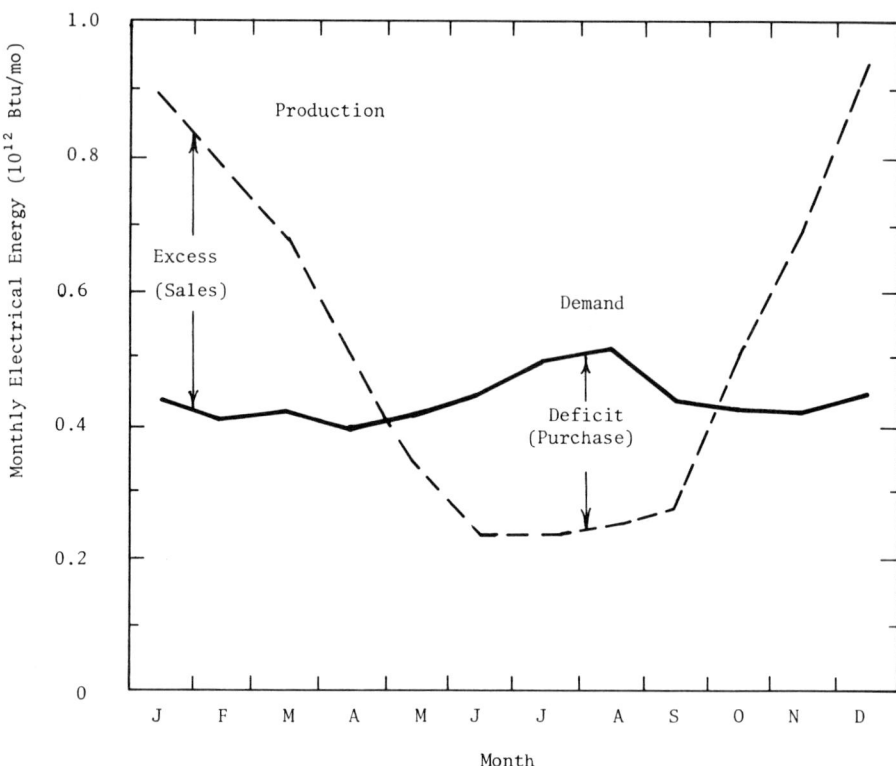

Fig. 8.6.4. Monthly electrical flows for cogeneration system.

proached. The equipment using compressed air would then run off of the air in the reservoirs. The area of management of loads at off-design conditions is only now receiving much attention and presents an opportunity for significant energy savings.

8.7. INDUSTRIAL COGENERATION

In Sections 8.4–8.6, cogeneration of heat and electricity was considered from the perspective of a utility supplying both outputs to its customers. An individual industrial plant which has a need for both heat and electricity is also a candidate for cogeneration. A system of the type described in Section 8.4.4 would probably be most suitable. In following this approach, the company would install a heating system which has the capability to produce electricity as a by-product. Additional needed electrical power or excess generated power would be purchased from or sold to the electrical grid.

There are many advantages to the company of such a cogeneration scheme. The overall conversion of fuel energy to useful energy is high (75–85 percent). Equally important, the system can produce electrical energy at lower cost than that from the utility. There then are strong economic incentives, especially in utility service areas in which demand charges on electrical consumption during certain periods such as weekdays, daytime, and/or summer are applied. The company can then operate its cogeneration system to supply a large amount of electricity during the demand periods, and then purchase electricity during the low cost off-peak periods. Such an approach is called "peak shaving," and financially benefits the individual company and reduces the national consumption of fuel to produce the same end products.

Another advantage of cogeneration to the individual plant is that since the plant requires a process heating system there is only the added cost of the power producing components. The investment is relatively low, about one-half that of a conventional power plant. The plant operation is not compromised in this scheme since the power grid is always there to supply electricity during critical periods. Finally, a cogeneration plant requires careful control, which in return requires knowledge and control of the plant processes. This will generally mean that a study of the plant operation is dictated, which in itself usually produces significant energy savings.

There are three basic systems suitable for industrial cogeneration. They employ steam turbines, gas turbines, or Diesel engines to produce electricity. All tie into the process heating system, which usually employs low temperature steam. Each will be discussed separately.

In the first type, the boiler pressure and temperature is raised and a steam turbine is installed in the line. Either extraction steam or that from the exhaust is used in the process. This configuration is most suitable to applications which require a large amount of process heating per unit electrical power produced.

The advantage of the steam turbine is that it readily fits into the process steam line. The fuel is that used in the boiler, and encompasses a wide range of gaseous, liquid, and solid fuels. It allows a reduction of the steam pressure and temperature to the levels required for the process. Newer units allow operation at lower pressures and increase the range of this approach. The main disadvantage of the steam turbine is the relatively high cost of the turbine and high pressure boiler. The power produced per unit of heat energy and the cost range are given in Table 8.7.1.

The second approach utilizes a gas turbine to produce electrical power. A "waste heat" boiler is installed in the exhaust to produce process steam. This system is most

TABLE 8.7.1
Properties of Industrial Cogeneration Systems

System	Electrical Power Process Heat (kWh/10^6 Btu)	Initial Cost ($/kW)
Steam turbine	50–100	300–600
Gas turbine	150–300	200–300
Diesel	300–500	500–800

suitable to situations requiring an intermediate ratio of process heating to electrical power. The advantages of the gas turbine are the relatively low costs of the units and the ability to burn gaseous and liquid fuels. The characteristics are given in Table 8.7.1.

The third option employs stationary Diesel engines to produce electricity. It has the ability to produce the highest electrical power output to process heat, but is most suitable to low power outputs (less than 40 kW). A waste heat boiler is used to recover energy in the exhaust and produce high temperature steam, while the engine cooling system is a source of low temperature heat. The Diesel engine is very reliable and has a long life. However, its cost is high (Table 8.7.1) and it requires distillate fuels.

The information contained in Table 8.7.1 shows that cogeneration systems can be used to meet a wide range of industrial applications. The power produced can be tailored to the process heating needs to significantly offset the electrical demands of the plant. The initial investment is less than that of conventional power plants, which helps reduce the cost to the company. Cogeneration has the potential to be advantageous on both energy and economic grounds for many companies.

8.8. SUMMARY

In this chapter the use of fuel to produce electricity is studied. Thermodynamically, the combustion of fuel is ineffective. In the combustion process itself the destruction of availability amounts to about 40 percent of the fuel availability. The combustion of fuel inherently limits the potential for producing electricity.

The performance of a power plant can be enhanced through simultaneous production of heat and electricity. The energy efficiency and availability effectiveness of combined plants can be 50 percent higher than those of single-output systems. Industries and utilities can both consider combined systems to reduce energy use and fuel costs.

SUGGESTED READING

G. J. Van Wylen and R. E. Sonntag, *Fundamentals of Classical Thermodynamics*, John Wiley & Sons, New York, 1976.

B. W. Wilkinson and R. W. Barnes, ed., *Cogeneration of Electricity and Useful Heat*, CRC Press, Boca Raton, FL, 1980.

R. A. Gaggioli, Proper Evaluation and Pricing of Energy, *Proc. Int. Conf. on Energy Use Mgmt. II*, Pergamon Press, Elmsford, NY, 1979, pp. 31-43.

PROBLEMS

8.1. An industrial plant currently produces steam for process heating. Currently, steam is produced at 350 psia and 600°F, and desuperheaters are used to reduce the temperature and pressure to saturated vapor at 300°F. (The desuperheater is a valve that mixes feed water with the steam to produce saturated vapor.) The water used is the condensate return from the process at 100°F. The process energy requirement for the plant is 20×10^6 Btu/hr for 8 hours per day and 5 days per week over the year. Currently, coal at $2.60/$10^6$ Btu is used in an 83 percent efficiency boiler.

The management is considering a cogeneration turbine to reduce purchased power costs. In this scheme, the turbine would be used to drop the steam pressure to 70 psia. A desuperheater valve would still be employed to bring the steam to saturation at 300°F. The turbine would cost about $120 per installed horse power, including the generator, and is available in 50-hp increments. It has an isentropic efficiency of 76 percent and is coupled to a 93 percent efficient generator. The electricity produced would displace purchased electric power at $0.04/kWh.

Determine if this is a good investment, and determine the annual savings (or costs). The management makes all purchases from its floating reserve, and will accept a 20 percent rate of return on energy investments and a 10-year life. Other company financial data are an income tax rate of 45 percent, property tax rate of 2 percent, an expected fuel inflation rate of 12 percent, and a general inflation rate of 10 percent. The turbine generator is expected to have a 2 percent maintenance rate each year and no salvage value. It is eligible for a 15 percent federal tax credit.

The current system flow diagram is (pumps are not shown):

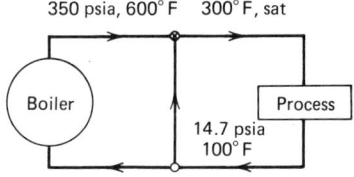

8.2. A steam power plant operates as a Rankine cycle with boiler pressure and temperature of 1400 psia and 1050°F, respectively, and a condensing temperature of 80°F. Boiler, turbine, and pump efficiencies are 0.85, 0.90, and 0.87, respectively. Coal is the fuel and combusted with 50 percent excess air which enters at 60°F. The condenser cooling water inlet temperature is 50°F and the effectiveness is 0.75 (see Chapter 10 for heat exchanger effectiveness.)

(a) Evaluate the availability destroyed in the various components of the system.

(b) Determine the reduction in availability destroyed for 10 percent improvement in each efficiency and in the condenser effectiveness.

(c) Determine the areas with the most opportunities for improvement.

■ 9 ■

AVAILABILITY OF FUEL FOR HEATING

9.1. THERMODYNAMIC CONCEPTS

The concept of availability has traditionally been applied to the production of useful work. In heating devices such as furnaces and boilers there is no useful work produced and all of the availability of the fuel is destroyed. Thus, it could appear that an availability analysis of a strictly heat producing device is irrelevant. However, this is not the case and in this chapter the value of such an analysis will be discussed.

The heat flow to a building that results from the combustion of fuel is given by the product of the fuel flow rate, fuel heating value, and furnace efficiency as

$$q = \eta_{\text{fur}} \dot{m}_{\text{fuel}} h_{rp} \qquad (9.1.1)$$

Although this heat flow does transfer availability because the building space is at a higher temperature than the environment, the fuel availability is ultimately destroyed. The heat flow from the furnace is transferred through the building walls to the environment. For the furnace and the building as a system, the heat flow is eventually transferred at the environmental temperature and has no availability.

The availability of the fuel used in a heating system becomes relevant when it is considered that the fuel could be used to produce work and this work then used to pump heat from the environment into the heated space. Such a system would require less fuel to produce the same heating effect as the heating system. Even though all of the fuel availability is destroyed in both situations, the total fuel requirement for the second system is less.

In order to evaluate the potential of fuel for heating only, an ideal furnace will be considered. This furnace will be ideal in the second law sense that availability is

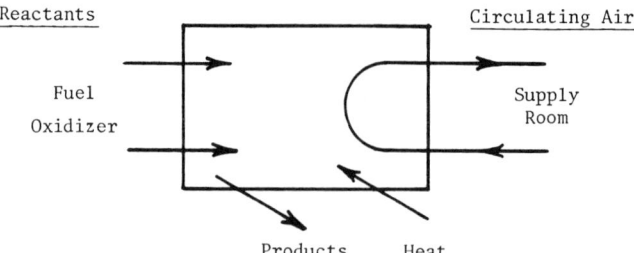

Fig. 9.1.1. Schematic of an ideal furnace.

not destroyed in the heating process. A schematic of the furnace operation and the availability flows are shown in Figs. 9.1.1 and 9.1.2. In this furnace the fuel and oxidizer are combined reversibly and the products leave at the pressure and temperature of the surroundings. The circulating room air is heated from the room discharge to the supply temperature. Heat is transferred into the furnace from the surroundings which are at a lower temperature. However, this heat flow is at the surroundings temperature and does not transfer availability.

The second law applied to this system yields

$$\dot{A}_r + \dot{A}_R - \dot{A}_s - \dot{A}_p = 0 \qquad (9.1.2)$$

This can be rearranged in terms of the change in the availability of the room circulating air. The fuel will be assumed to enter at the ambient temperature which is taken to equal the reference temperature. The products leave at the surroundings temperature to avoid availability destruction. Equation (9.1.2) can then be written as

$$(\dot{A}_s - \dot{A}_R) = \dot{m}_{\text{fuel}} a_{rp} \qquad (9.1.3)$$

where a_{rp} is the availability change of fuel and oxidizer to products at the temperature T_o.

The availability change of room air between the room and supply temperatures can be evaluated in terms of circulating air flow rate $\dot{m}_a$ using ideal gas relations.

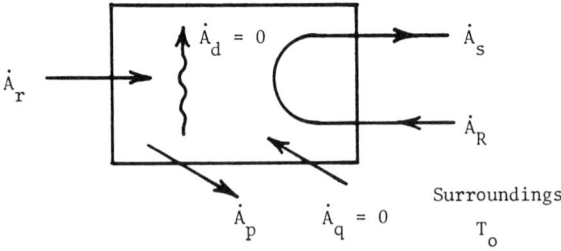

Fig. 9.1.2. Availability flows for an ideal furnace.

THERMODYNAMIC CONCEPTS

The pressures entering and leaving the furnace are equal, and Eq. (9.1.3) becomes

$$\dot{m}_a c_p \left[(T_s - T_R) - T_o \ln\left(\frac{T_s}{T_R}\right) \right] = \dot{m}_{\text{fuel}} a_{rp} \tag{9.1.4}$$

This equation really specifies the air flow rate required to avoid the destruction of availability in the process. The circulating air flow rate per unit mass of fuel input is given from Eq. (9.1.4) by

$$\frac{\dot{m}_a}{\dot{m}_{\text{fuel}}} = \frac{a_{rp}}{c_p [(T_s - T_R) - T_o \ln(T_s/T_R)]} \tag{9.1.5}$$

The heat flow from the ideal furnace can be evaluated from the circulating flow rate and room and delivery temperatures as

$$q = \dot{m}_a c_p (T_s - T_R) \tag{9.1.6}$$

Using Eq. (9.1.5), this can be written as

$$q = \dot{m}_{\text{fuel}} a_{rp} \bigg/ \left[1 - \frac{T_o}{(T_s - T_R)} \ln\left(\frac{T_s}{T_R}\right) \right] \tag{9.1.7}$$

The heating effect is greater than the product of fuel flow rate and availability of combustion. For the furnace both to be ideal and to satisfy the first law, heat must be transferred into the system from the surroundings. This heat flow has no availability since it is transferred in at the temperature of the surroundings.

The numerical values of the availability of combustion and heating value are essentially equal. Equation (9.1.7) can be used to define an effective furnace efficiency or overall coefficient of performance as

$$\text{COP}_o = \frac{q}{(\dot{m}_{\text{fuel}} h_{rp})} \tag{9.1.8}$$

Using Eq. (9.1.7), the overall COP_o is

$$\text{COP}_o = \left[1 - \frac{T_o}{(T_s - T_R)} \ln\left(\frac{T_s}{T_R}\right) \right]^{-1} \tag{9.1.9}$$

For example, if the surroundings are at 20°F, the room is at 65°F, and the delivery temperature is 120°F, the system COP_o is 7.7. This shows that 6.7 times as much heat entered the furnace from the atmosphere as did energy from the fuel. This furnace requires about 13 percent as much fuel as a conventional combustion furnace does to meet the same load. Clearly, if such a furnace could be developed it would be desirable.

The effect of supply temperature on COP_o is shown in Fig. 9.1.3. As the supply

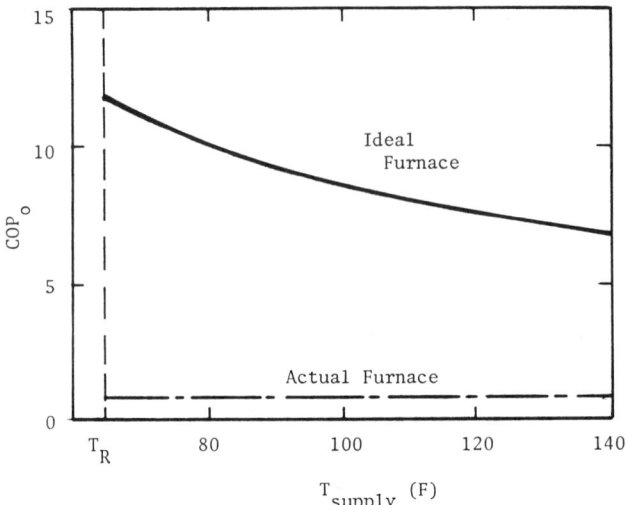

Fig. 9.1.3. Performance for ideal and actual furnaces.

temperature is reduced, the COP_o increases. This is because there is less work (availability) required to heat the air to a lower temperature. The maximum value of the COP_o occurs when the room and delivery temperatures are equal, and is given by

$$COP_o = \frac{1}{1 - T_o/T_R} \qquad (9.1.10)$$

The COP_o for an actual furnace, which is also its furnace efficiency, is also shown in Fig. 9.1.3. There is a significant difference between the two. This shows that the potential exists to considerably reduce the energy consumption for heating by designing a new furnace. Since this furnace would pump heat from the surroundings into the room, one component would have to be a heat pump. This heat pump would be driven by some power source, and a fuel cell-electric motor combination would be required. The system would be more complicated than the simple combustion furnace, but the fuel savings could make additional complexity cost effective.

9.2. SYSTEM PERFORMANCE

A schematic of an engine-heat pump combination which could approach an ideal furnace in operation is shown schematically in Fig. 9.2.1. The engine operates at a thermal efficiency of η_{th} and produces power in amount $\dot{W}$. This is related to the fuel flow by

$$\dot{W} = \eta_{th} \dot{m}_{fuel} h_{rp} \qquad (9.2.1)$$

SYSTEM PERFORMANCE

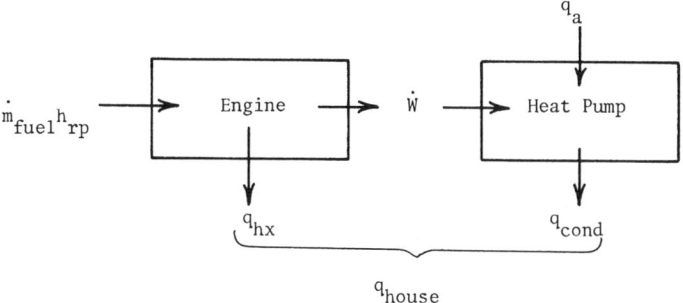

Fig. 9.2.1. Schematic of an engine–heat pump system.

The power produced is used to drive a heat pump operating between the environment and the building temperatures. The heating from the heat pump condenser is given in terms of the heat pump COP as

$$q_{cond} = \dot{W} COP_{hp} \qquad (9.2.2)$$

The heat rejected from the engine in the exhaust and cooling system is also used for space heating. The amount of the rejected energy is the difference between the fuel energy and power produced and can be written as

$$\dot{E}_{rej} = (1 - \eta_{th}) \dot{m}_{fuel} h_{rp} \qquad (9.2.3)$$

The exhaust products and cooling air or water would pass through a heat exchanger and transfer heat to the house. In terms of the heat exchanger effectiveness, this heat flow is given by

$$q_{hx} = \epsilon_{hx}(1 - \eta_{th}) \dot{m}_{fuel} h_{rp} \qquad (9.2.4)$$

The total heat transfer to the house is the sum of the heat exchanger and heat pump flows. Using Eqs. (9.2.1), (9.2.2), and (9.2.4) the heat flow to the house becomes

$$q_{house} = [(1 - \eta_{th}) \epsilon_{hx} + \eta_{th} COP_{hp}] \dot{m}_{fuel} h_{rp} \qquad (9.2.5)$$

The term in brackets is the equivalent furnace efficiency or coefficient of performance for the combined system. It relates the delivered heating to the fuel energy consumed. This overall COP_o is

$$COP_o = (1 - \eta_{th}) \epsilon_{hx} + \eta_{th} COP_{hp} \qquad (9.2.6)$$

The magnitude of COP_o depends on the heat exchanger effectiveness, the thermal efficiency of the engine, and the COP of the heat pump. An improved air-to-air heat pump would have a coefficient of performance of about 3. For an ideal engine such as a fuel cell, no availability is destroyed, all fuel energy is converted to work, and the thermal efficiency is unity. The heat exchanger is not needed and the combined system COP_o equals the heat pump COP. This upper limit on heat production from fuel is about three times greater than the fuel heating value and a factor of 4 better than conventional furnaces.

Existing internal combustion engines have thermal efficiencies of about 0.25. The exhaust heat flow is a major energy term, and it is important to have a high heat exchanger effectiveness. A reasonable value would be 0.8. Using a heat pump with a COP of 3, the overall COP_o would be 1.35. This is about twice existing furnace efficiencies.

The relation between the overall system COP_o and engine thermal efficiency is shown in Fig. 9.2.2. The effect of the heat exchanger effectiveness decreases and that of the heat pump COP increases as engine thermal efficiency increases. The range of efficiencies for conventional heating systems are shown for comparison. It is clear that significant energy savings can result from combined systems using existing technology.

The combined system COP is less than that of the heat pump, yet the combined system is more efficient in terms of resource energy use. The straight heat pump uses electricity generated in a power plant at a thermal efficiency of about 30 percent. The COP for the heat pump based on resource energy use is about 0.9. The combined system COP of 1.35 is also based on resource use, and thus the combined

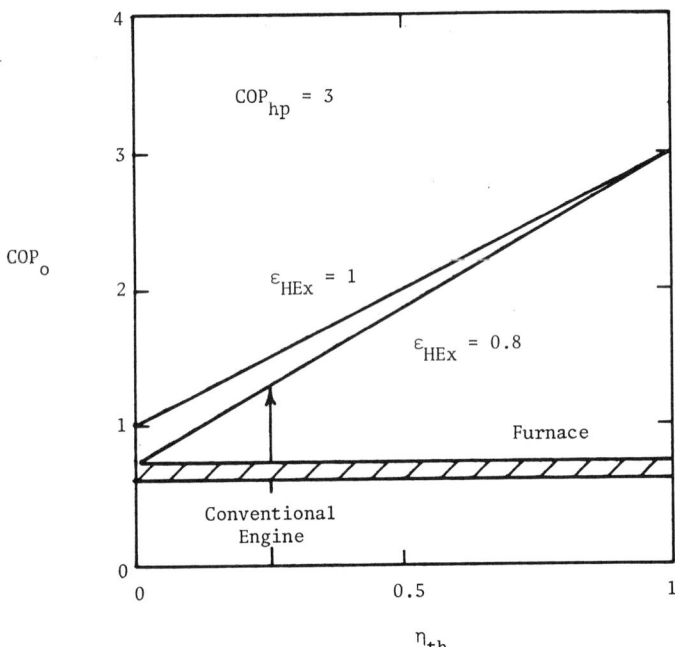

Fig. 9.2.2. Combined system efficiency as a function of engine efficiency.

system is 50 percent more efficient. A deterrent to it is that it uses a more scarce fuel such as gasoline or Diesel fuel rather than coal as in the power plant.

The energy reductions achieved by a combined system come at the expense of increased first cost. In order to estimate whether economic savings also result from a combined system, an example calculation will be performed. These systems are more suitable for housing complexes, and one with 40 living units will be chosen. It will be assumed that the total building load is 2×10^9 Btu/yr and the peak demand is 10^6 Btu/hr.

The conventional system will be a boiler costing $10,000 that burns natural gas with a price of $4.8/$10^6$ Btu at an efficiency of 0.8. The annual fuel cost is then $12,000. For commercial applications the values of P_1 and P_2 are 13.3 and 0.884, respectively. The life cycle costs are $168,000 and mainly reflect fuel costs.

The combined system will consist of a Diesel engine with a thermal efficiency of 0.25, a heat pump with a COP_H of 3, and a heat exchanger with an effectiveness of 0.8. The overall COP_O is 1.35, and at peak conditions a fuel energy flow of 740,000 Btu/hr is required. The engine produces 185,000 Btu/hr (73 hp) which, when coupled to the heat pump, delivers 560,000 Btu/hr. The remaining 440,000 Btu/hr is delivered by the exhaust heat exchanger. The heat pump capacity is about 45 tons. The engine size is relatively small by automotive standards, while the heat pump size is quite large.

The engine-heat exchanger first cost is assumed to be $5000, and the heat pump cost $15,000. The fuel used will be fuel oil at $0.90/gal, or $6.40/$10^6$ Btu. For the combined system efficiency of 1.35, the annual fuel cost is $9500. The life cycle costs are $148,000 and the life cycle savings over the conventional system is $20,000. The added investment in the combined system is paid off in 4 years.

There are several reasons why combined systems have not obtained widespread acceptance in residential applications. Small Diesel or gasoline engines are not readily available for stationary power usage. The number of hours of operation per year are fairly large (2000-4000). An engine capable of going 100,000 miles has a life of only 5000 hours based on an average speed of 20 mph. Motorcycle or lawnmower scale engines are not suitable. Small reliable Diesel engines need to be developed. Noise is also a problem. For these reasons, combined system development has been limited to large (multifamily or commercial) installations which use large Diesel or gas turbine engines.

9.3. GAS-FIRED HEAT PUMPS

9.3.1. System Description

A proposed combined system for residential applications is the gas-fired heat pump. Natural gas is used as the fuel for a small engine. The engine power is used to drive a vapor compression heat pump. The heat rejection from the engine and the heat transfer from the condenser are both used to heat the space. In concept, the system is similar to those using internal combustion engines. The difference lies in the power cycle employed and the use of natural gas.

One of the most promising gas-fired heat pump systems is based on the Stirling cycle. This cycle has the potential to approach the Carnot value of thermal efficiency, and is probably better suited to the small power requirements of residential applications than is a gasoline or Diesel engine. In one such system currently under development, a free piston engine is used to directly couple the engine and heat pump compressor to eliminate rotating shafts. This cycle will be evaluated in this section.

A schematic of the system is shown in Fig. 9.3.1. There are four different circuits with four different working fluids. The combustion system uses air and has a combustion chamber, a high temperature heat exchanger, and a heat recovery heat exchanger. It is similar to a forced draft furnace, but requires a heat exchanger capable of withstanding the high temperatures of the working fluid in the Stirling cycle. The energy in the exhaust leaving this heat exchanger is used to preheat the air entering the combustion chamber which increases the furnace efficiency.

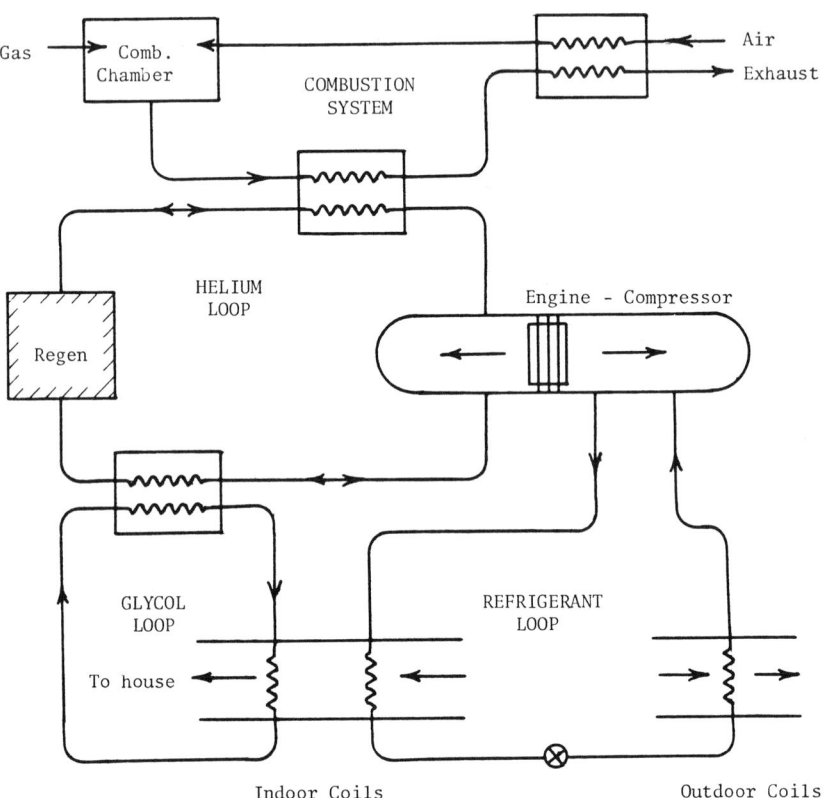

Fig. 9.3.1. Schematic of a proposed gas-fired heat pump.

The Stirling cycle engine consists of a free piston expander-compressor, a high temperature heat exchanger for heat addition, a low temperature exchanger for heat rejection, and a regenerator. The working fluid is helium. The cycle processes are shown schematically in Fig. 9.3.2, and consist of isothermal heat addition (1-2), constant volume expansion (2-3) in which the fluid is cooled by heat transfer into the regenerator material, isothermal heat rejection (3-4), and constant volume heat addition from the regenerator. The cycle thermal efficiency approaches that of the Carnot cycle since external heat addition and heat rejection are at constant temperature.

To accomplish these processes, the engine consists of two interacting piston-cylinder combinations. The helium flows cyclically back and forth through the exchangers and the regenerator. The regenerator consists of a large block of porous material. Glycol is pumped through the heat rejection exchanger to transfer this energy to the heating coils of the house.

The engine drives a double-acting freon compressor. The remaining components are the standard heat pump condenser, evaporator, and expansion valve. The entire system is envisioned to mount in a package of about 3 ft on a side.

The design operating temperatures for this cycle are shown in Fig. 9.3.3. The heat transfer from the combustion chamber is at a high temperature to minimize availability destruction and maximize cycle thermal efficiency. However, as discussed in Chapter 8, the combustion process has inherently reduced the potential for producing work. The heat rejection from the engine is at a relatively low temperature and used to heat air at 120°F.

The energy transfers as a percentage of the inlet fuel energy are shown in parentheses. The combustion system has a relatively high overall efficiency due to the economizer heat exchanger. Of the 79 percent of the energy transferred to the Stirling engine, 24 percent is transferred as work. The cycle thermal efficiency is 0.30, compared to the Carnot efficiency of 0.63 for these same conditions. There is

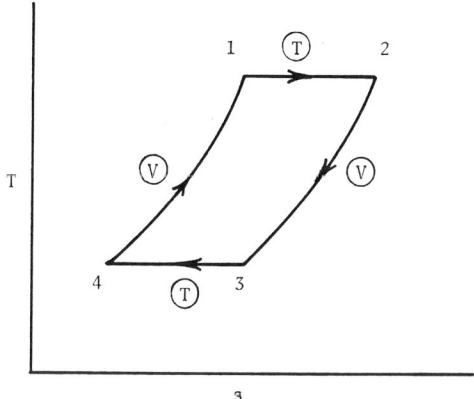

Fig. 9.3.2. Process diagram for Stirling cycle.

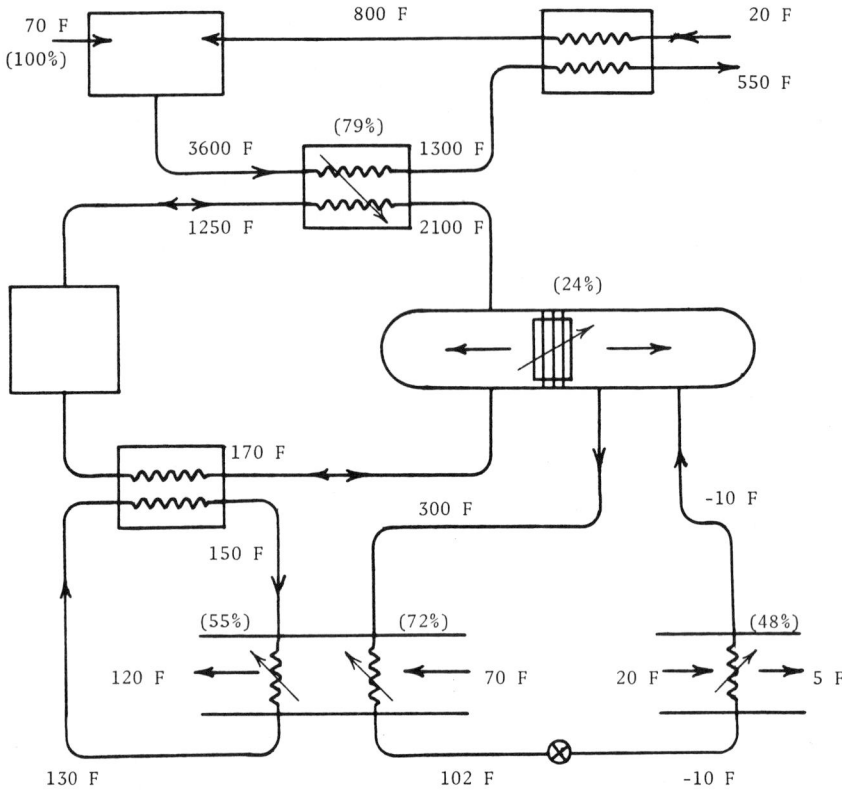

Fig. 9.3.3. Temperature and energy flows as a percentage of energy input for gas-fired heat pump.

potential for improvement here. The Stirling cycle delivers 55 percent of the fuel energy to the house. The heat pump work is 24 percent of the fuel energy. With a COP of 3, this allows the heat pump to deliver another 72 percent of the fuel energy to the house.

The overall COP_o is 1.27. Although this is not as high as the example of a conventional internal combustion engine (Section 9.2), the cycle uses a relatively more available and lower priced fuel. The engine thermal efficiency is about one-half the Carnot value, and, hopefully, can be improved by further development.

9.3.2. Overall Performance

The gas-fired heat pump can be represented as a furnace which supplies heat to an engine. All of the engine heat rejection is used for heat. The engine power drives a heat pump to pump heat into the house. The house heat flow can then be calculated as

$$q_{house} = q_{rej} + q_{cond} \tag{9.3.1}$$

GAS-FIRED HEAT PUMPS

In terms of the cycle thermal efficiency, heat addition from the furnace, furnace efficiency, and fuel energy, the heat rejected from the engine is given by

$$q_{rej} = (1 - \eta_{th}) \eta_{fur} \dot{m}_{fuel} h_{rp} \tag{9.3.2}$$

Similarly, the condenser heat flow is given in terms of the heat pump COP, cycle thermal efficiency, furnace efficiency, and fuel energy as

$$q_{cond} = \eta_{th} COP_{hp} \eta_{fur} \dot{m}_{fuel} h_{rp} \tag{9.3.3}$$

The house heat flow is then

$$q_{house} = (1 - \eta_{th} + \eta_{th} COP_{hp}) \eta_{fur} \dot{m}_{fuel} h_{rp} \tag{9.3.4}$$

The overall COP_o is

$$COP_o = (1 - \eta_{th} + \eta_{th} COP_{hp}) \eta_{fur} \tag{9.3.5}$$

This expression is similar to Eq. (9.2.6). The difference lies in the use of an external combustion circuit for the Stirling cycle. This expression also shows the importance of the furnace efficiency. The temperatures given an Fig. 9.3.3 show that the furnace efficiency is somewhat limited by the preheating of the combustion air, which means that the exhaust is at a relatively high temperature. If this energy were transferred to the house through another heat exchanger, the cycle COP_o could be improved. Using an exhaust heat exchanger with an effectiveness of ϵ_{hx} would increase the COP_o to

$$COP_o = (1 - \eta_{th} + \eta_{th} COP_{hp}) \eta_{fur} + (1 - \eta_{fur}) \epsilon_{hx} \tag{9.3.6}$$

For the conditions of Fig. 9.3.3, the use of an 75 percent effectiveness heat exchanger would increase COP_o from 1.27 to 1.43, and would probably be cost effective.

9.3.3. Economics

The components of gas-fired heat pumps are not now commercially available, and the costs are unknown. However, the price at which a gas fired heat pump is cost competitive can be estimated. The life cycle costs of any heating device are given by

$$LCC = P_1 C_F \frac{q_{house}}{COP_o} + P_2 C_E \tag{9.3.7}$$

The life cycle savings of an alternative (a) over a conventional system (b) is the difference in these costs. The cost of the alternative at which the savings equal zero is the maximum price at which the system can sell for, and is given by

$$C_{E,a} = \frac{(C_E P_2)_b}{P_{2,a}} + q_{house} \frac{P_1}{P_{2,a}} \left[\left(\frac{C_F}{COP_o}\right)_b - \left(\frac{C_F}{COP_o}\right)_a \right] \quad (9.3.8)$$

The fuel present worth factor P_1 is probably the same for the two alternatives, but the first cost factor P_2 may not be. A gas-fired heat pump may be eligible for federal energy tax credits while the conventional furnace is not.

The gas-fired heat pump will be compared to conventional and intermittent combustion furnaces. For the house of Chapter 3, the heating load is 46.3×10^6 Btu/yr, and the cost of natural gas is $3.50/10^6$ Btu. The economic factor P_1 is 22.17, while the values of P_2 are 1.174 and 0.974 for the conventional furnace and two alternatives, respectively. The conventional furnace with an efficiency of 0.7 costs $1200.

The price at which the gas-fired heat pump competes economically with the conventional furnace is about $4000. In order for it to save money over the 95 percent efficient intermittent combustion furnace, it can cost only $1300 more. It appears that gas-fired heat pumps must be developed to cost about $4000 in order to save both money and energy.

9.4. SUMMARY

The thermodynamic potential of fuel to supply heat has been studied in this chapter. The combustion of fuel to supply only heat is ineffective in a thermodynamic sense, and a system that produces work internally to drive a heat pump is more effective. A reversible engine-heat pump combination has an overall COP_o, or furnace efficiency, up to 10 times that of the conventional furnace. This potential would be approached using a power source such as a fuel cell.

The overall COP_o for engine-heat pump combinations using existing or developing components is in the range of 1.3-1.5. These systems significantly reduce energy use over conventional heating systems. The price must be in the vicinity of $4000 for these to save money also. The development of these devices could have a significant impact on residential and commercial energy usage.

SUGGESTED READING

G. J. Van Wylen and R. E. Sonntag, *Fundamentals of Classical Thermodynamics*, John Wiley & Sons, New York, 1976.

L. A. Sarkes, J. A. Nicholls, and M. S. Menzer, Gas Fired Heat Pumps: An Emerging Technology, *ASHRAE Journal*, March 1977, p. 36.

PROBLEMS

9.1. Determine the cost effectiveness of a system such as depicted in Fig. 9.2.1 for a housing complex. The annual loads and peak loads are

	Annual	Peak
Heating	6×10^9 Btu	3×10^6 Btu/hr
Cooling	2×10^9 Btu	2×10^6 Btu/hr
Electricity	10^6 kWh	500 kW

9.2. Evaluate the availability destroyed in each component and process of the system in Figure 9.3.3

· 10 ·

ENERGY USE IN INDUSTRY

10.1. OVERVIEW

The use of energy in the industrial sector is diverse. There are many different industries involved, and each one has processes that are specific to it. It is more difficult to generalize use here than in other sectors. Overall, the industrial sector accounts for about 40 percent of the total U.S. use. The seven major industries listed in Table 10.1.1 consume about three-quarters of the energy used in the industrial sector. Electricity is widely used, and amounts to one-quarter of the total use in industry. Coal, natural gas, and fuel oil supply three-quarters of the energy.

The energy required to produce various materials is a guide as to energy intensive industries. In Table 10.1.2 the energy costs per pound of product are given for a variety of materials. This does not include the manufacturing operations required to make the finished product. The values are determined from the total energy input of the industry divided by the total output and reflect many overhead operations.

Metals are highly energy intensive due to the nature of the raw material and the number of processes involved. Recycling of metals such as aluminum reduces use considerably. Building and packaging materials are relatively low in intensiveness, although insulating materials are quite high. Plastics are also very energy intensive. The energy required to produce a pound of material represents between $\frac{1}{2}$ and 5 lb of coal, or 0.05-0.5 gal of fuel oil. The energy required to produce a product is significant.

An interesting aspect of the energy costs for materials is the calculation of energy payback periods for insulation. For the example house of Section 3.6, $3\frac{1}{2}$ inches of insulation produced energy savings of 43.7×10^6 Btu/yr (Table 3.6.4). The energy cost of insulating the 1348 ft^2 of wall area is about 10×10^6 Btu. Thus wall insulation will save as much energy as it costs to produce it in 0.2 year. Double-

TABLE 10.1.1
Use of Energy by Industry Type

Industry	Percent of U.S. Total
Primary metals	8.7
Chemical	8.2
Petroleum refining	4.7
Food processing	2.2
Paper	2.1
Concrete and glass	2.0
Manufacturing	3.2
	31.1

pane windows have an energy cost of about 100,000 Btu/ft^2, or, for the example house, a total energy cost of 21×10^6 Btu. The savings for double-pane windows are 26.1×10^6 Btu/yr, and so the energy payback period is about 1 year. This demonstrates that insulation has a rapid energy as well as economic payback period.

Energy is mainly used to produce heat in industries, with direct heating of products accounting for about 28 percent of the total. Process steam, which is used both for indirect heating and space conditioning, comprises 40 percent of the total. Direct electric drive for process equipment accounts for 20 percent. In many industries, refrigeration is an important use. The two thermal uses, heating and cooling, have many common characteristics among the different industries and will be studied in some detail.

TABLE 10.1.2
Energy Cost of Various Materials

Material	Heat (Btu/lb$_m$)	Electricity (kWh/lb$_m$)	Total[a] (Btu/lb$_m$)
Aluminum	10,000	15	60,000[b,c]
Aluminum scrap	4,500	–	4,500[b]
Copper	16,000	0.5	60,000[b]
Steel	12,000	0.3	30,000[b]
Steel scrap	6,000	0.3	24,000[b]
Concrete	3,500	0.2	5,500
Glass	8,000	–	8,000
Paper	10,000	0.6	16,000
Recycled paper	4,000	0.6	10,000
Plastics	50,000	0.9	63,000
3½-in. insulation	7,000 Btu/ft^2		
6-in. insulation	9,400 Btu/ft^2		

[a] Includes resource energy required to produce electricity.
[b] Includes costs to produce finished stock.
[c] Electricity content is for hydroelectric production.

The temperatures required for industrial processes vary over a wide range. About 50 percent of the total energy requirements are below about 1400°F, and 25 percent below about 500°F. About 40 percent of the requirements are at high temperatures of 2500–3000°F. The spectrum of required process temperatures indicates that it is possible to "reuse" heat. The waste heat from high temperature processes is potentially useful for low temperature applications.

Conventional boilers and furnaces are able to supply heat at virtually any temperature at the same efficiency and cost. There is little increase in heat loss if appropriate insulation measures are taken. Thus, for heating situations, there are large potential savings in "waste" heat utilization where the heat from combustion of fuel is initially used at high temperature and subsequently transferred to lower temperature applications.

Refrigeration requirements are more specialized than heating requirements, and occur in fewer industries. However, refrigeration systems commonly use electrically driven compressors, and thus costs and resource energy use are high. The compressor work requirement is a strong function of evaporator and condensing temperature, and thus there is a premium on careful matching of supply and process temperatures. Thus, when refrigeration is employed, there is a need for careful economic consideration of component sizes and insulation.

Although energy use in the industrial sector is diverse, certain fundamental heat transfer and thermodynamic principles are common to all processes. It is the intent of this section to consider several basic problems which can serve as examples for more complex analyses of industrial processes.

The economic selection of insulation will be considered first. This differs from the application to residential and commercial buildings in that the reduction in first cost of the heating or cooling system may be considered and it can have an effect on the optimum level. Next, waste heat recovery systems will be studied. These will include heat exchangers for recovering high temperature heat and heat pumps for recovering low temperature heat. The two aspects will be to determine first, the conditions under which the best heat recovery system is cost effective, and second, the optimum system for the application. Refrigeration systems will be evaluated to show the interplay between the various pumping and heat transfer costs. Finally, the costs and energy savings through recycling will be studied.

10.2. ECONOMIC OPTIMUM INSULATION LEVELS

The economic optimum level of insulation was considered in Chapters 2 and 3 where the costs included the first cost of insulation and the heating fuel savings. In an industrial application, additional complexities may exist. For a given process, added insulation reduces the heating requirement the heating plant must meet. Since the heating plant can be closely sized when process requirements are known, a smaller plant can then be selected. There are then savings in plant first cost. This is especially important when a new plant or addition is being considered.

Heating or refrigeration piping may run through conditioned spaces. For example, a steam line may run through a cold storage room. Heat loss from the steam line then has a cost associated with the fuel for heating and also the fuel for cooling the conditioned space. Insulation saves both heating and cooling costs and reduces both the steam and refrigeration plant size. These factors need to be included in determining the economic optimum insulation level.

The total life cycle cost of insulation for a given application is given by the usual relation

$$\text{LCC} = P_1 \text{ (fuel costs)} + P_2 \text{ (first costs)} \qquad (10.2.1)$$

The fuel costs are written in terms of the price of fuel C_F, the annual heat loss Q, and the efficiency of the heating or cooling system η_o.

$$\text{fuel cost} = \frac{C_F Q}{\eta_o} \qquad (10.2.2)$$

If the heat loss is to or from an unconditioned space (e.g., the ambient), the fuel price is the cost of heating if only heating is involved or the cost of cooling if only cooling is involved. However, if the heat flow is to a refrigerated space the fuel cost is the sum of the refrigeration and heating costs.

The annual heat loss can be represented by the thermal resistance of the insulation and the operating conditions as follows for walls

$$Q = \frac{A \, \Delta T_o N}{R} \qquad (10.2.3)$$

and for pipes as

$$Q = \frac{L \, \Delta T_o N}{R} \qquad (10.2.4)$$

where A and L are the area and length of walls and pipes, respectively. ΔT_o is the average temperature difference for heat flow and N is the annual hours of operation. The product $\Delta T_o N$ is the number of "degree hours" for the system.

The first costs are the sum of the insulation and the cost of the heating and/or refrigeration plant. These two costs can be expressed in terms of their component prices. The price of insulation is the sum of the material (and fixed costs) and will be written as

$$\text{first cost of insulation} = C_I R A + C_{\text{fix}} A \qquad (10.2.5)$$

where C_I is the installed cost per unit R (e.g., \$/(hr ft^2 °F/Btu) for wall insulation). For many industrial applications, the cost of labor and materials add about equally to the total. For some materials, and for pipe insulation, the cost C_I is also a function of the resistance value. The fixed costs account for setting up the installation equipment, transportation, and fixed labor costs.

The plant costs are those necessary to provide a heating (or cooling) system big enough to meet the design load. This is the heat transfer rate at design conditions, and is related to the design temperature difference. For walls, the design heat flow q_{des} is expressed as

$$q_{des} = \frac{A \Delta T_{des}}{R} \qquad (10.2.6)$$

and for pipes as

$$q_{des} = \frac{L \Delta T_{des}}{R} \qquad (10.2.7)$$

where ΔT_{des} is the temperature difference used to size the system.

The first costs are given in terms of the cost of the system per unit capacity and the design heat flow as

$$\text{first cost of plant} = C_p q_{des} \qquad (10.7.8)$$

where C_p is, for example, the cost in \$/(Btu/hr) of installed capacity for a heating system. As with the fuel costs, the plant cost would be the sum of the heating and cooling plant costs if the heat loss was into a cooled space.

For a wall, the life cycle cost becomes

$$\text{LCC} = \frac{P_1 C_F}{\eta_o} \frac{A \Delta T_o N}{R} + P_2 (C_I R A + C_p \frac{A \Delta T_{des}}{R} + C_{fix} A) \qquad (10.2.9)$$

The minimum cost is found by differentiating Eq. (10.2.9) with respect to the thermal resistance and equating it to zero. The area cancels out and, in a similar manner, the length would not be included for a pipe application. The derivative of the fixed costs is zero and thus this cost is not relevant to determining the optimum thickness. The optimum insulation R value for walls and pipes is given by

$$R_{opt} = \left[\frac{P_1 C_F \Delta T_o N/\eta_o + P_2 C_p \Delta T_{des}}{P_2 C_I} \right]^{1/2} \qquad (10.2.10)$$

Equation (10.2.10) specifies the optimum R value of the insulation. From manufacturer's information, the thickness of insulation can then be determined. It is more convenient to calculate the R value first and then determine the thickness next rather than solve for thickness directly.

10.2.1. Example

A wall separating a cold food locker from a heated workroom is to be insulated using fiberglass insulation. The locker is maintained at 20°F continuously, and the room is at 70°F. It is assumed that the room temperature is maintained by a heating system during winter and by heat from steam pipes in summer. Thus, there is a heating fuel cost the entire year. If this were not the case, the heat loss could be divided into winter and summer values. The refrigeration system removes all of the heat that enters the locker and thus operates continuously.

Heating is provided by a coal-fired boiler with an 85 percent efficiency. The coal price is $2/10^6$ Btu. The cost of fuel for refrigeration is the cost of electricity to power the compressors. The electrical cost is 4.5¢/kWh, and the refrigeration system COP is 2. The combined, delivered fuel cost is $9/10^6$ Btu.

The first cost of steam generator is taken as $0.04/(Btu/hr), which is representative of large coal-fired boilers. The first cost of the refrigeration system is about four times as large, or $0.15/(Btu/hr). These first costs are incremental costs representing the additional cost of added capacity to meet the heat loss. The installed cost per square foot of insulation is $0.30/in. The R value is 3/in., and thus the cost per unit area and R value is $0.10/(R ft^2)$. The design temperature difference for sizing the heating and cooling systems is 50°F, which in this case is the same as the operating temperature difference.

The values of P_1 and P_2 are taken for a 20-year life as 13.30 and 1.084, respectively. Insulation is not considered as eligible for tax credits in new applications. Using Eq. (10.2.10), the optimum insulation R value becomes

$$R_{opt} = 24 \text{ hr ft}^2 \text{ °F/Btu}$$

The corresponding life cycle cost would be $5.18/ft^2$. The effect of including plant costs is to increase the level of insulation. Neglecting the savings in plant size due to installing insulation would reduce the optimum R value to 22.4. In a new plant, the cost of the heating or cooling system should be included. For a retrofit application where the system already exists, the plant cost should not be included since the insulation saves only fuel costs. However, the insulation in this case may be eligible for tax credits.

10.3. HEAT EXCHANGERS FOR WASTE HEAT RECLAMATION

Heat exchangers are widely used for waste heat reclamation. In a typical application, heat is transferred from a hotter waste fluid stream to a cooler process stream. The desired result is that the process stream is increased in temperature. This heat transfer then replaces a primary energy source such as fuel oil in a heating application. Alternatively, the waste stream could be a cold stream and used to precool a process stream that would eventually be refrigerated. This would save on refrigeration system energy.

This is shown schematically in Fig. 10.3.1. The waste stream discharges to the surroundings at a high temperature. The process stream is heated from its inlet temperature to an outlet temperature using heat from a boiler. With a waste heat recov-

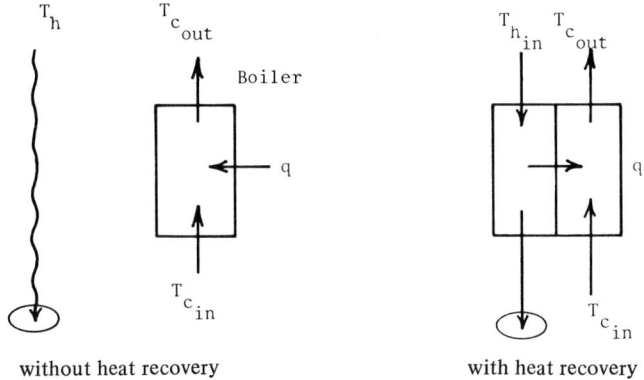

Fig. 10.3.1. Process operation with and without heat recovery.

ery heat exchanger, the hot stream is used to heat the cold stream to the desired temperature. The heat flow from the boiler is displaced by the heat exchanger.

It is always technically feasible to recover heat using a waste-heat heat exchanger. However, it is not always cost effective to do so. The value of the energy saved must offset the first costs of equipment. There are two questions that need to be addressed in waste heat recovery. First, it must be determined whether any heat exchanger is cost effective. Second, it a heat exchanger is cost effective, the optimum size heat exchanger must be determined.

There are some general requirements necessary for waste heat recovery to be cost effective. Both the waste and process streams must be simultaneously available for many hours throughout the year. Large temperature differences between the streams yield large energy savings and high rates of heat transfer. This reduces the size and costs of the exchanger. Flow rates of both streams should be comparable so that temperature changes of each stream are similar. This effectively uses the available waste energy.

The costs associated with the heat exchanger are the first costs of the heat exchanger and the heating system and the pump work required to pump fluid through the unit. The heat exchanger saves fuel energy, and this saving must be determined. In this section, the thermal performance of heat exchangers will be reviewed to determine the fuel savings. The cost of heat exchangers will then be considered and the economics of waste heat recovery demonstrated.

10.3.1. Thermal Performance

The thermal performance of heat exchangers can be represented by the effectiveness-number of transfer units ($\epsilon - N_{tu}$) approach. The effectiveness ϵ is a measure of how closely the exchanger approaches ideal behavior, and is defined as

$$\epsilon = \frac{q_{actual}}{q_{maximum}} \qquad (10.3.1)$$

where q_{actual} is the actual amount of heat transferred. It is given in terms of the energy transferred from the hot to the cold stream by

$$q_{act} = (\dot{m}c_p)_h (T_{h_{in}} - T_{h_{out}}) = (\dot{m}c_p)_c (T_{c_{out}} - T_{c_{in}}) \qquad (10.3.2)$$

It is convenient to write Eq. (10.3.2) in terms of the capacitance rates defined as

$$C = \dot{m}c_p \qquad (10.3.3)$$

Equation (10.3.2) becomes

$$q_{act} = C_h(T_{h_{in}} - T_{h_{out}}) = C_c(T_{c_{out}} - T_{c_{in}}) \qquad (10.3.4)$$

The maximum heat transfer is the amount transferred in an ideal exchanger in which the fluid with the *minimum* capacitance rate is changed from the cold fluid inlet temperature to the hot fluid inlet temperature. This is given by

$$q_{max} = C_{min}(T_{h_{in}} - T_{c_{in}}) \qquad (10.3.5)$$

[Note that if q_{max} were defined using C_{max}, the first law expression Eq. (10.3.4) would specify that the fluid with the minimum capacitance rate would leave either hotter than $T_{h_{in}}$ or cooler than $T_{c_{in}}$. This would violate the second law.]

The effectiveness is a function of flow arrangement (e.g., counterflow, crossflow, etc.), the capacitance rate ratio C_{min}/C_{max}, and the number of transfer units, N_{tu}. The N_{tu} is defined as

$$N_{tu} = \frac{AU}{C_{min}} \qquad (10.3.6)$$

where A is the surface area of the exchanger. The conductance U is a heat transfer parameter that characterizes the exchanger. Often manufacturers will report the heat transfer rate for a given application, and the design engineer must calculate the conductance from this information and expressions for heat exchanger performance.

A plot of effectiveness as a function of N_{tu} for four common flow arrangements in heat exchangers is shown in Fig. 10.3.2. The effectiveness asymptotically approaches a limit as N_{tu}, or heat transfer area, increases. For counter- and crossflow exchangers, the effectiveness approaches unity. For parallel flow exchangers, the asymptote depends on the capacitance rate ratio. The counterflow exchanger always gives the best performance (highest effectiveness) for given values of N_{tu} and C_{min}/C_{max}.

The rate of heat transfer can be written in terms of the effectiveness by combining Eqs. (10.3.1) and (10.3.5) as

$$q_{act} = \epsilon C_{min}(T_{h_{in}} - T_{c_{in}}) \qquad (10.3.7)$$

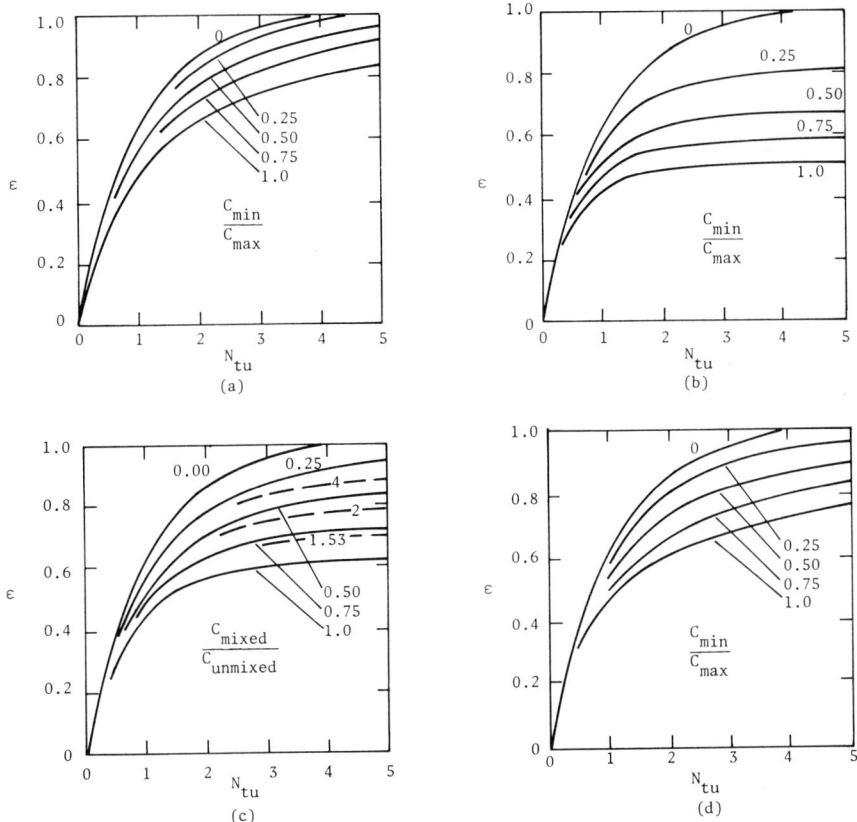

Fig. 10.3.2. Heat transfer effectiveness as a function of number of transfer units and capacity rate ratio. (*a*) Counterflow exchanger. (*b*) Parallel-flow exchanger. (*c*) Crossflow exchanger with one fluid mixed. (*d*) Crossflow exchanger with both fluids mixed.

The annual energy transfer is then the instantaneous rate times the annual number of hours N that the two streams are flowing together through the exchanger.

$$Q = N\epsilon C_{min}(T_{h_{in}} - T_{c_{in}}) \qquad (10.3.8)$$

For a particular manufacturer's line of heat exchangers and a given application, Eq. (10.3.8) represents a relation between the total energy transfer and the heat exchanger area. As the area increases, the effectiveness increases and total heat transfer increases. This is shown schematically in Fig. 10.3.3. The total heat transfer approaches a maximum value as surface area becomes large and is given by

$$Q_{max} = NC_{min}(T_{h_{in}} - T_{c_{in}}) \qquad (10.3.9)$$

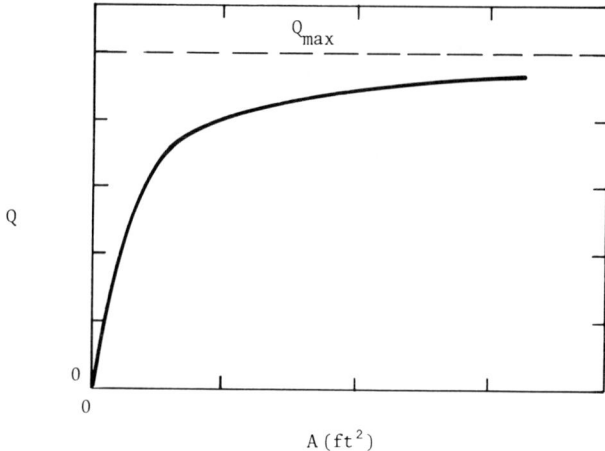

Fig. 10.3.3. Annual heat transfer as a function of heat exchanger surface area.

Thus there is a point of diminishing returns at which the added area does not produce commensurate energy savings. The heat exchanger area is the parameter that will be varied in order to optimize the exchanger.

10.3.2. Economic Performance

The economic performance of a waste heat recovery system is most readily accounted for using a life cycle savings approach. The costs of operation without a waste heat exchanger are only fuel costs. With an exchanger, the costs include the first cost of the unit and the pumping power costs. The savings are the differences in the two costs, and will be maximized in order to determine the optimum exchanger size.

The heating fuel cost for a system without a heat exchanger is the cost of heating the desired stream to the same outlet temperature as the heat exchanger does. The life cycle cost without an exchanger is

$$\text{LCC}_{no\,hx} = \frac{P_1 C_F Q}{\eta_{fur}} \quad (10.3.10)$$

The cost of the system with a heat exchanger is the sum of the first cost of the unit and the pumping work required to force the fluids through the exchanger. The first cost includes a base cost for those items needed for all exchangers regardless of size such as supports, piping, headers, and so on. There is also an area dependent cost that includes the cost of manufacturing the heat exchanger surfaces. The installation cost of a heat exchanger is a function of labor and material costs, and these too can be separated into base and area dependent costs. The total first cost can be represented as

$$\text{first cost} = C_o + C_A A \quad (10.3.11)$$

Equation (10.3.11) represents the cost of many different types of heat exchangers fairly well, with the constants differing for each type. Several quotations from suppliers for heat exchangers of different size may be needed to determine the values C_o and C_A.

For liquid-to-liquid exchangers, the pump work term is usually negligible. However, when a gas is one of the fluids, this term may be significant. The fan power would increase proportional to the exchanger size or surface area, and fan power would be supplied by electricity. The life cycle costs of the exchanger are the first costs and pumping power costs, and are given by

$$\text{LCC}_{hx} = P_2(C_o + C_A A) + \frac{P_1 C_{\text{elec}} W A}{\eta_{\text{pump}}} \qquad (10.3.12)$$

where C_{elec} is the delivered electrical cost and W is the annual pump work per unit surface area. Heat exchangers are designed to transfer heat efficiently, which means that the heat flow is usually large compared to the pump or fan work. In the remaining development the pump work will be assumed to be negligible.

The life cycle savings are the difference in costs between the situation with a boiler and that with an exchanger. Using Eq. (10.3.12), this becomes

$$\text{LCS} = \frac{P_1 C_F \epsilon Q_{\max}}{\eta_{\text{fur}}} - P_2(C_o + C_A A) \qquad (10.3.13)$$

10.3.3. Break-Even Fuel Cost C_F^*

The general relationship between life cycle savings and heat exchanger area is illustrated in Fig. 10.3.4 for three different values of fuel cost C_F. For all fuel costs, life cycle savings are negative for small areas. In general, the savings increase with area, reach a maximum, and then decrease. Beyond the maximum the cost of added heat exchanger area does not produce proportionate savings. The area $A°$ is the area that

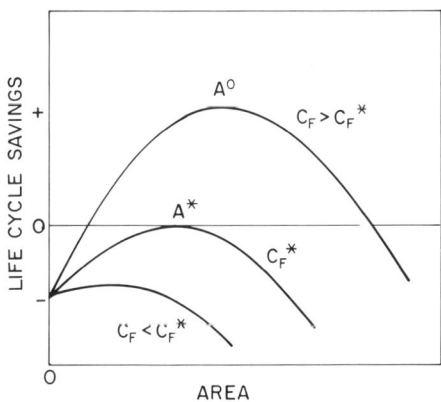

Fig. 10.3.4. Life cycle savings as a function of heat exchanger area.

produces maximum savings and thus it specifies the optimum heat exchanger for the application.

There is one value of fuel cost C_F^* for which the optimum exchanger with area A^* yields zero life cycle savings. At this fuel cost, there is no advantage of either the conventional heating system or the heat exchanger over the other. If the actual fuel cost is less than C_F^*, it is cheaper to purchase fuel. If it is greater, a heat exchanger would produce savings. Thus, the value of C_F^* relative to the actual fuel price can be used to determine whether any heat exchanger is cost effective.

The conditions on the break-even fuel cost are that the life cycle savings are zero and also that the savings are a maximum. This latter condition specifies that the derivative of savings with respect to area is zero. From Eq. (10.3.13),

$$\frac{\partial \text{LCS}}{\partial A} = \frac{P_1 C_F Q_{\max}}{\eta_{\text{fur}}} \frac{\partial \epsilon}{\partial A} - P_2 C_A = 0 \tag{10.3.14}$$

At the optimum area, denoted by the superscript $^\circ$, the condition becomes

$$\left(\frac{\partial \epsilon}{\partial A}\right)^\circ = \frac{P_2 C_A \eta_{\text{fur}}}{P_1 C_F Q_{\max}} \tag{10.3.15}$$

The effectiveness of heat exchangers is given in terms of $N_{tu}, C_{\min}/C_{\max}$, and flow arrangement as shown in Fig. 10.3.2. Equation (10.3.15) can be rewritten in terms of these parameters as

$$\left(\frac{\partial \epsilon}{\partial N_{tu}}\right)^\circ = \left(\frac{P_2 C_A}{P_1 C_F}\right)\left(\frac{\eta_{\text{fur}} C_{\min}}{Q_{\max} U}\right) \tag{10.3.16}$$

The other condition on the break-even fuel cost is that the savings are zero. From Eq. (10.3.13), this condition can be written in terms of the group on the right-hand side of Eq. (10.3.16) as

$$\left(\frac{P_2 C_A}{P_1 C_F}\right)\left(\frac{\eta_{\text{fur}} C_{\min}}{Q_{\max} U}\right) = \frac{\epsilon}{(C_o U/(C_A C_{\min}) + N_{tu})} \tag{10.3.17}$$

The break-even conditions are found by combining Eqs. (10.3.16) and (10.3.17). The break-even condition, denoted by the superscript *, is given by

$$\left(\frac{\partial \epsilon}{\partial N_{tu}}\right)^* = \frac{\epsilon^*}{(C_o U/(C_A C_{\min}) + N_{tu}^*)} \tag{10.3.18}$$

This condition is evaluated for the best type of heat exchanger possible. This is a counterflow exchanger with $C_{\min}/C_{\max}$ equal to zero. For this exchanger, the effectiveness is given by

$$\epsilon^* = 1 - e^{-N_{tu}^*} \tag{10.3.19}$$

Equation (10.3.19) is used to eliminate ϵ^* in Eq. (10.3.18). The resulting break-even condition can be written as

$$N_{tu}^* + \frac{C_o U}{C_A C_{min}} + 1 = e^{N_{tu}^*} \qquad (10.3.20)$$

This equation is trancendential in N_{tu}^*, and in order to facilitate calculations, N_{tu}^* is plotted against the group $(C_o U)/(C_A C_{min})$ in Fig. 10.3.5. The corresponding area A^* is evaluated from N_{tu}^* by

$$A^* = \frac{N_{tu}^* C_{min}}{U} \qquad (10.3.21)$$

The break-even fuel cost can now be determined from Eq. (10.3.13) as

$$C_F^* = \frac{P_2(C_o + C_A A^*) \eta_{fur}}{P_1 \epsilon^* Q_{max}} \qquad (10.3.22)$$

This break-even fuel cost represents the minimum fuel price that would allow the best heat exchanger to save money. If fuel prices are less than C_F^*, no heat exchanger will be economic. If fuel prices are greater than C_F^*, a counterflow heat exchanger will save money. This provides a criterion as to whether a heat exchanger should be considered further in a given application.

10.3.4. Optimum Heat Exchanger Size and Maximum Life Cycle Savings

If the actual fuel cost is greater than the break-even fuel cost, then there is an area for which the life cycle savings are a maximum. The optimum conditions are given

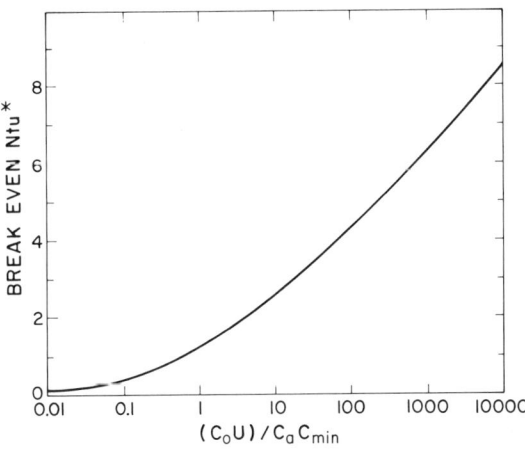

Fig. 10.3.5. Relation for break-even N_{tu}^*.

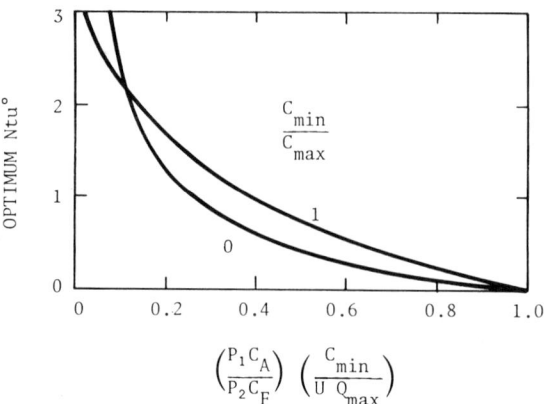

Fig. 10.3.6. Relation for optimum N_{tu}°.

by Eq. (10.3.16). To facilitate the calculation, the value of $(\partial \epsilon / \partial N_{tu})^\circ$ was calculated for counterflow heat exchangers. This graphical relation is given in Fig. 10.3.6 with ϵ° as a function of the right-hand side of Eq. (10.3.16). The assumption that all exchangers can be treated as a counterflow unit is reasonable for crossflow exchangers with both fluids mixed. For parallel flow and crossflow exchangers with one fluid mixed, this assumption underpredicts the exchanger size and savings. However, these latter two types are not commonly employed because of their low effectiveness.

Figure 10.3.6 allows evaluation of the optimum N_{tu}°. The optimum area is given by

$$A^\circ = \frac{N_{tu} C_{\min}}{U} \qquad (10.3.23)$$

The optimum effectiveness is found from the appropriate graph in Fig. 10.3.2. The maximum savings are then computed from Eq. (10.3.13) as

$$\text{LSC}^\circ = \frac{P_1 C_F \epsilon^\circ Q_{\max}}{\eta_{\text{fur}}} - P_2 (C_o + C_A A^\circ) \qquad (10.3.24)$$

10.3.5. Worksheet for Heat Exchanger Evaluation

A worksheet has been developed for evaluating these quantities for a waste heat recovery application, and is given in Table 10.3.1. The entries follow the equations developed in Sections 10.3.1 through 10.3.4. Also included in the worksheet is a section that allows evaluation of nonoptimum exchangers. Often exchangers are built only in discrete sizes, and the optimum exchanger must be chosen from one of those available.

TABLE 10.3.1
Worksheet for Heat Exchanger Evaluation

Data Required

Thermal Parameters

Waste stream:
- flow rate ($\dot{m}$)   _____ lb$_m$/hr
- specific heat (c_p)   _____ Btu/lb$_m$ °F
- capacitance rate ($C = \dot{m}c_p$)   _____ Btu/hr °F
- inlet temperature ($T_{w\,in}$)   _____ °F

Process stream:
- flow rate ($\dot{m}$)   _____ lb$_m$/hr
- specific heat (c_p)   _____ Btu/lb$_m$ °F
- capacitance rate ($C = \dot{m}c_p$)   _____ Btu/hr °F
- inlet temperature ($T_{p\,in}$)   _____ °F

Number of coincident hours per year (N)   _____ hr
Minimum capacitance rate (C_{min})   _____ Btu/hr °F
Inlet temperature difference ($T_{w\,in} - T_{p\,in}$)   _____ °F
Maximum heat transfer [Q_{max}, Eq. (10.3.9)]   _____ Btu

Heat Exchanger Parameters
- conductance (U)   _____ Btu/hr ft^2 °F
- base cost (C_o)   _____ $
- area dependent cost (C_A)   _____ $/ft^2

Economic Parameters
- displaced fuel price (C_F)   _____ $/Btu
- furnace efficiency   _____
- present worth of fuel costs (P_1)   _____
- present worth of owning costs (P_2)   _____

Break-even Fuel Cost C_F^*

$(C_o U)/(C_A C_{min})$ _____
Number of transfer units, N_{tu}^* (Fig. 10.3.5)   _____
Optimum area A^* [Eq. (10.3.21)]   _____ ft^2
Optimum effectiveness ϵ^* [Eq. (10.3.19)]   _____
Break-even fuel cost C_F^* [Eq. (10.3.22)]   _____ $/Btu

Is C_F greater than C_F^*?
- _____ If yes, heat exchanger is cost effective.
- _____ If no, heat exchanger is not cost effective.

(*Continued*)

TABLE 10.3.1 (Continued)

Optimum Area and Maximum Life Cycle Savings

$\left(\dfrac{P_2 C_A}{P_1 C_F}\right)\left(\dfrac{\eta_{fur} C_{min}}{Q_{max} U}\right)$ _____

C_{min}/C_{max} _____

Optimum number of transfer units N_{tu}° (Fig. 10.3.6) _____

Optimum effectiveness ϵ° (Fig. 10.3.2) _____

Optimum area A° [Eq. (10.3.21)] _____ ft²

Life cycle fuel savings $[P_1 C_F \epsilon^{\circ} Q_{max}/\eta_{fur}]$ _____ $

First costs $[P_2(C_o + C_A A^{\circ})]$ _____ $

Life cycle savings (LCS) _____ $

Nonoptimum Heat Exchangers Area

Area A	_____ ft²	_____ ft²	_____ ft²
N_{tu}	_____	_____	_____
ϵ (Fig. 10.3.2)	_____	_____	_____
Life cycle fuel costs	_____ $	_____ $	_____ $
Life cycle first costs	_____ $	_____ $	_____ $
Life cycle savings	_____ $	_____ $	_____ $

10.3.6. Example of Heat Exchanger Evaluation

As an example of the procedure developed in this section, it will be assumed that there is a waste stream of water flowing at 1000 gpm and 110°F, and it has been proposed that this stream could be used to preheat water entering the plant at 500 gpm and 70°F. It has also been estimated that the waste stream is available for 100 hr/yr. From this information the thermal parameters are entered on the worksheet for both waste and process streams.

The heat exchanger type is a counterflow shell and tube exchanger. Cost and performance information on the heat exchanger obtained from the manufacturer are tabulated in Table 10.3.2. An average value of the conductance (U) of 850 Btu/hr ft² °F will be used. A plot of cost as a function of area yields values of C_o and C_A of $4480 and $7.55/ft², respectively.

The cost of fuel is taken to be that of coal at a price of $2.55/10⁶ Btu used in a boiler of 85 percent efficiency. The economic parameters are taken for a 10-year life and an acceptable rate of return of 15 percent. The fuel inflation rate is taken to be 12 percent and the corporate tax rate is 45 percent. A tax credit of 20 percent is available for this heat recovery system. The values of P_1 and P_2 are then 4.26 and 0.80, respectively.

The break-even fuel cost is calculated to $1.59/10⁶ Btu. As this is less than the displaced fuel cost of $2.55/10⁶ Btu, the heat exchanger will be cost effective.

The optimum area is determined to be 589 ft². The corresponding life cycle savings are $2827. The heat exchanger is a good investment.

HEAT EXCHANGERS FOR WASTE HEAT RECLAMATION

TABLE 10.3.2
Heat Exchanger Parameters

Heat Exchanger Model	Surface Area (ft^2)	Cost ($)	Conductance (Btu/hr ft^2 °F)
1	158	5,650	857
2	312	6,850	853
3	475	8,050	850
4	622	9,250	848
5	790	10,400	847

Table 10.3.2 implies that only certain size exchangers may be purchased. The optimum heat exchanger area of 589 ft^2 is bracketed by models 3 and 4. The life cycle savings of each of these is calculated as shown on the worksheet (Table 10.3.3). It is seen that both of the life cycle savings are very close to the optimum. Near the optimum size, the selection is not critical, and many alternatives will yield close to the optimum value.

The calculations for the filled in worksheet are shown below.

① $Q_{max} = 100 \text{ hr } (250,300 \text{ Btu/hr °F}) \, 40°F = 1.00 \times 10^9 \text{ Btu/yr}$

② $\dfrac{C_o U}{C_A C_{min}} = \dfrac{\$4480(850 \text{ Btu/hr ft}^2 \text{ °F})}{\$7.55/\text{ft}^2(250,300 \text{ Btu/hr °F})} = 2.02$

③ $A^* = \dfrac{1.5(250,300 \text{ Btu/hr °F})}{(850 \text{ Btu/hr ft}^2 \text{ °F})} = 442 \text{ ft}^2$

④ $\epsilon = 1 - e^{-1.5} = 0.78$

⑤ $C_F^* = \dfrac{0.80(7.55 \, \$/\text{ft}^2)(442 + 4480/7.55) \text{ ft}^2 \, (0.85)}{4.26(0.78)(1.0 \times 10^9 \text{ Btu})}$

$= 1.59 \times 10^{-6} \, \$/\text{Btu}$

⑥ $\dfrac{P_2 C_A \eta_{fur} C_{min}}{P_1 C_F Q_{max} U} = \dfrac{(0.80)(7.55)(0.85)(250300)}{4.26(2.55 \times 10^{-6})(1.00 \times 10^9)(850)} = 0.14$

⑦ $\dfrac{C_{min}}{C_{max}} = \dfrac{(250,300)}{(500,000)} = 0.5$

⑧ $A° = \dfrac{2.0(250,300)}{(850)} = 589 \text{ ft}^2$

⑨ $P_1 C_F \epsilon° Q_{max} = 4.26(2.55 \times 10^{-6})(0.78)(1.0 \times 10^9)/0.85 = \$9968$

⑩ $P_2(C_o + C_A A°) = 0.8[4480 + 7.55(589)] = \$7141$

⑪ $\text{LCS} = 9968 - 7141 = \$2827$

TABLE 10.3.3
Worksheet for Heat Exchanger Evaluation

Data Required

Thermal Parameters

Waste stream:
- flow rate ($\dot{m}$)   500,600 lb_m/hr
- specific heat (c_p)   1.0 Btu/lb_m °F
- capacitance rate ($C = \dot{m}c_p$)   500,600 Btu/hr °F
- inlet temperature ($T_{w\,in}$)   110 °F

Process stream:
- flow rate ($\dot{m}$)   250,300 lb_m/hr
- specific heat (c_p)   1.0 Btu/lb_m °F
- capacitance rate ($C = \dot{m}c_p$)   250,300 Btu/hr °F
- inlet temperature ($T_{p\,in}$)   70 °F

Number of coincident hours per year (N)   100 hr
Minimum capacitance rate (C_{min})   250,300 Btu/hr °F
Inlet temperature difference ($T_{w\,in} - T_{p\,in}$)   40 °F

(1) Maximum heat transfer [Q_{max}, Eq. (10.3.9)] 1.0×10^9 Btu

Heat Exchanger Parameters
- conductance (U)   850 Btu/hr ft² °F
- base cost (C_o)   4480 $
- area dependent cost (C_A)   7.55 $/ft²

Economic Parameters
- displaced fuel price (C_F) 2.55×10^{-6} $/Btu
- furnace efficiency   0.85
- present worth of fuel costs (P_1)   4.26
- present worth of owning costs (P_2)   0.80

Break-even Fuel Cost C_F^*

(2) $(C_o U)/(C_A C_{min})$ 2.02

Number of transfer units, N_{tu}^* (Fig. 10.3.5)   1.5

(3) Optimum area A^* [Eq. (10.3.21)]   442 ft²

(4) Optimum effectiveness ϵ^* [Eq. (10.3.19)]   0.78

(5) Break-even fuel cost C_F^* [Eq. (10.3.22)] 1.59×10^{-6} $/Btu

Is C_F greater than C_F^*?

     _X_ If yes, heat exchanger is cost effective.

     ___ If no, heat exchanger is not cost effective.

HEAT EXCHANGERS FOR WASTE HEAT RECLAMATION

TABLE 10.3.3 (Continued)

Optimum Area and Maximum Life Cycle Savings

(6)	$\left(\dfrac{P_2 C_A}{P_1 C_F}\right)\left(\dfrac{\eta_{fur} C_{min}}{Q_{max} U}\right)$	0.14
(7)	C_{min}/C_{max}	0.5
	Optimum number of transfer units N_{tu}^o (Fig. 10.3.6)	2.2
	Optimum effectiveness ϵ^o (Fig. 10.3.2)	0.78
(8)	Optimum area A^o [Eq. (10.3.21)]	589 ft²
(9)	Life cycle fuel savings $[P_1 C_F \epsilon^o Q_{max}/\eta_{fur}]$	9968 $
(10)	First costs $[P_2(C_o + C_A A^o)]$	7141 $
(11)	Life cycle savings (LCS)	2827 $

Nonoptimum Heat Exchangers Area

Area A	475	ft²	622	ft²	ft²
N_{tu}	1.6		2.1		
ϵ (Fig. 10.3.2)	0.72		0.77		
Life cycle fuel costs	9201	$	9841	$	$
Life cycle first costs	6453	$	7340	$	$
Life cycle savings	2748	$	2501	$	$

10.3.7. Effect of Fuel Price on Heat Exchanger Selection

The selection of the optimum heat exchanger depends on many economic parameters, including fuel price. The life cycle savings for the heat exchanger example are plotted in Fig. 10.3.7 as a function of heat exchanger area for three different fuel costs. The lowest corresponds to coal at $2/10⁶ Btu, and the intermediate value corresponds to oil at $4/10⁶ Btu. The highest value is for electricity at $6/10⁶ Btu, and would represent savings if the heat exchanger were to reclaim waste "cold" and displace refrigeration energy.

The life cycle savings increase as the fuel price increases, as expected. The optimum area also increases as fuel price increases but not in direct proportion to fuel cost. For coal, the optimum area is 500 ft², for oil it is 850 ft², and for electricity it is 1100 ft². As is usually the case, the region of optimum savings is fairly broad. Although the fuel savings with oil are about 10 times that with coal, a heat exchanger which is optimum for a coal-fired system will provide 93 percent of the savings for an oil-fueled system as does the optimum exchanger. If the optimum exchanger for coal is used when electricity is the fuel, the savings are about 90 percent of the optimum. Near the optimum value, the savings are not sensitive to exchanger size. This mitigates the need to know fuel inflation rates accurately.

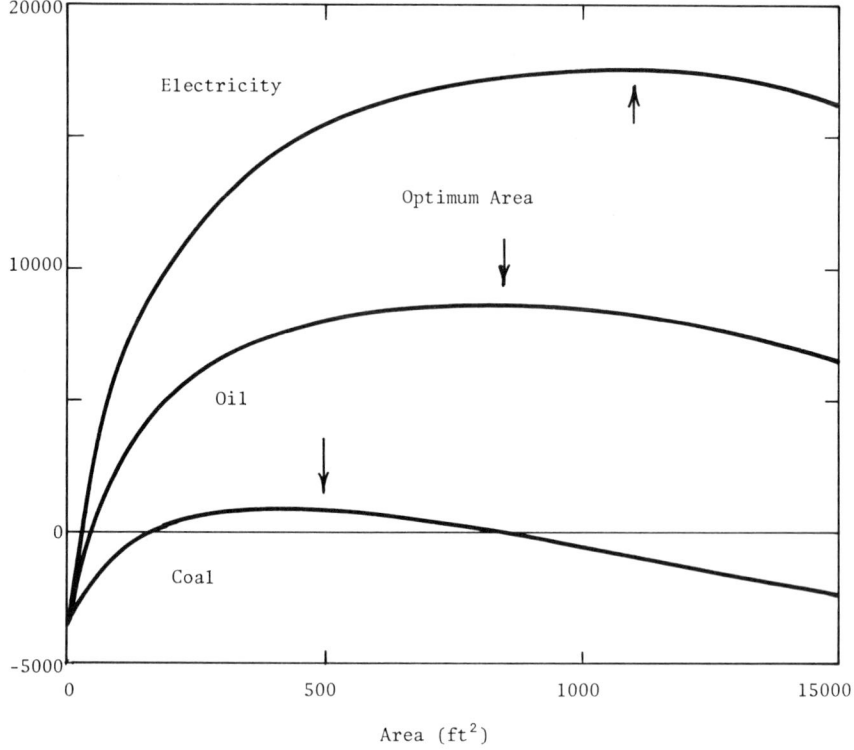

Fig. 10.3.7. Life cycle savings as function of heat exchanger area.

10.4. HEAT PUMPS FOR WASTE HEAT RECOVERY

Heat pumps have the potential for reclaiming energy from a relatively cold waste stream and pumping this energy to a higher temperature process stream. The heat pump extracts the energy from the waste stream by lowering its temperature. Through the work supplied to the heat pump compressor, energy is delivered to the process stream. As with a heat exchanger for waste heat recovery, it is always technically feasible to use a heat pump to recover waste heat. However, it may not be economically feasible. The value of the heating fuel energy saved must be greater than the sum of first and operating costs of the heat pump.

In this section, the thermal and economic performance of heat pumps for waste heat recovery will be determined. The development will parallel that for heat pumps used for heating as discussed in Chapter 7. A criterion for considering a heat pump for a given application will be developed first. The optimum size heat pump for an application will then be determined, and the life cycle savings calculated.

10.4.1. Thermal Performance

The energy delivered by the heat pump to the process stream is the capacity. The heat pump coefficient of performance is the ratio of this delivered energy to the work input $\dot{W}$.

$$\text{COP} = \frac{q_{htg}}{\dot{W}} \tag{10.4.1}$$

The value of COP depends on the temperature difference between the waste and process streams, and ranges from 2 to 6 for commercially available heat pumps used for waste heat recovery.

When used in a waste heat recovery application, the energy delivered equals the product of the process stream flow rate, specific heat, and the difference between the outlet and inlet process stream temperatures.

$$q_{htg} = \dot{m} c_p (T_o - T_i)_p \tag{10.4.2}$$

The purchased energy required to deliver this heat is the compressor work. On an annual basis the work is the capacity times the number of hours per year N that the waste and process streams are simultaneously available divided by the heat pump COP. The annual cost of this energy is the annual work times the cost of electricity used to drive the compressor and is

$$C_{comp} = \frac{C_{elec} q_{htg} N}{\text{COP}} \tag{10.4.3}$$

The heat pump displaces an amount of heating energy equal to the capacity of the heat pump times the number of hours it is used. The fuel that must be supplied equals the heating energy divided by the furnace or boiler efficiency of the device that would be used in the absence of the heat pump. The cost of fuel displaced is the fuel energy times the cost of fuel and is given by

$$C_{htg} = \frac{C_F q_{htg} N}{\eta_{fur}} \tag{10.4.4}$$

The development in Section 7.1 may also be used to determine the conditions under which an industrial heat pump is energy effective. From Eq. (7.1.8), the heat pump will save resource energy if

$$\text{COP} > \frac{\eta_{fur}}{\eta_{pp}} \tag{7.1.8}$$

Industrial boilers have furnace efficiencies of about 0.85, and power plant efficiencies are about 0.33. For a heavy pump to save energy, the COP must be greater than

2.4. This is higher than that for residential applications due to the use of more efficient furnaces in industry.

10.4.2. Economic Performance

A heat pump is cost effective if the life cycle savings, which are the difference in life cycle costs betwen using a conventional fuel and using a heat pump, are positive. For the conventional boiler or furnace, the life cycle costs are those of the fuel only and are given by

$$\text{LCC}_{htg} = \frac{P_1 C_F q_{htg} N}{\eta_{fur}} \tag{10.4.5}$$

The life cycle costs of the heat pump include the purchase and installation cost plus the cost of electrical work supplied:

$$\text{LCC}_{hp} = \frac{P_1 C_{elec} q_{htg} N}{\text{COP}} + P_2 \text{ (first)} \tag{10.4.6}$$

The life cycle savings are the difference between the conventional heating and the heat pump costs, and are given by

$$\text{LCS} = P_1 q_{htg} N \left[\frac{C_F}{\eta_{fur}} - \frac{C_{elec}}{\text{COP}} \right] - P_2 \text{ (first)} \tag{10.4.7}$$

A preliminary estimate of the economic feasibility of a heat pump can be made as for a residential application. The life cycle savings will always be negative if the term in brackets in Eq. (10.4.7) is less than zero. Thus, for a heat pump to be cost effective, the COP must satisfy the following criteria.

$$\text{COP} > \frac{C_{elec} \eta_{fur}}{C_F} \tag{10.4.8}$$

This provides a readily determined criteria as to whether to consider a heat pump on economic grounds. Even if the COP is greater than this value, a heat pump may not be cost effective unless the hours of operation and the heat delivered are sufficiently large so that the first cost is relatively small. For an industrial application using coal at $2/10^6$ Btu in a boiler of 85 percent efficiency and purchasing electricity at \$0.03/kWh, the COP must be greater than 3.7. However, if fuel oil at 80¢/gal (\$5.70/10^6 Btu) is used, the COP has to be greater than about 1.3. The economic viability depends strongly on the fuel currently used for heating.

Performance curves for a heat pump designed for industrial applications are shown in Fig. 10.4.1. These are representative of Westinghouse Templifier heat pumps which are used for transferring heat between liquids. The graph indicates that a wide variety of inlet and outlet conditions can be accommodated. The two-

HEAT PUMPS FOR WASTE HEAT RECOVERY

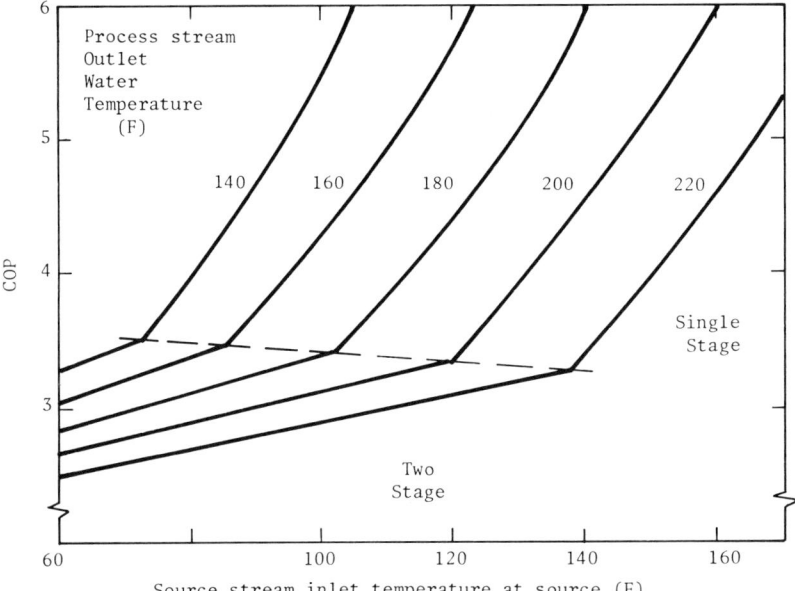

Fig. 10.4.1. Performance of industrial heat pumps.

stage units have two compressors with intercooling between stages and are more efficient than single-stage units. These units are designed to be operated at constant source and process temperatures, and are sized to meet a specific load. In this manner, they are different from residential heat pumps which are designed to operate over a wide range of source temperatures.

The performance curves can be plotted against the temperature difference between source inlet and process outlet temperatures to yield a ready determination of the economic and energy effectiveness. Figure 10.4.2 presents the results of Fig.

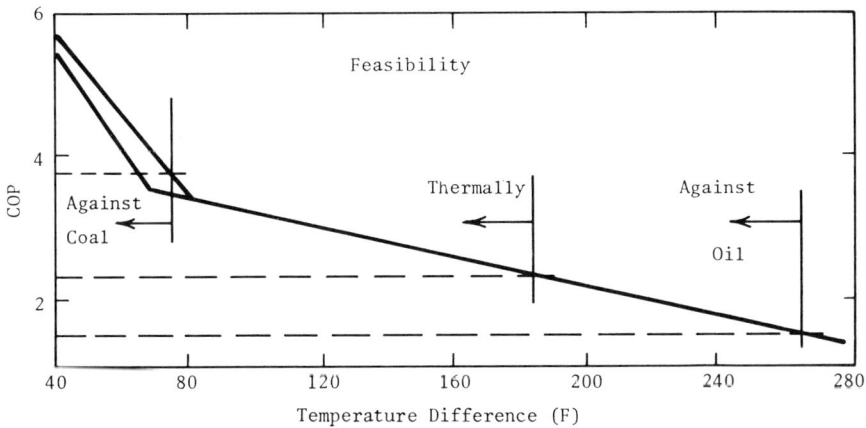

Fig. 10.4.2. Performance as a function of temperature difference.

10.4.1 in this manner. Based on the criteria for energy, heat pumps are viable for temperature differences below about 200°F. However, if coal is the conventional fuel, heat pumps will save money only for low temperature differences. When oil is the conventional fuel, the heat pump can save money up to very large temperature differences. At levels above 200°F, heat pumps use more energy but save money.

The optimum size heat pump for a given application depends on its first cost as well as fuel costs. The purchase price of industrial sized heat pumps is about $50/ton. This is considerably lower than that for a residential application due to much larger sizes which bring the unit price down.

10.4.3. Worksheet

A worksheet has been developed for evaluating the economic viability of a heat pump, and is given as Table 10.4.1. The entries follow the development given in Sections 10.4.1 and 10.4.2. First, the application parameters are entered. Then, economic data for the application and the heat pump are entered. The preliminary evaluation based on COP is made. Finally, if the heat pump could be cost effective, the life cycle savings of the various heat pumps are determined.

10.4.4. Example of Heat Pump Evaluation

As an example of the procedure it will be assumed that a process stream of water at 100 gpm is needed at 130°F. A waste stream at 90°F is available. Over the course of the year, there are 1200 hours of plant operation during which both the process and waste streams are flowing. There are several heat pumps available that range in heating capacity from 20 to 60 tons. Three units will be considered and their costs, including installation, are given on the worksheet (Table 10.4.2). The nominal COP of all units is the same.

The cost of conventional fuel is taken to be that of fuel oil at 80¢/gal, used in a boiler with an 85 percent efficiency. The cost of electricity is 3¢/kWh. The present worth factors P_1 and P_2 for a 10-year life are 4.26 and 0.8, respectively.

The preliminary estimate shows that the COP must be greater than 1.3. Since the COP of 4.66 for each unit is greater than this, all of the heat pumps are less costly in terms of fuel use than conventional boilers. The next consideration is whether the fuel savings are greater than the first costs.

The life cycle fuel savings are computed for the three heat pumps. The first costs are then subtracted from the fuel savings to determine the life cycle savings.

Only the two largest heat pumps are cost effective. While the smallest heat pump does save fuel, the savings are not sufficient to balance the first costs. Of the two larger heat pumps, the largest capacity unit saves the most energy and money. The largest unit, however, may not always be the most cost effective. In some applications, a large capacity unit may either cool the waste stream down or heat the process stream up to such values that the capacity and COP are reduced. In this example, the temperature changes are in the range of 10-20°F and within the manufacturers' performance specifications.

HEAT PUMPS FOR WASTE HEAT RECOVERY

TABLE 10.4.1
Worksheet for Heat Pump Evaluation

Application Parameters

Process stream: Flow rate ($\dot{m}$)	_____	lb_m/hr
Specific heat (c_p)	_____	Btu/lb_m °F
Outlet temperature	_____	°F
Waste stream: Inlet temperature	_____	°F
Number of coincident hours (N)	_____	hr

Heat Pump Parameters

Capacity (Btu/hr)	_____	_____	_____
COP	_____	_____	_____
First cost, $	_____	_____	_____

Economic Parameters

Conventional fuel cost (C_F)	_____	$/Btu
Conventional heating efficiency (η_{fur})	_____	
Electricity cost (C_{elec})	_____	$/Btu
Present worth of fuel costs (P_1)	_____	
Present worth of owning costs (P_2)	_____	

Preliminary Feasibility

$C_{elec}\eta_{fur}/C_F$	_____	_____	_____
Is heat pump feasible? YES	_____	_____	_____
NO	_____	_____	_____

Life Cycle Savings

Life cycle fuel savings	$ _____	_____	_____	_____
Life cycle first costs	$ _____	_____	_____	_____
LCS, $	$ _____	_____	_____	_____

The calculations for the filled in worksheet are shown below.

① $\quad \dfrac{C_E \eta_{fur}}{C_F} = \dfrac{(8.8 \times 10^{-6})(0.85)}{(5.7 \times 10^{-6})} = 1.3$

② $\quad P_1 q_{htg} N \left[\dfrac{C_F}{\eta_{fur}} - \dfrac{C_E}{COP} \right]$

$\quad = (4.26)(488{,}000)(1200) \left[\dfrac{5.70 \times 10^{-6}}{0.85} - \dfrac{8.80 \times 10^{-6}}{4.60} \right] = \$12{,}018$

③ $\quad P_2$ (first) $= 0.8(15{,}900) = \$12{,}720$

④ $\quad$ LCS $= 12{,}018 - 12{,}720 = -\$702$

TABLE 10.4.2
Worksheet for Heat Pump Evaluation

Application Parameters

Process stream: Flow rate ($\dot{m}$)	50,060	lb_m/hr
Specific heat (c_p)	1.0	$Btu/lb_m\,°F$
Outlet temperature	130	°F
Waste stream: Inlet temperature	90	°F
Number of coincident hours (N)	1200	hr

Heat Pump Parameters

Capacity (Btu/hr)	486,000	732,000	976,000
COP	4.66	4.66	4.66
First cost, $	15,900	18,700	21,600

Economic Parameters

Conventional fuel cost (C_F)	5.7×10^{-6}	$/Btu
Conventional heating efficiency (η_{fur})	0.85	
Electricity cost (C_{elec})	8.8×10^{-6}	$/Btu
Present worth of fuel costs (P_1)	4.26	
Present worth of owning costs (P_2)	0.80	

Preliminary Feasibility

① $C_{elec}\eta_{fur}/C_F$		1.3	1.3	1.3
Is heat pump feasible?	YES	X	X	X
	NO			

Life Cycle Savings

② Life cycle fuel savings	$	12,018	18,026	24,035
③ Life cycle first costs	$	12,720	14,960	17,280
④ LCS, $	$	−702	3,066	6,755

10.5. INDUSTRIAL REFRIGERATION

The energy requirements for refrigeration are significant in the food processing and cryogenics industries. The food industry is characterized by many small individual plants which are often quite old. Traditional engineering methods have been used in designing and selecting the refrigeration equipment, which means that the lowest first cost was the usual criteria. The recent energy situation has created the need to include energy costs in the selection of refrigeration components.

A common situation in the design or alteration of a refrigeration system in a plant is the selection of a refrigeration coil (heat exchanger) to provide a given amount of cooling. The load might be that generated by cooling of a product such as freezing vegetables or that required to maintain a space such as a frozen meat locker at a low temperature. The wall and pipe insulations should be selected to minimize life cycle costs (Section 10.2). The refrigeration coil should also be chosen to minimize both first and energy costs. For a given situation, there are many possible coils available from a given manufacturer that will meet the given load, but the first and energy costs will vary dramatically between them. The plant engineer is best qualified to make the selection based on his particular economic situation. The selection process is complicated, and the basic concepts and relations will be laid out in this section.

A schematic of an industrial refrigeration plant is given in Fig. 10.5.1. The refrigerant is pumped by the compressor to the condenser, through the expansion valve, and then into the cooling coil (evaporator). The room air is blown over the coils, cooled, and returned to the room. In selecting the heat exchanger, the relevant energy terms are the room cooling load q_R, the fan power $\dot{W}_{fan}$, and the compressor power $\dot{W}_{comp}$. The goal is to select a heat exchanger to minimize the life cycle costs associated with these energy forms and the first cost of the exchanger.

The independent variables in the selection of a given line of heat exchangers are the heat transfer surface area and the frontal, or flow, area. As the surface area of an exchanger increases, the effectiveness increases and the temperature of the refrigerant increases and approaches that of the room. Increased refrigerant temperatures in the evaporator increase refrigeration system COP and reduce compressor power. For a given exchanger frontal area, increased surface area means the flow length increases, which raises the pressure drop and fan power. The frontal area may be increased to reduce flow velocity and, consequently, the pressure drop and fan power. However, the surface conductance may be lowered, which means the effectiveness drops. These interrelated effects complicate the heat exchanger selection procedure.

The cooling load is the sum of the room load and that generated by the fan-motor combination. All of the electric energy consumed by the fan motor must be

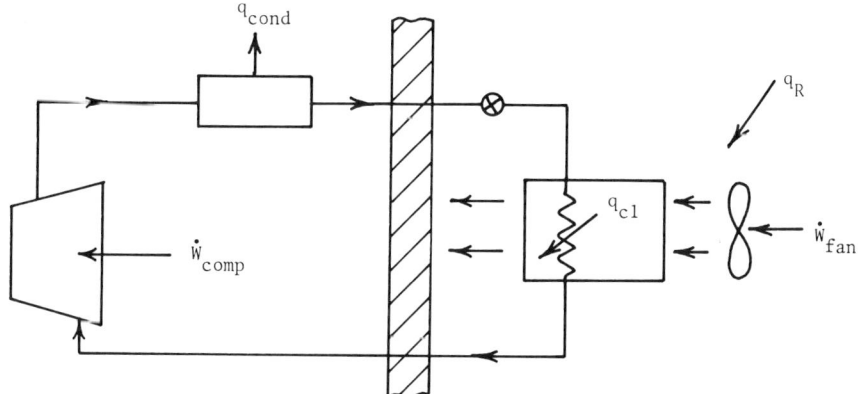

Fig. 10.5.1. Schematic of an industrial refrigeration process.

removed through the cooling coils. The cooling load is then

$$q_{cl} = q_R + \dot{W}_{fan} \tag{10.5.1}$$

The room load is determined from the heat flow through walls, infiltration due to door openings and closings, cooling of the product, and people and equipment in the room.

The fan power is given in terms of the pressure drop and air flow rate through the exchanger by

$$\dot{W}_{fan} = \frac{\dot{m} \Delta P}{\rho \eta_{fan}} \tag{10.5.2}$$

where the fan efficiency accounts for both fan motor and blade efficiencies. The pressure drop across the exchanger is related to the velocity and the exchanger flow length by

$$\Delta P = cLV^2 \tag{10.5.3}$$

where the constant c includes entrance and exit losses and flow friction. Manufacturers' data are needed to determine the exact relation. The flow velocity and flow rate are related to the exchanger frontal area by

$$\dot{m} = \rho A_f V \tag{10.5.4}$$

Comparisons of Eqs. (10.5.2)–(10.5.4) show that fan power depends strongly on flow velocity. It is important to have low velocities to minimize the effect of fan power on both direct and indirect purchased fuel costs.

The cooling load is given in terms of the heat exchanger effectiveness as

$$q_{cl} = \epsilon C_{min}(T_R - T_{evap}) \tag{10.5.5}$$

where the air flow is the minimum capacitance rate fluid. For an evaporator, the effectiveness is given by

$$\epsilon = 1 - e^{-N_{tu}} \tag{10.5.6}$$

The N_{tu} is given in terms of surface area and conductance by

$$N_{tu} = \frac{AU}{C_{min}} \tag{10.5.7}$$

The surface area is related to the flow length and frontal area through

$$A = \beta A_f L \tag{10.5.8}$$

The area density (β) is the heat transfer surface area per unit volume. It is a function of surface geometry (fin spacing, tube spacing, etc.), and depends on the particular line of exchangers under consideration.

The conductance is a function of flow velocity. It is usually proportional to velocity raised to an exponent between 0.5 and 0.8. Again, manufacturers' data must be consulted for specific values.

The compressor power required to meet the entire cooling load is given in terms of the refrigeration system COP by

$$\dot{W}_{comp} = \frac{q_{cl}}{COP} \qquad (10.5.9)$$

For industrial refrigeration systems, the COP depends on both condenser and evaporator temperatures. For a given condenser temperature, the COP is of the form

$$COP = ae^{b(T_{evap} - T_o)} \qquad (10.5.10)$$

where T_o is a reference temperature. For nominal ammonia systems with water cooled condensers, the typical values of a and b might be 3 and 0.02/°F for a value of the reference temperature T_o of 0°F. The COP is seen to be a strong function of evaporator temperature.

The life cycle costs are the sum of fuel and first costs. The fuel costs are the sum of compressor and fan costs, and their life cycle value is given by

$$\text{fuel costs} = P_1 C_E \left(\frac{q_R + \dot{W}_{fan}}{COP} + \dot{W}_{fan} \right) \qquad (10.5.11)$$

It is seen that fan power has both a direct cost on purchased electricity for the fan and an indirect cost on purchased compressor power.

The first costs of the exchanger depend on area. The size also affects the fan size selection, and this must be included. The exchanger costs are then given by an equation of the form

$$(\text{first cost})_{hx} = C_o + C_A A \qquad (10.3.11)$$

The life cycle costs become

$$LCC = P_1 C_E \left(\frac{q_R + \dot{W}_{fan}}{COP} + \dot{W} \right) + P_2 (C_o + C_A A) \qquad (10.5.12)$$

This expression is to be minimized with respect to exchanger area. The subsidiary relations for the effects of frontal area and length on COP and fan power need to be included. The optimization is complicated and best done using computer techniques.

The effect of area on the various energy and cost terms are shown schematically

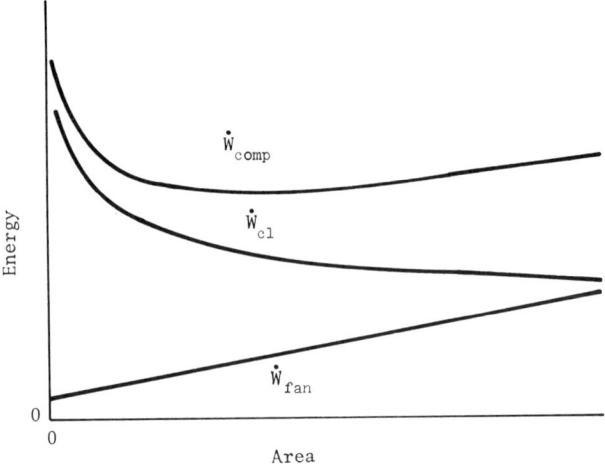

Fig. 10.5.2. Energy use of compressor and that due to cooling load and fan as a function of area.

in Figs. 10.5.2 and 10.5.3. These figures are more or less to scale for typical industrial systems. As shown in Fig. 10.5.2, the component of compressor work due to the room cooling load drops rapidly as exchanger area increases and approaches an asymptotic value. The increasing effectiveness with area allows the refrigerant temperature to approach that of the room. In contrast, fan power increases directly as exchanger area increases. The curves are realistic in that the fan power can be as large as that required to meet the cooling load. The total compressor power is the sum of these two components.

The life cycle costs are shown in Fig. 10.5.3. The compressor costs decrease with area and then increase in parallel with the energy curve of Fig. 10.5.2. The fan

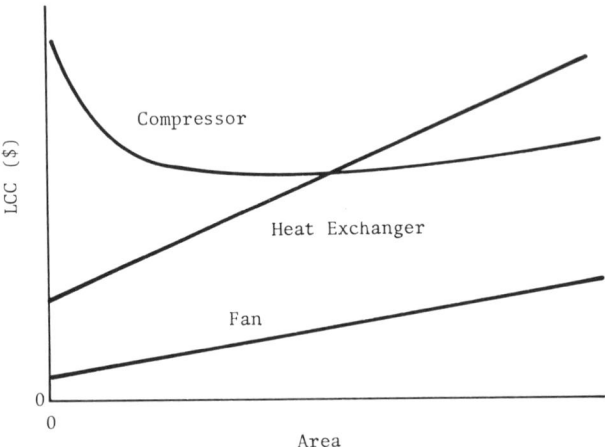

Fig. 10.5.3. Life cycle costs of refrigeration components.

power and heat exchanger first costs increase with area. These curves are also realistic in that energy costs are comparable to that for the exchanger.

The optimum selection procedure has been performed for a wide variety of situations. In general, heat exchangers are now selected which have larger areas and higher effectiveness than those previously installed. This allows refrigerant temperatures to be within about 5°F of the leaving air temperature. The exchanger geometries are selected to have larger frontal areas and lower velocities (by up to 50 percent) than previous exchangers. Energy costs are significantly reduced (up to one-third) from the earlier units. These selections demonstrate the importance of energy considerations in sizing equipment.

10.6. RECYCLING

The energy cost of various materials was given in Section 10.1. These costs are those to produce a basic stock ready for manufacturing from a raw material. The stock undergoes further processing and refinement to produce a finished product. Recycling of the used or worn-out finished product can eliminate many of the steps in the process and has the potential for significant energy savings. There are also benefits in reducing waste and the amount of virgin raw material required.

The production of automobiles from iron ore will be considered to illustrate the costs and benefits of recycling. Many steps are involved and each has certain energy requirements. Recycling will eliminate some, but not all, steps and significantly reduce energy requirements. This example will point out what is required in making such an analysis for another process.

The flow diagram for the production of the iron and steel in an automobile is shown in Fig. 10.6.1. The energy requirements for these processes per ton of finished product are shown to the right of the process. In the mining process, taconite ore is taken from the ground and then processed to increase the iron concentration. This ore is then transported by ships to the steel plant.

There are two major processes inside the steel plant. In the first, iron ore is reduced to pig iron in the blast furnace. This is a high temperature process and very energy intensive. The pig iron is then transported to arc, open hearth, and oxygen furnaces where the pig iron is converted into iron and steel stock. Each furnace can accommodate a mixture of raw and recycled material. It will be at this point in the flow stream that recycled automobiles will be used.

The metal stock is then sent to the steel mill where the basic forming operations of rolling, forging, and stamping are carried out. These processes consume the greatest portion of energy use, but cannot be eliminated by recycling. The component parts are shipped to the manufacturing plant where the finished automobile is assembled. Finally, the automobile is shipped to the dealer.

The total energy requirements per ton of finished automobile are 96×10^6 Btu. An average sized automobile has about 1.5 tons of iron and steel in it, and so about 144×10^6 Btu of energy are used to make one automobile. At an average energy price of $\$5/10^6$ Btu, the cost of energy to make an automobile is about \$720. This

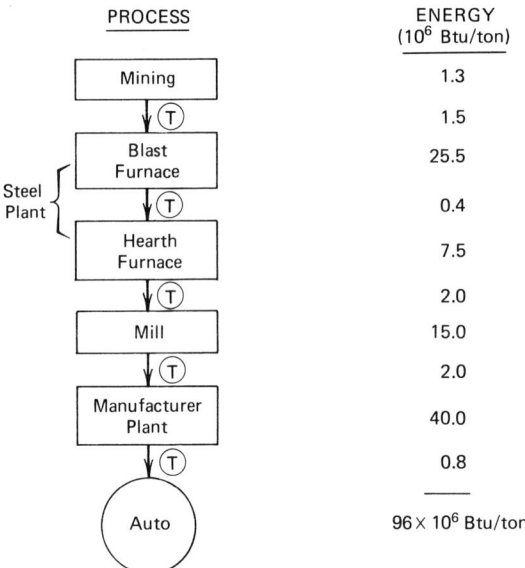

Fig. 10.6.1. Flow diagram for automobile manufacture.

is less than 10 percent of the total price, and shows that direct energy costs are a relatively minor portion of total costs.

Energy costs are significant, though, in each of the plants and processes. The biggest consumption occurs in the manufacturing operation. Measures that reduce energy use here have a potentially large payoff, and so manufacturing methods themselves bear close study. The blast furnace operations are second in energy use. Here the potential for heat recovery is large due to the amount of heating required. The transportation requirements add up to 6.7×10^6 Btu/ton, or only about 7 percent of the total consumption. This is due to the use of highly efficient ships and trains for much of the transportation. For each operation, then, there are possible energy reductions.

The recycling steps involve reprocessing the used automobile and introducing it into the manufacturing flow stream. The flow diagram for the manufacturing process with recycling is shown in Fig. 10.6.2. The used automobile is taken to a salvage yard where it is stripped and then crushed. The crushed car is taken to a scrap processor which mechanically converts the car into metal scrap and removes most of the plastic, aluminum, and other impurities. This scrap is transported, usually by rail, to the steel mill where the scrap can then be reused.

There are limits to the amount of scrap that is employed. The scrap has been oxidized, it contains impurities, and some of the iron originally had other metals and carbon added to it. In addition, not all automobiles are recycled. As a result, some fresh iron is required. The maximum amount of scrap that can be recycled is around two-thirds of a ton per finished ton in an automobile.

The energy requirements for the recycling process are shown on the right side of

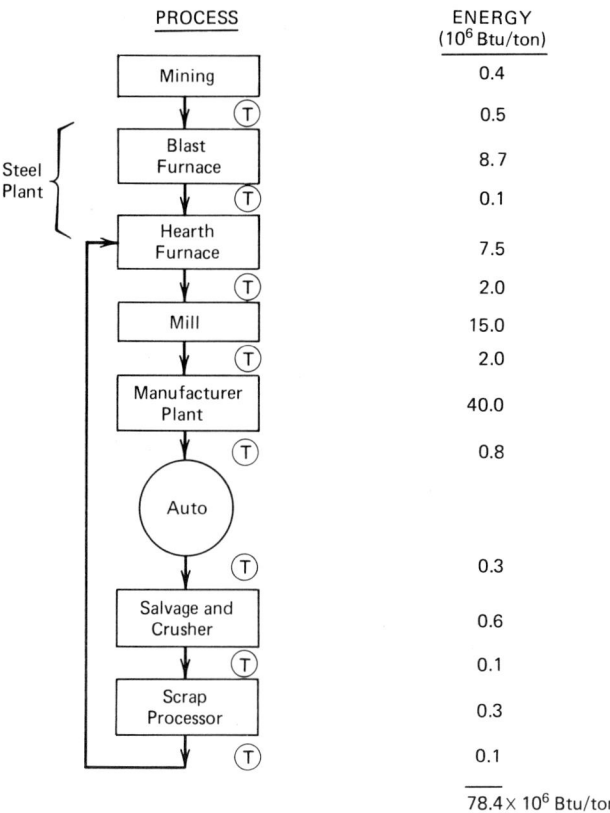

Fig. 10.6.2. Flow diagram for automobile manufacture with recycling.

Fig. 10.6.2. Recycling reduces the energy requirements for ore processing by 19×10^6 Btu/ton, and adds 1.4×10^6 Btu/ton for processing the scrap automobile. The net result is that energy requirements are reduced about 18 percent. The dollar value of the energy savings are about $130.

These energy savings represent only 1–2 percent of the final purchase price of the automobile, and are relatively small in comparison. However, they may be large in comparison to the profit. The manufacturer's profit on his sales are on the order of 10 percent, and thus the energy savings represent 10–20 percent of his profit. In general, a dollars worth of energy savings are equivalent to 100 dollars' worth of sales, and from this point of view energy savings are quite attractive.

The overall effect on the United States of recycling automobiles could be significant. If two million new automobiles are recycled annually the total energy savings would be about 50×10^{12} Btu/yr. This represents the heating energy requirements for about 0.5 million homes. These energy savings also create industrial changes. The mining and steel industry requirements are reduced, which may affect employment and the economy in this industry, while increasing the need for reprocessing facilities. All of these factors are important from the national viewpoint.

10.7. SUMMARY

The energy use in the industrial sector has been discussed in this chapter. Two main topics were covered. The first deals with the energy costs of materials and the energy savings due to recycling. The second deals with the energy use in heating and cooling and the economic optima for insulation, heat exchangers, and heat pump. There are specific examples of the use of energy in industry, which reflect the fact that the industrial sector is very diverse. Each industry itself has many specific processes which would benefit from an energy analysis. The techniques presented in this chapter should serve as a guide to conducting these analyses. The economic value of energy savings can be large compared to the profits from sales, and this enhances energy conserving techniques in industry.

SUGGESTED READING

W. K. Foell et al., *Industrial Waste Heat Recovery*, Report to Argonne National Laboratories from Resource Management Associates, Madison, WI, June 1980.

W. M. Kays and A. L. London, *Compact Heat Exchangers*, McGraw-Hill Book Co., New York.

J. M. Calm, *Recovery of Wasted Heat with Centralized and Distributed Heat Pump Systems*, ASME paper 78-WA/HT-631918.

B. Sternlicht, Capturing Energy from Industrial Waste Heat, *Mechanical Engineering*, August 1978.

D. A. Reay (ed.), *Industrial Energy Conservation*, A. Wheaton and Co., Exeter, UK, 1977.

PROBLEMS

10.1. Wall insulation is to be used in a building with 5000 heating degree days. Determine the optimum level of insulation considering the energy to manufacture the insulation and the heating energy. Heat is supplied by:
  (a) Electric resistance heat.
  (b) A natural gas furnace.

10.2. For the example of Section 10.3.6, assume that the waste stream is at $90°F$ at a flow of 250 gpm, and that there are 400 coincident hours of operation. Determine the optimum heat exchanger size.

10.3. Repeat Problem 10.2 using fuel oil as the furnace fuel.

10.4. For the example of Section 10.4.4, assume that the desired process stream outlet temperature is $110°F$, and compute the optimum heat pump savings.

PROBLEMS

10.5. For the example of Section 10.4.4, assume that the conventional fuel is coal, and compute optimum heat pump savings.

10.6. A refrigeration system is being designed for a large food locker. The cooling load is 100,000 Btu/hr throughout the year. Ammonia is the refrigerant, and the constants in the COP equation [Eq. (10.5.10)] are $3, 0.02/°F$, and $0°F$ for a, b, and T_o, respectively. The heat exchanger conductance is 15 Btu/hr ft² °F. Heat exchanger costs are a base cost of $10,000 and an area dependent cost of $1.50/ft². The fan power may be neglected. Determine the optimum heat exchanger area and refrigerant temperature for this application.

▪ 11 ▪

TRANSPORTATION ENERGY USE

11.1. OVERVIEW

The energy used for passenger and freight transportation accounts for about 25 percent of the total U.S. energy used. The fuel source is essentially all liquid petroleum and thus transportation consumes about 50 percent of the total petroleum supplies. The distribution of energy use by major transportation mode is given in Table 11.1.1. The automobile is the single largest energy consumer. Including the approximately 7 percent of the aircraft energy used for passenger travel, the passenger transportation energy consumption amounts to about two-thirds of the total.

The growth in transportation usage has paralleled that of the other major sectors. Within the sector, though, there have been dramatic shifts. Railroad usage has dropped eightfold since 1950, and aircraft usage has increased five times. Trucking has become the dominant mode of freight transportation, and has doubled over the last 15 years.

The energy intensiveness for various passenger and freight transportation modes is given in Tables 11.1.2 and 11.1.3. For passenger transport, the values represent typical, current usage. It is surprising that for urban trips, the common vehicular modes are fairly close in intensiveness. For intercity transit, the most popular mode, the airplane is the most intensive, while trains, which are better in energy use by a factor of 7, are disappearing as an option. The intensiveness for all modes is strongly dependent on the number of passengers. For example, although the automobile and bus intensivenesses are close for present loads, a filled urban bus (60 passengers) uses about 50 percent less energy than does a full automobile. Shifts to car pools, buses, and bicycles can significantly reduce energy consumption.

The energy intensiveness of freight transport as presented in Table 11.1.3 shows that trucking consumes a disproportionate amount of energy. The more efficient

TABLE 11.1.1
Transportation Energy Use

Mode	Percentage of Total
Automobile	53
Truck	26
Aircraft	8
Waterways	3
Pipelines	6
Railroads	3
Bus	1

modes (waterways, railroads, and pipelines) provide almost 80 percent of the total ton-miles at slightly over one-third the total energy cost. The use of the airplane is currently quite small. Pipelines and waterways are limited to certain materials and locations. Railroads, in contrast, can deliver goods to virtually any location at about one-half the energy cost of trucks. As with passenger transport, railroad usage is rapidly decreasing both in routes traveled and efficiency of transport.

The survey of energy use in the transportation sector shows that about three-quarters of the energy is consumed by on-road vehicles. About 72 percent of this amount is due to automobiles. In the next section, some use characteristics for the automobile will be discussed. The physical mechanisms which determine the power requirements for automotive and truck energy use will be presented, and alternatives that reduce engine power and fuel usage will be evaluated. The engine characteristics and their interaction with the vehicle will be presented. Although many changes in design can only be implemented by automobile and truck manufacturers, the development will provide some insight in the reasons for gasoline consumption.

TABLE 11.1.2
Energy Intensiveness of Passenger Transport

Vehicle	mpg	No. of Passengers	Passenger-mile / gal	Btu / Passenger-mile
Motorcycle	30	1	30	4100
Automobile	20	2	40	3100
Urban bus	4	12	48	2900
Urban train	1	50	50	2800
Jet airplane	0.12	250	30	4200
Intercity bus	6	20	120	1200
Intercity train	0.6	400	240	600
Walking	—	1	—	500[a]
Bicycle	—	1	—	300[a]

[a]Based on energy content of food.

TABLE 11.1.3
Energy Intensiveness of Freight Transport

Vehicle	Percentage of Energy	Percentage of Ton-miles	Btu/ton-mile
Jet airplane	2	0.2	12,000
Truck	61	24	2,500
Waterway	8	16	500
Railroad	19	36	500
Pipeline	10	24	400

11.2. AUTOMOBILE TRANSPORTATION

11.2.1. Use Profile

Automobiles consume about one-quarter of all the liquid petroleum fuel used in the United States. The uses, average trip lengths, and average occupancy for the family automobile are given in Table 11.2.1. These are from national surveys and may not represent changes that have occurred in the last few years.

Table 11.2.1 shows that most of the automobile uses occur for relatively short trips. Business-related uses such as commuting and shopping account for 61 percent of the total and many of these trips could be shifted to more efficient mass transit systems. The often critized long distant "vacation trip" accounts for a surprisingly low percent of total miles (2 percent) and is relatively efficient due to higher occupancy (about 3.3 people).

A further breakdown shows that 55 percent of the trips are less than 5 miles in length and account for 55 percent of the miles driven. Fifty-five percent of the trips have only one occupant. The overall average trip length is 9 miles, and the overall occupant level is two people. The average car makes a total of about 1400 trips annually, or about four trips per day. The automobile has become the mode of transportation for short trips, and usually carries few people per vehicle.

Another facet of the automobile usage is that the average number of vehicles per

TABLE 11.2.1
Automobile Use Profile

Function	Percentage of Miles	Trip Length (miles)	No. of Occupants
Earning a living	41	10	1.4
Recreational	34	9	2.5
Family business	20	6	2.0
Educational	5	5	2.5
Average		9	2.0

home is about 1.6. This means that roughly one-half of the homes have one car. A family that is purchasing only one car has to decide whether to purchase on the basis of average use or the maximum expected load. It appears that the decision is to buy one car that can accommodate the longest trips with the most people. As a result, the large vehicle is used inefficiently much of the time. Owning two cars, one large and one small, may seem to be a more efficient solution. However, there appears to be a fundamental law that two cars are driven twice as many miles as one car, and any savings are thus illusionary.

11.2.2. Historical Trends

The growth in automobile transportation over the period 1950 to present has been dramatic. The total number of automobiles has doubled, and the total number of surfaced roads has increased 50 percent. The interstate highway system has been added. As a result, the total number of miles driven has tripled.

The growth in automobile usage initially occurred with automobiles that became progressively less efficient. The average fuel economy over the years is shown in Fig. 11.2.1. Both the average and new car mileages are shown. There was a slow decrease in economy prior to 1974 as cars became progressively larger and more powerful. Legislation was then introduced setting targets for 1980 and 1985, and new car mileage is improving and will probably reach those goals.

New car fuel economy does not directly translate into improved overall mileage. The number of automobiles replaced every year is about 10 percent, and about one-half the automobiles on the road are over 5 years old. Thus, total fuel economy lags new car economy significantly.

The EPA mileage figures for some selected automobiles are given in Table 11.2.2. These cars were selected as being representative of different size classes. There is a

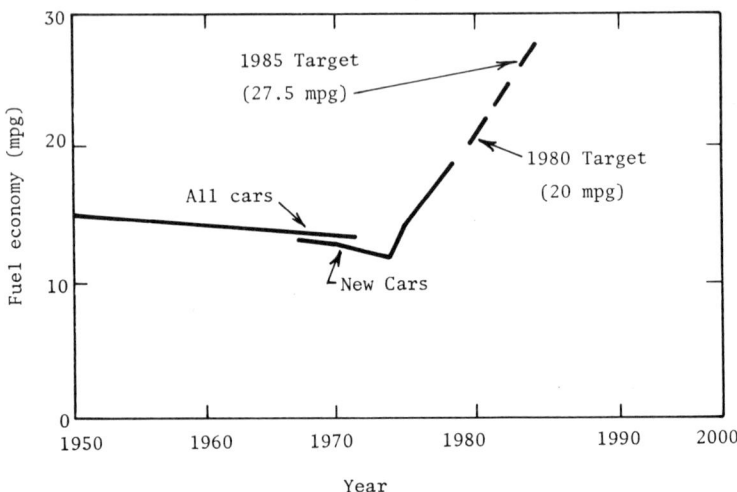

Fig. 11.2.1. Variation in automobile fuel economy over time.

TABLE 11.2.2
Gasoline Mileage of Selected Automobiles

	1977 Weight Class (lb$_f$)	Mileage (mpg)			
		1976	1977	1978	1979
Subcompact	2200				
Datsun		33	34	40	35
Toyota		29	32	39	31
Compact	4000				
Nova		20	21	21	19
Granada		18	21	24	18
Midsize	4800				
Malibu				24	22
Diplomat				20	18
Large	5500				
Mercury		14	16	16	15
Chrysler				15	14
Highest					
VW Rabbit (Diesel)				45	41
Lowest					
Chrysler				12	12

large effect of size on mileage, with larger automobiles having significantly poorer mileage than smaller ones. There has been a gradual increase in mileage over time reflecting changes in design. There is also a large difference between the lowest and highest fuel economy.

The sudden drop in 1979 mileage is due to a change in reporting. Prior to 1979, the EPA mileage figure that appeared to best represent the actual use was a combination of city and highway driving. Recently, it has been found that the previous "city" value best reflects overall useage, and this number is reported in the 1979 ratings. Although these mileage values are based on controlled dynamometer tests instead of road tests, one-half of the drivers report that their average fuel economy is within 2 mpg of the EPA figure. However, 10 percent report that their average is more than 5 mpg below the estimated value. As a result, these figures are only of use in representing the changes required to achieve better overall use of fuel.

The improvements in the total, or fleet, economy result in part from a shift from larger to smaller cars. The percentage of standard automobiles sold has dropped from about 60 percent of the market in 1960 to about 25 percent in 1975. The number of intermediate and compact cars has increased proportionately.

11.3. VEHICLE POWER REQUIREMENTS

The energy consumed by a given vehicle is a function of many variables, some of which are technical while others are social or legislative. In this section, those variables that are technical and that can be improved through engineering to reduce fuel consumption will be discussed.

The total amount of fuel used by a vehicle in a given year is a function of the miles driven and average fuel economy, and given by

$$\text{annual fuel use} = \frac{\text{miles driven}}{\text{mpg}} \qquad (11.3.1)$$

The miles driven depend on the function of the vehicle and the driving habits of the driver. The vehicle fuel economy is an average over all of the miles driven. At any one driving condition, the instantaneous mileage can be evaluated as

$$\text{mpg} = \frac{\text{velocity}}{(\text{specific fuel consumption})(\text{power})} \qquad (11.3.2)$$

The vehicle velocity V depends on function, driver's style, and legislation (speed limits). The specific fuel consumption is the fuel flow rate per unit power output and is primarily a function of engine design. It depends upon such factors as number of cylinders, compression ratio, fuel type, and displacement. The power required at any operating condition is determined from the friction, aerodynamic, gravitational, and inertial forces acting on the vehicle. The power requirement can be determined from

$$\dot{W} = (F_{fr} + F_{ae} + F_{gr} + F_{ir}) V \qquad (11.3.3)$$

The forces are a function of chassis design, the road, and vehicle speed.

The technical factors that influence vehicle mileage are then the engine design, which influences specific fuel consumption, and chassis design, which influences the forces on the vehicle. These latter factors will be discussed first, and the engine characteristics second. The objective will be to evaluate those that can readily be altered to improve fuel economy.

11.3.1. Frictional Forces

The frictional force F_{fr} acting on a vehicle is given by the product of its weight W and a coefficient of rolling friction k.

$$F_{fr} = kW \qquad (11.3.4)$$

The coefficient of rolling friction includes both the frictional effects at the interface between the tire and the road and the friction in the wheel and axle bearings. The friction between tire and road depends on both surface and the construction of the tire. To some extent, the overall coefficient depends on vehicle speed. Some nominal values for various road surfaces and standard belted tires are given in Table 11.3.1.

VEHICLE POWER REQUIREMENTS

TABLE 11.3.1
Coefficient of Rolling Friction

Road Surface	0 mph	55 mph
Concrete	0.015	0.02
Dirt	0.1	0.11
Sand	0.3	0.31

From Table 11.3.1, it is seen that road construction has a strong effect on the value of k. Further, a rough surfaced road increases k by 10-30 percent, while a road wet with rain or snow causes a 10 percent increase in k. These parameters are not under the control of the automobile designer.

Increased inflation pressure reduces contact area and k accordingly. For concrete roads, a 25 percent increase in pressure reduces k by 5-10 percent. Work is done on a tire as it flexes during rotation, and tire construction can change the corresponding value of k. Radial tires have a 6-10 percent lower value than do tires with bias construction. A bald tire has a value of k about one-half that of one with tread; this obviously reduces energy consumption at the expense of safety. Overall, then, a properly inflated radial tire could reduce k from 0.015 to 0.0125, or a 20 percent drop in frictional force and power.

Vehicle weight can also be reduced to lower rolling friction. Vehicle weight has dropped as size is reduced. Currently, the steel, iron, and other metals comprise about three-quarters of the total weight. Plastics are commonly used for trim, body panels, and some structural members, and save about 70 percent in weight over metal components. They have a potential for many more functions. There are however, trade-offs in reducing weight for fuel economy and providing safety for occupants.

Since the frictional force and power depend directly on vehicle weight, reductions in weight reduce energy consumption in direct proportion. As an example of the role of the frictional force in the overall energy usage, two automobiles representative of small and large classes and a truck will be considered. The forces and horsepower requirements at 55 mph for these are given in Table 11.3.2 for a value of k of 0.015. It is seen that the horsepower requirements are low compared to installed engine size for the automobiles. For a truck, the frictional horsepower requirements are significant.

TABLE 11.3.2
Frictional Force and Horsepower Requirements

	Force	Power at 55 mph
Subcompact (2200 lb$_f$)	33 lb$_f$	4.8 hp
Large (5500 lb$_f$)	82.5 lb$_f$	12.1 hp
Truck (20 tons)	600 lb$_f$	88.0 hp

11.3.2. Aerodynamic Forces

The aerodynamic force F_{ae} acting on a vehicle depends on the area normal to the direction of motion (frontal area, A_f), air density ρ, vehicle velocity V, and a drag coefficient C_d that reflects the shape of the vehicle. This relation is given by

$$F_{ae} = \frac{C_d A_f \rho V^2}{2} \qquad (11.3.5)$$

The velocity in Eq. (11.3.5) is really the relative velocity between the air and the vehicle. However, for speeds at which aerodynamic forces become significant and for usual wind speeds (less than 5 mph), it is sufficiently accurate to use the vehicle velocity directly in Eq. (11.3.5).

The aerodynamic drag is due mostly to form drag as opposed to skin friction. The air flow separates from the vehicle surfaces and forms a low pressure wake behind the vehicle. The air flow patterns are shown schematically in Fig. 11.3.1 for a typical car and a streamlined one.

For both vehicles, the air flow over the front is relatively smooth and flows along the surface. For the typical automobile, the flow separates from the roof and sides at about the point of maximum cross-sectional area. The pressure in the wake region is about equal to the atmospheric static pressure, while the pressure on the

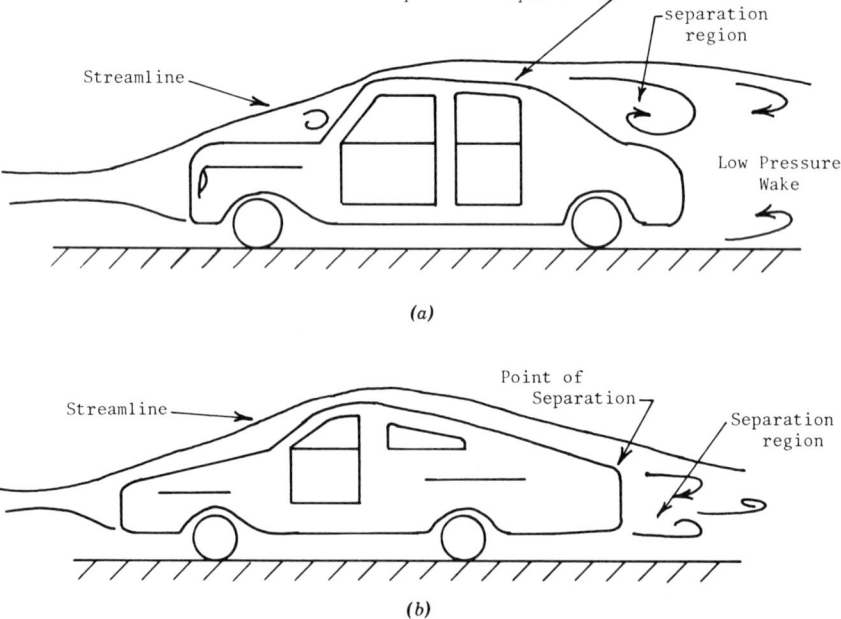

Fig. 11.3.1. Schematic of air flow over vehicles. (*a*) Typical automobile. (*b*) Streamlined automobile.

VEHICLE POWER REQUIREMENTS

front is on the order of the stagnation pressure (dynamic pressure plus static pressure). This pressure difference acting over the frontal area of the car accounts for most of the aerodynamic drag.

The effect of streamlining is to reduce the size of the wake region. By slanting the back, the flow remains attached to the surface until the abrupt truncation at the rear of the car. The reduced size of the low pressure wake reduces the net force for the given frontal area, which lowers C_d. The slant angle is critical and fairly shallow. Flow separation occurs for angles greater than about 15 degrees with respect to the direction of motion. For this reason, most streamlining currently found on automobiles is cosmetic.

The drag coefficient C_d accounts for the variation in pressures over the surface. Values of C_d are given in Table 11.3.3 for various vehicular shapes. Most vehicles have value of drag coefficient between that for a sphere and for a flat plate normal to the flow. Motorcycles have high drag coefficients due to projections such as handlebars, mirrors, foot pedals, and limbs. The potential for improvement is indicated by the tear drop shape, which has a drag coefficient one-fifth to one-tenth that of current vehicles.

The frontal area is the projected area normal to the direction of motion. For automobiles, this may be estimated from the vehicle height H and center-to-center distance between wheels W, by

$$A_f = 0.9HW \qquad (11.3.6)$$

The power required to overcome aerodynamic forces is obtained from combining Eqs. (11.3.3) and (11.3.5). The aerodynamic power goes up as the cube of vehicle velocity. In order to illustrate this effect of velocity and to compare it to the friction power, the example cars and the truck of Section 11.3.1 will again be considered. The smaller car has a frontal area of 18 ft^2, the larger one an area of 27 ft^2, and the truck a frontal area of 80 ft^2. The cars have a drag coefficient of 0.45 while that of the truck is 1.0. The frictional, aerodynamic, and total horsepower are shown for each in Fig. 11.3.2.

TABLE 11.3.3
Drag Coefficients for Vehicular Shapes

Shape	C_d
Passenger automobile	0.4–0.5
Convertible (top down)	0.6–0.65
Bus	0.7–0.9
Truck	0.8–1.0
Motorcycle	1.8
Teardrop shape	0.1
Sphere	0.45
Flat plate (normal to flow)	1.0

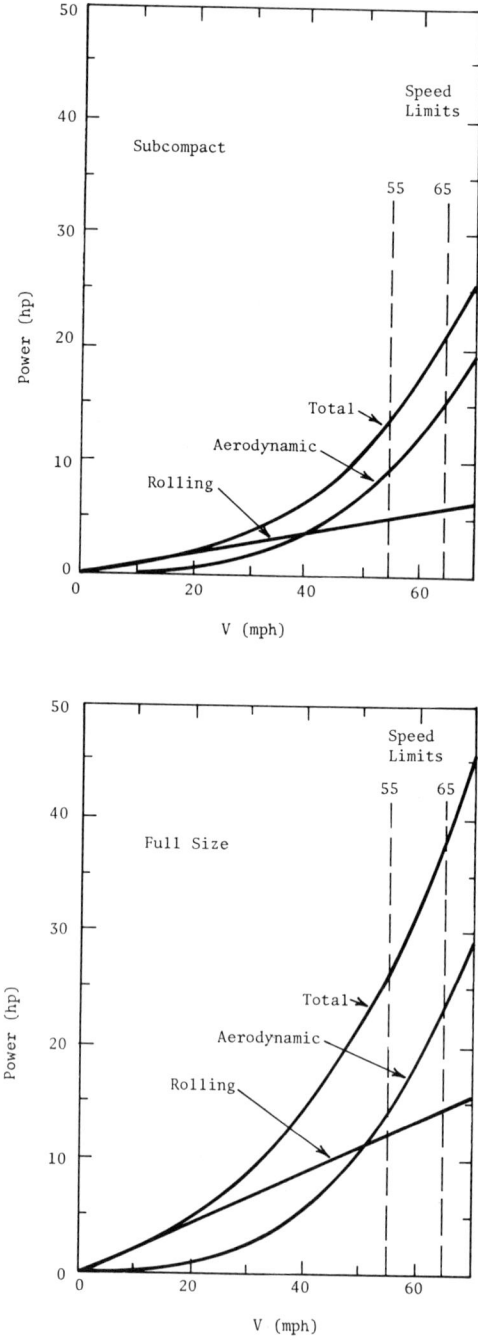

Fig. 11.3.2. Horsepower requirements to overcome aerodynamic and rolling friction as a function of speed.

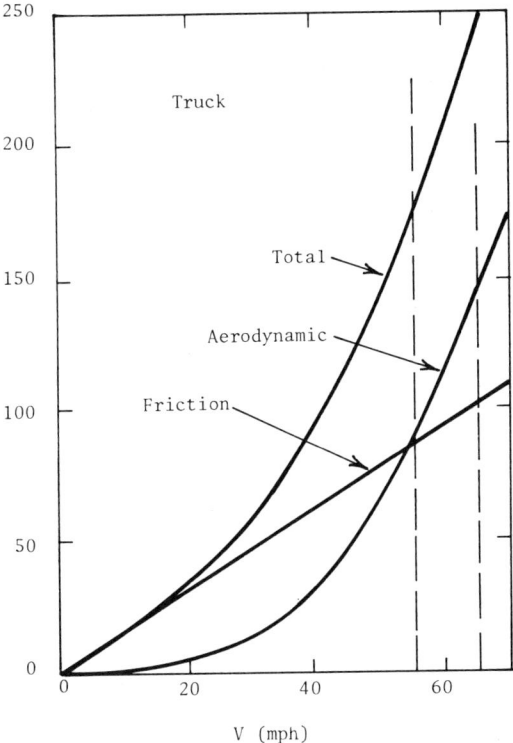

Fig. 11.3.2. (*Continued*)

There are several conclusions that can be drawn from this figure. First, horsepower requirements to overcome the automotive design forces are generally small compared to installed engine power. The engine size for the cars is in the range of 60–250 hp, and is four to eight times the power requirements for aerodynamic and rolling friction only. As will be discussed, other considerations such as hills and acceleration increase engine power requirements. At a steady speed on a flat road, the engine is operating at a part load of about 10–25 percent of maximum. Engines thus the need to be designed for good fuel economy at these low power levels.

For the small car, aerodynamic forces are more important than for the large car. At 55 mph, the total horsepower per unit weight for the compact car is 50 percent higher than that of the full size vehicle. This is due to the relatively large volume of the compact car needed in order to accommodate passengers. The horsepower per unit weight of the truck is about equal to that for the full size automobile, and rolling and aerodynamic friction are in about the same proportions.

Various design changes can be affected to reduce drag coefficients. The elimination of outside mirrors, radio antennae, door handles, hood ornaments, and so on, reduces the flow disruptions. The undercarriage may be streamlined through the addition of pans. For automobiles, a reduction in C_d to a value of 0.35 is probably technically feasible at no sacrifice of function.

Trucks and buses can also benefit from streamlining. Rounding the front edge of the vehicle prevents separation of the air on the top, and reduces the drag coefficient by about 25 percent. For a truck, this is achieved at the expense of a reduction in cargo space. Fairings placed on the cab of a truck and trailer combination serve the same purpose and can reduce the drag coefficient to 0.55–0.6. Since trucks operate mainly at highway speeds, the resultant energy and economic savings are significant for these improvements, and are undoubtedly cost effective as evidenced by their current use.

11.3.3. Gravitational Forces

The power required for vehicle travel is affected by hills on the road. As shown in Fig. 11.3.3, there is a component of vehicle weight in the direction of motion, and this force must be overcome. This force F_g is given in terms of the slope of the road θ as

$$F_{gr} = W \sin \theta \qquad (11.3.6)$$

The slope of a roadbed is usually given in terms of the grade, which is the rise of the road per unit horizontal distance. For example, the maximum grade on the interstate system is 4 percent, which is a 4-ft rise in 100 ft of distance. The slope of the road is then 2.3 degrees.

Gravitational forces can significantly increase power requirements. From Eq. (11.3.4), it is seen that both the rolling friction coefficient k and the sine of the slope are multiplied by the weight. The rolling coefficient is on the order of 0.015, while the value of $\sin \theta$ for a 4 percent grade is 0.04. Thus, the force introduced by a moderate grade can be considerably greater than that introduced by rolling friction. For a full size car at 55 mph on a 4 percent grade, the additional power requirement for the hill is 32 hp. As seen from Fig. 11.3.2, this would increase the total power from 27 to 59 hp. The installed size of the automobile engine is usually large enough to accommodate these expected grades. For the truck, the increased power requirement is 235 hp for a total of about 400 hp. This may be larger than the installed engine size, which slows the truck down on grades.

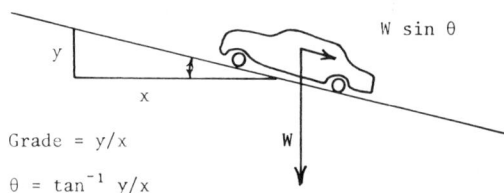

Fig. 11.3.3. Gravitational forces acting on a vehicle.

11.3.4. Inertial Forces

The inertial force F_{ir} introduced by vehicle acceleration has two components. One is the component due to acceleration of the entire vehicle in the direction of motion. The other is due to the rotational acceleration of the motor, drive train, and tires. For ease of representation, these two components are combined into one equation as

$$F_{ir} = \gamma m \frac{dV}{dt} \qquad (11.3.7)$$

where m is the vehicle mass and γ is a factor that accounts for both linear and angular acceleration. The value of γ is between 1.1 and 1.4, depending on gear ratio.

The power to accelerate the vehicle is given by

$$P_{ir} = F_{ir} V \qquad (11.3.8)$$

Combining Eqs. (11.3.7) and (11.3.8) yields

$$P_{ir} = \gamma m V \frac{dV}{dt} \qquad (11.3.9)$$

This can be reformulated as

$$P_{ir} = \gamma \frac{d}{dt}\left(\frac{mV^2}{2}\right) \qquad (11.3.10)$$

Equation (11.3.10) shows that the power required depends on the change in kinetic energy, which depends on the square of vehicle velocity. For a given change in speed, more power is required at a high speed than a low one.

For example, the power required to accelerate the automobiles from 10 to 20 mph in 5 seconds is compared to that for the same acceleration from 45 to 55 mph in Table 11.3.4. At low speeds, these power requirements are about five times the rolling and aerodynamic friction power, while at high speeds they are about double those power levels. The power requirements to accelerate the truck over the same

TABLE 11.3.4
Power Requirements for Vehicle Acceleration

	10 to 20 mph	45 to 55 mph
Subcompact	10 hp	32 hp
Full size	24 hp	80 hp
Truck	48 hp	160 hp

TABLE 11.3.5
Design Requirements at 55 mph

	Subcompact	Full Size	Truck
Friction	5	12	88
Aerodynamic	9	14	89
Grade (4% grade)	14	32	235
Acceleration (45 to 55 mph)	20	50	160
Design power	48 hp	108 hp	572 hp

speed ranges but in 15 seconds are also given. These power levels are quite significant. Installed engine size must be large enough to overcome the expected inertial forces also.

11.3.5. Summary of Power Requirements for a Vehicle

The forces acting on a vehicle and the resulting power requirements show that in terms of vehicle chassis design, the weight and the aerodynamic shape are important. By careful design, the engineer is able to reduce power requirements for these two components. However, gravitational forces are outside the scope of vehicle design, but significantly influence the selection of installed engine power. The need, or desire, to accelerate the vehicle also increases the installed engine size. For these reasons, installed engine power is often several times larger than that required to overcome friction alone. Engines thus operate at part load most of the time, and must be designed to be efficient at these loadings.

The role of each of these components in engine sizing is illustrated in Table 11.3.5. The design load is assumed to be that required to drive at the current speed limit (55 mph) up a 4 percent grade (maximum allowable on the interstate system). The design acceleration value is from 45 to 55 mph in 8 seconds for the automobile and 15 seconds for the truck. The design horsepower is seen to be mainly to overcome grades and to accelerate. These factors are outside the range of the designer. Further, a factor of safety must always be included to account for out-of-tune engines or unexpected road conditions.

11.4. ENGINE CHARACTERISTICS

11.4.1. Thermodynamic Relations

The engine characteristics have a significant effect on the fuel economy of the automobile. The engine type (Diesel or spark ignition), compression ratio, and the relation between fuel consumption, engine speed, and horsepower are all relevant. In this section, the basic thermodynamic relations for ideal engine cycles will be discussed first, and the performance relations next.

The air-standard Otto and Diesel cycles are closed system idealizations of the actual spark ignition and Diesel cycle engines. They are helpful in showing the role of compression ratio and in comparing the two cycles. The process representation for both are shown in Fig. 11.4.1. The Otto cycle processes consist of isentropic compression (1-2), constant volume heat addition (2-3), isentropic expansion (3-4), and constant volume heat rejection (4-1). The thermal efficiency of the cycle depends only on the compression ratio

$$r_v = \frac{V_1}{V_2} \tag{11.4.1}$$

and is given by

$$\eta_{th} = 1 - \frac{1}{r_v^{k-1}} \tag{11.4.2}$$

where k is the specific heat ratio for air (1.4).

The Diesel cycle processes are isentropic compression (1-2), constant pressure heat addition (2-3), isentropic expansion (3-4), and constant volume heat rejection (4-1). The thermal efficiency depends on both compression and temperature ratios, and is given by

$$\eta_{th} = 1 - \frac{(T_4/T_1 - 1)}{(T_3/T_2 - 1) k r_v^{k-1}} \tag{11.4.3}$$

There are several conclusions that can be drawn from these simplified cycle relations. First, the thermal efficiency of both cycles increases as compression ratio increases. From Eq. (11.4.2), the thermal efficiency of the Otto cycle at a compression ratio of 10:1, which is representative of values in the 1960-1970 era, is 60 per-

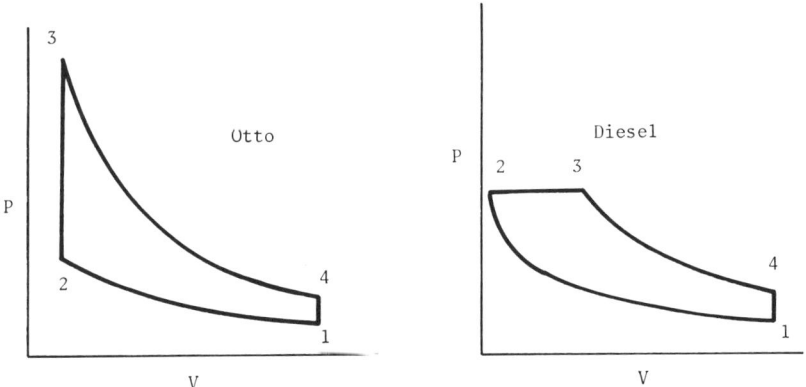

Fig. 11.4.1. Air standard Otto and Diesel cycle representation.

cent. Current compression ratios are about 8:1, and the corresponding thermal efficiency has dropped 7 percent to 56 percent. The lowered compression ratios result from pollution legislation which has reduced the lead content in gasoline with a consequent lowering of temperature after compression to prevent detonation (knock).

The Diesel cycle has a lower thermal efficiency for the same compression ratio since the temperature at which heat is added (2-3) is lower. However, the Diesel cycle in practice can operate at a higher compression ratio than the Otto cycle. This is because both fuel and air are compressed in the Otto cycle and the limit is that mixture may detonate (knock) if compressed to too high a temperature. In the Diesel cycle, only the air is compressed and the fuel is sprayed into the cylinder. As a result, Diesel cycles can operate at compression ratios of up to 20:1 with significantly higher thermal efficiencies than the Otto cycle. In practice, then, the Diesel cycle is the more efficient one.

11.4.2. Engine Characteristics

The second aspect of the engine characteristics is the relation between fuel consumption, engine speed, and power. A representative engine map for a modern four-cylinder engine is given in Fig. 11.4.2. It is seen that at low power outputs, the specific fuel consumption is quite high. This is due to the large fraction of the engine output that is used to overcome bearing friction, water pump, fan, alternator, and fan belt windage. As the power output is increased, the fuel consumption per unit

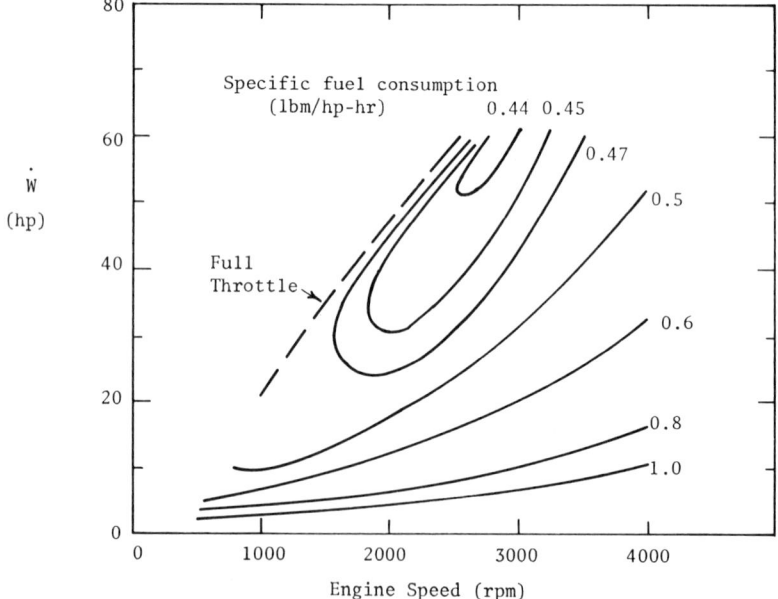

Fig. 11.4.2. Relation between power, engine speed, and fuel consumption.

horsepower output decreases. There is a minimum specific fuel consumption for any engine speed, and this value decreases as power increases. A present best fuel consumption value is about 0.4 lb_m/hp hr. However, there is a limit on engine speed in that high power output cannot be obtained at low engine speeds.

The engine map can be converted to a vehicle performance map to show the relation between fuel consumption, road speed, and power. In high gear, the ratio of engine to road speed is 40 rpm per mile per hour. There is about 5.5 lb_m/gal of gasoline. The power can be normalized by the peak power to extend the map to an engine of any displacement or number of cylinders. The resulting map is shown in Fig. 11.4.3 in which the fuel consumption is plotted against road speed for various values of fractional power output. The engine is seen to be relatively inefficient at low power outputs. The highest efficiencies are at high speeds and power outputs, which rarely occur.

The vehicle performance map can be combined with the vehicle power requirements (Fig. 11.3.2) to determine the fuel economy (mpg) as a function of road speed at steady conditions. The fuel economy relates to the engine power and specific fuel consumption by Eq. (11.3.2)

$$(\text{mpg}) = \frac{V(\text{mph})}{\text{sfc} \cdot (\text{gal/hp} \cdot \text{hr}) P(\text{hp})} \qquad (11.3.2)$$

The fuel economy will be estimated for the subcompact and full size automobile examples. It will be assumed that the installed engine sizes are 60 and 150 hp, respectively. By comparing Fig. 11.3.2 with Fig. 11.4.3, it can be seen that the relative power requirements are low. For most of the driving speed range, the power requirements are on the order of 10 percent of engine capacity, and fuel consumption is relatively high.

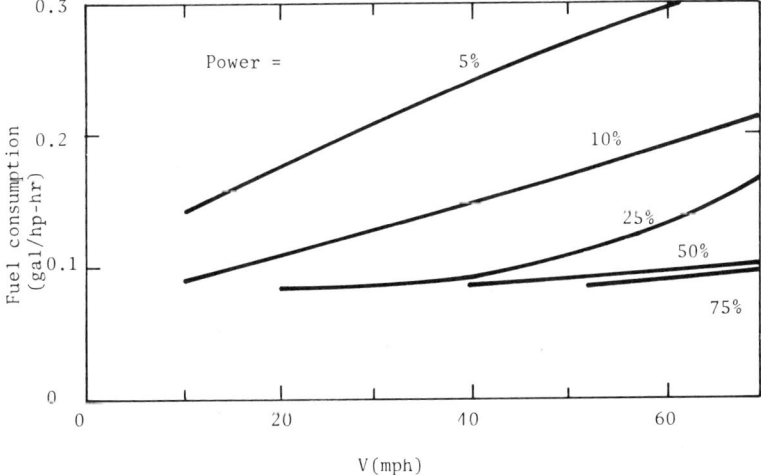

Fig. 11.4.3. Relation between fuel consumption, road speed, and percent power.

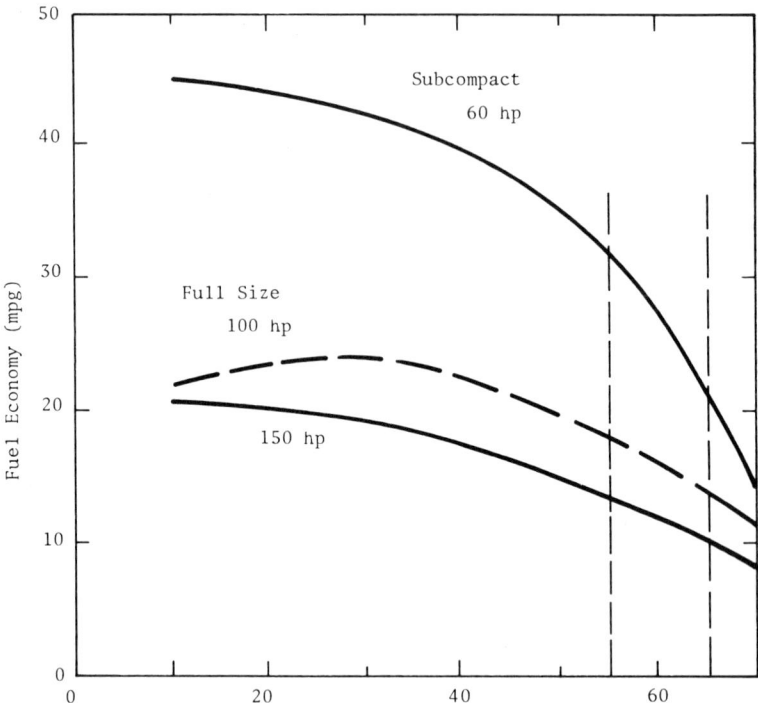

Fig. 11.4.4. Fuel economy as a function of road speed.

The steady state fuel economy as a function of road speed is shown in Fig. 11.4.4 for the two vehicles. As expected, the fuel economy decreases as speed increases, but the curve is relatively flat up to about 30-40 mph. There is a drastic drop in fuel economy in the 55-65 mph range.

The reason for this behavior is that at low values of road speed, the power is essentially directly proportional to vehicle speed. Thus fuel economy is almost inversely proportional to specific fuel consumption which, from Fig. 11.4.2, is seen to decrease as speed and power increase. At higher road speeds, aerodynamic forces are important and power increases faster than speed. Specific fuel consumption continues to decrease, but at a relatively low rate. The fuel economy thus decreases.

The fuel economy at 55 mph is about 45 percent better than that at 65 mph for the small car and 25 percent better for the large car. The larger car has inherently poorer economy due to the larger power requirement even though the relative engine performance is the same. This demonstrates the importance of lowering power requirements through weight reductions and improved aerodynamic design.

The effect of installed engine size on fuel economy is also shown in Fig. 11.4.4. It is assumed that a 100-hp engine is put in the full size automobile. This increases the relative power requirement by a factor of 2 and increases fuel economy by 15-20 percent. However, this engine size may be unsatisfactory in this large auto-

horsepower output decreases. There is a minimum specific fuel consumption for any engine speed, and this value decreases as power increases. A present best fuel consumption value is about 0.4 lb_m/hp hr. However, there is a limit on engine speed in that high power output cannot be obtained at low engine speeds.

The engine map can be converted to a vehicle performance map to show the relation between fuel consumption, road speed, and power. In high gear, the ratio of engine to road speed is 40 rpm per mile per hour. There is about 5.5 lb_m/gal of gasoline. The power can be normalized by the peak power to extend the map to an engine of any displacement or number of cylinders. The resulting map is shown in Fig. 11.4.3 in which the fuel consumption is plotted against road speed for various values of fractional power output. The engine is seen to be relatively inefficient at low power outputs. The highest efficiencies are at high speeds and power outputs, which rarely occur.

The vehicle performance map can be combined with the vehicle power requirements (Fig. 11.3.2) to determine the fuel economy (mpg) as a function of road speed at steady conditions. The fuel economy relates to the engine power and specific fuel consumption by Eq. (11.3.2)

$$(\text{mpg}) = \frac{V(\text{mph})}{\text{sfc} \cdot (\text{gal/hp} \cdot \text{hr}) P(\text{hp})} \qquad (11.3.2)$$

The fuel economy will be estimated for the subcompact and full size automobile examples. It will be assumed that the installed engine sizes are 60 and 150 hp, respectively. By comparing Fig. 11.3.2 with Fig. 11.4.3, it can be seen that the relative power requirements are low. For most of the driving speed range, the power requirements are on the order of 10 percent of engine capacity, and fuel consumption is relatively high.

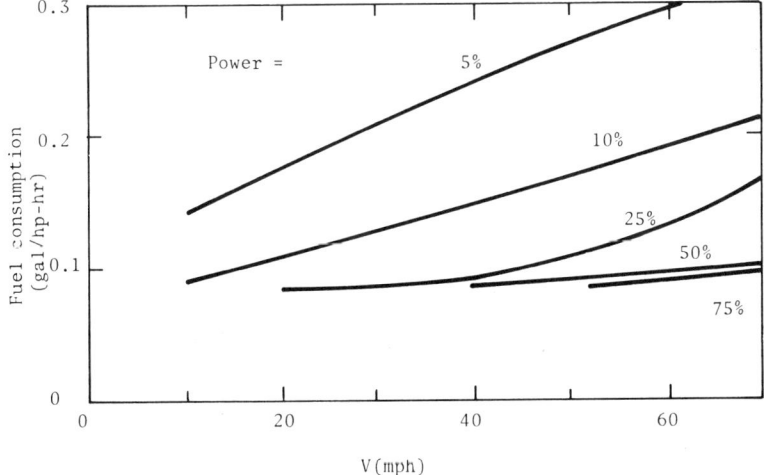

Fig. 11.4.3. Relation between fuel consumption, road speed, and percent power.

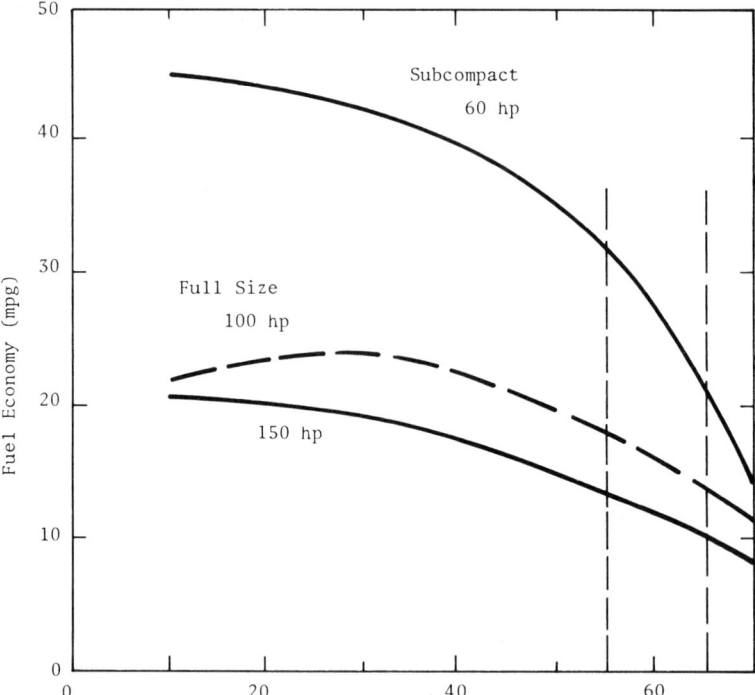

Fig. 11.4.4. Fuel economy as a function of road speed.

The steady state fuel economy as a function of road speed is shown in Fig. 11.4.4 for the two vehicles. As expected, the fuel economy decreases as speed increases, but the curve is relatively flat up to about 30–40 mph. There is a drastic drop in fuel economy in the 55–65 mph range.

The reason for this behavior is that at low values of road speed, the power is essentially directly proportional to vehicle speed. Thus fuel economy is almost inversely proportional to specific fuel consumption which, from Fig. 11.4.2, is seen to decrease as speed and power increase. At higher road speeds, aerodynamic forces are important and power increases faster than speed. Specific fuel consumption continues to decrease, but at a relatively low rate. The fuel economy thus decreases.

The fuel economy at 55 mph is about 45 percent better than that at 65 mph for the small car and 25 percent better for the large car. The larger car has inherently poorer economy due to the larger power requirement even though the relative engine performance is the same. This demonstrates the importance of lowering power requirements through weight reductions and improved aerodynamic design.

The effect of installed engine size on fuel economy is also shown in Fig. 11.4.4. It is assumed that a 100-hp engine is put in the full size automobile. This increases the relative power requirement by a factor of 2 and increases fuel economy by 15–20 percent. However, this engine size may be unsatisfactory in this large automobile when acceleration and hills are taken into consideration. This does demonstrate the importance of matching engine and vehicle size for good fuel economy. It also shows that the technique developed by General Motors of changing the number of active cylinders in a V-8 from 8 to 4 at light loads significantly increases fuel economy. This has the effect of switching to a smaller installed engine size.

The transmission gear ratio can also affect fuel economy. For example, a transmission with overdrive might reduce engine speed by 30 percent at the same power output. For the compact automobile traveling at 55 mph with a power requirement of 15 hp, overdrive would reduce engine speed from 2200 rpm to 1700 rpm. From Fig. 11.4.2, this would reduce the specific fuel consumption by 15 percent, and increase fuel economy a like amount. An overdrive option is probably a cost-effective addition for an automobile which is driven at highway speeds a large fraction of the time.

11.5. DISTRIBUTION OF ENERGY USE FOR AUTOMOBILES

The distribution of energy consumption for a typical automobile has been estimated for the EPA composite cycle and is given in Table 11.5.1. One-half of the use is due to frictional effects, and this may be reduced by careful chassis design. Design of the mechanical components (transmission, drive train) and accessories (power equipment) can influence about 21 percent of the use. The remaining 29 percent is due to driver and road interaction, and the need for the engine to be on even if no power is required. Changes in driving style to reduce acceleration can reduce these losses somewhat.

The effect of various design changes on fuel consumption will be assessed for the two example automobiles. The various parameters are given in Table 11.5.2 for the base vehicle design. It is assumed that each vehicle travels 15,000 miles/yr, and that gasoline costs are $1.40/gal.

Several design changes will be considered. The resulting improvements in fuel usage will be estimated based on the distribution of energy uses given in Table 11.5.1. For example, a 10 percent weight reduction reduces rolling friction 10 percent, but, as rolling friction accounts for only 25 percent of total use, the total fuel consumption is reduced only 2.5 percent.

TABLE 11.5.1
Automobile Distribution of Energy Usage

Rolling friction	25%
Aerodynamic drag	25%
Coasting and idle	16%
Braking	13%
Transmission losses	12%
Accessories (AC, power steering, etc.)	9%
	100%

TABLE 11.5.2
Base Parameters for Automobile Energy Use

	Subcompact	Full Size
Weight (lb$_f$)	2200	5500
Frontal area (ft^2)	18	27
Rolling friction	0.015	0.015
Aerodynamic drag C_d	0.45	0.45
Fuel economy (mpg)	34	15
gal/yr	441	1000
$/yr	617	1400

The various conservation measures considered are:

1. A 15 percent reduction in rolling friction due to reduced weight and better tires.
2. An improvement in aerodynamic styling to reduce C_d to 0.35.
3. An improvement in engine performance to reduce fuel consumption 10 percent.

The resulting effects of each measure taken singly are given in Table 11.5.3. Each one plays a significant part in reducing fuel consumption. Taken together, they can lower fuel usage by about 20 percent.

The assumed improvements in tire and aerodynamic design are about as large as current technology suggests. The limits on vehicle weight reduction are more open-ended. It appears that further improvements must come about through engine and mechanical component design.

Improvements in the engine and drive train also have a large potential. Engine modifications such as Diesel instead of spark ignition engines and stratified charge

TABLE 11.5.3
Effect of Conservation Measures on Fuel Consumption

	15% Weight Reduction	Reduce C_d to 0.35	Improve sfc 10%	All Measures
	Subcompact			
mpg	35.3	36.4	37.0	40.6
Gallons saved/yr	15	29	36	72
$ saved/yr	$21	$41	$ 50	$100
	Full Size			
mpg	15.6	16.0	16.3	17.9
Gallons saved/yr	39	63	80	163
$ saved/yr	$55	$88	$112	$228

engines which can run lean instead of rich could increase fuel economy by 15-25 percent. Reductions in engine size (maximum horsepower) allow the engine to run more fully loaded and at a higher efficiency. Improved transmission design and possibly a return to manual four-speed units reduce fuel use. Taken together, these modifications might improve automobile mileage by 40 percent. Through technical changes, then, a 50-60 percent improvement in fuel economy at little or no loss in performance is feasible. However, the dollar value of the savings are relatively small. The largest annual savings amount to only 1-2 percent of the purchase price, or only 25 to 50¢/day.

The fuel economy of trucks can be improved through some of the same measures as for automobiles. Radial tires can improve fuel economy by up to 10 percent, and a more aerodynamic design could add another 5 percent. Improved engines and fans could reduce fuel usage by 20 percent. For tractor-trailer combinations, the aerodynamic drag of an added trailer is very low. A double trailer has 5 percent better mileage than two single trailers, while a triple trailer is 25 percent better than three single units. Considering the large fuel use in the trucking industry, these techniques can produce significant overall savings.

11.6. THE EFFECT OF SPEED LIMITS

In 1974, speed limits on highways were reduced from 65 to 55 mph in order to save fuel. This decision has been very controversial, and both the freight industry and the general public have opposed it to varying degrees. If the speed limits are obeyed, there are energy savings, but these may be accompanied by economic losses.

For passenger automobiles, the difference in fuel usage between driving 65 and 55 mph is about 30 percent (Section 11.4.2). The number of miles driven at highway speeds might be 20 percent, at most, for an average automobile. A typical driver, then, would realize a maximum reduction of 30 percent of 20 percent, or about 6 percent, on fuel consumption by reducing his speed. For an average fuel economy of 20 mpg and 15,000 miles/yr, his annual economic savings would be about $60. This amounts to 2¢/mile at highway speeds which is small compared to other costs. For the driver, this can also be thought of as saving $60 in fuel cost at the expenditure of an increase of a total of 8 hours in trip time. This amounts to savings at a rate of about $8 per hour. These economic incentives do not appear to be large enough for widespread, willing voluntary compliance.

The fuel savings and economic costs for freight transport are significantly different. Long distance trucking is almost entirely at highway speeds, the fuel costs per mile are higher, and there is the labor cost of the driver involved. A 30 percent reduction in fuel use might produce savings of 14¢/mile, or about $8/hr at highway speeds. The truck driver's wage alone is probably two to four times this amount, the cost of owning the truck is significant, and the cash value of the load is large. It is undoubtedly uneconomical for the trucking industry to save fuel by reducing speeds.

On the other hand, lower speed limits do produce important overall savings in petroleum usage. Passenger automobiles are 55 percent of the transportation energy, and transportation accounts for 50 percent of the total petroleum usage. A 6 percent reduction in passenger car usage reduces total petroleum usage by a maximum of 1.7 percent. Truck savings would contribute about a maximum of 1.1 percent. The maximum combined savings in petroleum usage would be on the order of 3 percent, and is certainly significant in view of our present petroleum deficits.

11.7. SUMMARY

The transportation sector accounts for about 50 percent of the total oil used in the United States. Of this, automobile and truck use amount to three-quarters of the total. Thus, technical improvements in vehicle energy use have a significant effect on both overall energy consumption and oil imports. Most improvements, however, are made by the automotive companies. The individual owner has control only over his driving habits and his initial vehicle selection.

At low speeds the rolling friction components are the major factors in energy use. Thus low weight and radial tires with proper pressure are beneficial. Aerodynamic friction becomes significant at highway speeds and here streamlining is valuable. The closest match between the horsepower output of the engine and the vehicle power requirements yields the best fuel economy. High compression spark ignition and Diesel engines are inherently lower in fuel use. All of these fuel economy factors have to be weighed against safety and air pollution requirements. Through technical changes, there is the possibility of significantly reduced petroleum usage at no loss in function.

SUGGESTED READING

Transportation Energy Conservation Data Book, ORNL 5493, Oak Ridge National Laboratory, Oak Ridge, TN, Edition 2, 1977 and Edition 3, 1979.

A. C. Mascy and R. L. Paullin, *Transportation Vehicle Energy Intensities*, DOT-TST-13-79-1, June 1974.

J. R. Pierce, The Fuel Consumption of Automobiles, *Scientific America*, 1975, p. 34.

EPA Gas Mileage Guide, Environmental Protection Agency, Washington, DC, Annual.

G. Sovran, T. Morel, and W. T. Mason (eds.), *Aerodynamic Drag Mechanisms of Bluff Bodies and Road Vehicles*, Plenum Press, New York, 1978.

PROBLEMS

11.1. An automobile weighs 2500 lb$_f$, has a frontal area of 20 ft^2, a drag coefficient of 0.43, and an installed engine size of 100 hp. The annual number of hours driven is 600 and distributed over the following speeds.

PROBLEMS

Average Speed	Percent Time
20 mph	10%
30 mph	30%
40 mph	40%
50 mph	20%

Estimate the annual fuel consumption and cost.

An overdrive options costs $250.00. Would this be a cost-effective option?

11.2. A truck fully loaded weighs 30 tons, has frontal area of 80 ft^2, a drag coefficient of 1.0, and an installed engine size of 350 hp. The annual number of hours driven is 2400, and distributed over the following speeds.

Average Speed	Percent Time
25 mph	10%
35 mph	10%
45 mph	30%
55 mph	50%

Estimate the annual fuel consumption and cost.

(a) A streamling cowl is available that reduces the drag coefficient to 0.65 and costs $1200.00. Is this a cost-effective option?

(b) Driving 60 mph on the highway allows more trips and increases the truck profit by $5000.00. The annual cost of speeding tickets is $150.00. Is driving 60 mph cost effective?

· 12 ·

RENEWABLE ENERGY SOURCES

The previous chapters have been concerned with more effective use of current energy sources. These present sources, which are mainly coal, petroleum, and natural gas, are depletable and will eventually run out. In this chapter the use of renewable sources will be briefly discussed. The major alternatives at present are bioconversion, wind, and solar.

For each of these the technical problems involved in utilizing the source will be discussed first. The economics will then be considered. Finally, an attempt will be made to assess the overall role of these sources in the energy situation. In these times of change the assessment of the economics is especially difficult. It is hoped that the methods of analysis will be helpful and lend some guidance as to the potentials of each source.

12.1. BIOCONVERSION

Bioconversion is the name given to the process by which the sun's energy is converted to food and energy in plants. The term is now applied to the production of fuel sources on a renewable basis, although all of our current depletable sources are the result of this process. The problem today is to produce fuels of high quality and in sufficient quantity to reduce or eliminate our dependency on coal, oil, and natural gas.

The photosynthetic process uses the sun's energy in the visible spectrum. This energy is absorbed in the chlorophyll in the plant. There it provides the energy to convert carbon dioxide and water into glucose and oxygen. The glucose is used in growth to provide the plant structure. This structure and the chemical energy bound up in it eventually becomes the fuel for our use.

12.1.1. Energy Farming

The three major types of plantings that can be grown to produce fuels are annual energy crops (grains and grasses), silviculture energy farms (trees), and aquatic energy farms. The land based crops have had extensive development over the years for both food and fuel, while aquatic systems are in the developmental stages.

Annual energy crops for energy could be grown on either prime agricultural land, which might displace food crops, or marginal land that is not now under cultivation. Promising crops at present are sorghum, sugarcane, sunflowers, and similar grasses. Both gaseous and liquid fuels can be obtained from these crops. Synthetic natural gas (methane) is produced by anaerobic digestion of the plant. Alcohol can be produced by fermentation and oil by extraction processes. The processes are known and small scale pilot plants have been built.

In silviculture, rapid growing trees are cultivated and harvested. The rotation period is on the order of 5-10 years to maximize yield. Suitable species are eucalyptus, sycamore, cottonwood, poplars, and pines. The wood can be used directly as a solid fuel in industrial or power plant boilers, or it can be made into charcoal for use in power plants and steel mills. Liquid fuels such as alcohol and oils as well as low Btu gas can be produced. As with annual crops, the techniques are established and well known.

Aquatic energy farms can be either land or ocean based. Two land based crops that have been proposed are water hyacinth and microalgae. Kelp appears to be the best ocean grown crop, and experimental plantations have been built. Water hyacinth grows all too readily in parts of the United States and can be harvested easily while harvesting microalgae and kelp is more difficult. A major problem is the large amount of water present in the plants which has to be removed. Synthetic natural gas would be the major fuel produced. This approach is in the development and feasibility stage.

The expected yields in terms of material and energy for these three types are given in Table 12.1.1. The energy content is that of the crop and does not include any conversion losses. The energy produced per acre varies, but is on the order of the amount of energy required to heat one to four homes. The ratio of input energy to that in the crop is also given, and shows that 5-10 percent of the output is required for fertilizer, irrigation, and equipment to plant and harvest the crop.

TABLE 12.1.1
Bioconversion Energy Supplies

Type	Annual Productivity		Energy Ratio	Cost ($/$10^6$ Btu)
	Tons/acre	10^6 Btu/acre		
Annual	5-20	200-300	1:20	1-2
Silviculture	1-8	40-150	1:10	1-4
Aquatic	15-25	200-500	1:10	1-2

The estimated costs per unit of energy output are also given, and show that the fuels are competitive with current sources. This also shows that the costs to process and transport the crop must be on the order of $1-$2/10^6 Btu for the fuel to be competitive. Wood fuels have been used in industrial plants located near the forest, and are competitive. Anaerobic digestion has been used to produce methane in sewage treatment plants at low cost. It does appear that fuels produced from bioconversion techniques can be competitive.

The area of land to produce energy crops has several implications. First, there is the trade-off between food needs and fuel needs. Given the current world situation, it may not be politically desirable to change from food to fuel production. The amount of land currently under cultivation in the United States is 400×10^6 acres. It is estimated that there is a potential to increase this by 25 to 100×10^6 acres, but this is, of course, marginal land. If the average productivity were 200×10^6 Btu/acre yr, this land could produce between 5 and 20×10^{15} Btu/hr. This amounts to 6-25 percent of the total U.S. consumption. Bioconversion has the potential to contribute significantly to renewable energy supplies.

12.2. WOOD

Wood appears to be attractive for home heating in many sections of the country. It is sometimes quite inexpensive, it may be stored at the site of consumption, and can be burned at a rate to match the load. In many rural and suburban areas, wood burning furnaces, stoves, and fire place accessories have found a ready market.

The heating value of dry wood is in the range of 7000-8000 Btu/lb_m. Water in the wood represents a loss in that it is vaporized during combustion and adds no heat to the house. This loss can amount to 10-20 percent of the heating value. Wood is often sold by the cord, which is 128 ft^3 of stacked wood, but is only about 80 ft^3 of actual wood and weighs 2 tons. The heating value of wood on the basis by which it is usually purchased is

$$HV = 24 \times 10^6 \text{ Btu/cord}$$

or

$$HV = 12 \times 10^6 \text{ Btu/ton}$$

The energy from burning wood that is delivered to the home depends on the efficiency of the equipment used. This efficiency is difficult to estimate, and representative values are given in Table 12.2.1 for common types of wood burning equipment.

The open fireplace commonly found in most homes is very inefficient from an energy point of view. While it is aesthetically pleasing and may warm the occupants near it, it often increases the energy consumption of the furnace. The draft created by the fire creates a large flow of warm room air up the chimney and this exfiltration loss may be greater than the radiant heating from the fire. In addition, the

TABLE 12.2.1
Efficiencies of Wood Burning Equipment

Type	Efficiency
Open fireplace	−10−+10%
Fireplace with doors, liner, and grate, etc.	10−20%
Stove	30−60%
Furnace	50−80%

damper must be left open when the fire is going out, and there is a continued flow of warm air up the chimney with no gain from the fire.

The addition of glass doors reduces the exfiltration both when the fire is burning and when it is being extinguished. A liner that allows room air to circulate around the warm combustion box increases heat transfer to the room. Grates made of hollow tubes also increase heat flow to the room. Depending on the modifications, the efficiency may increase to as high as 20 percent.

Stoves and wood burning furnaces are less aesthetic options, but increase the amount of energy obtained from the wood. The more complex units with controlled draft for slow combustion, internal circulating ducts to cool down the combustion products, and a forced draft heat exchanger have highest efficiencies. As Table 12.2.1 shows, it is possible to burn wood at an efficiency comparable to a natural gas or oil furnace.

The cost of wood varies tremendously between localities. It is, obviously, less expensive in the rural areas where it is plentiful, and more expensive in urban areas where it is not. A cost of \$100/cord or \$50/ton is a representative value, although costs undoubtably vary by a factor of 2 over the United States. The cost of wood heat is then about $\$4/10^6$ Btu, which is comparable to natural gas.

The amount of wood required for a home is large, and the cost comparable or in excess of conventional fuels. The volume of wood required and its cost are given in Table 12.2.2 for a home requiring 50×10^6 Btu/yr. These costs are contrasted with those for natural gas and fuel oil. It is essential to have a high efficiency wood furnace if one is to depend on wood for heating. Otherwise, the volume and cost make wood heating unattractive.

TABLE 12.2.2
Costs of Home Heating

Fuel	Device and Efficiency	Volume	Annual Cost
Wood	Fireplace, 10%	40 cords	\$2000
	Stove, 40%	10 cords	\$ 500
	Furnace, 70%	6 cords	\$ 285
Natural gas	Furance, 70%	−	\$ 200
Oil	Furnace, 70%	−	\$ 460

These costs are based on purchasing wood, and it is, of course, cheaper if one cuts his own wood from his wood lot, although the costs of equipment, transportation, and time may be significant. There are also additional costs for the home owner who burns wood. Insurance premiums may rise if wood burning equipment is installed in a house since chimney and roof fires are more likely. Chimneys will build up a cresote layer and require cleaning annually. There are also societal costs in that wood is a dirty fuel unless combusted in a high performance furnace. Air quality has deteriorated in localities with many wood fires, and legislation has been enacted to regulate the amount of wood burned.

Wood is a renewable fuel, but it requires significant land area for growth. In a northern area, about one-half cord of hardwood per acre of land can be produced on a renewable basis. Intensive farming using fertilizer and fast growing trees may increase the yield to one to two cords per acre. Thus, each moderate size home that converts to wood requires about 6 acres of land. This use competes with that for farming, recreation, mining, housing, and so forth. For example, in Wisconsin there are 1.3×10^6 homes which would require a total of 10×10^6 acres of land if all converted to wood for home heating. This is one-third of the total land area and equals that currently farmed. A switch to wood would undoubtedly require a conversion of productive land area.

In summary, wood as a home heating fuel is not a viable option for the nation. For those rural homes with a ready supply, wood may be an attractive option. For the urban or suburban home dweller, wood can yield some heat if burned in an improved fireplace.

12.3. WIND

Wind is an attractive energy source for many localities. It is a form of solar energy in that wind currents are created by differential solar heating of the ground. The thermal differences set up wind currents on a global scale. There are also local effects such as mountains, hills, and valleys which augment winds.

The energy content of the wind is its kinetic energy. The total amount of energy is the product of flow rate and kinetic energy

$$\dot{E}_{wd} = \frac{\dot{m} V_{wd}^2}{2} \qquad (12.3.1)$$

or, using the continuity expression for flow rate

$$\dot{E}_{wd} = \frac{\rho A_{wd} V_{wd}^3}{2} \qquad (12.3.2)$$

The energy varies as the cube of the wind speed. This places a premium on those locations with high wind speeds.

The energy that may be extracted from the wind is less than the kinetic energy.

If all of the kinetic energy were removed from the wind by a wind machine, the wind velocity would be zero and the wind could not exit the machine. The maximum power that may be obtained from the wind using a wind machine is 8/27, or 0.59, of the kinetic energy. This maximum power from an ideal wind machine is then

$$\dot{W}_{ideal} = \frac{0.59 \rho A_{wd} V_{wd}^3}{2} \tag{12.3.3}$$

Thus can also be expressed as the power density in terms of wind speed in miles per hour as

$$\frac{\dot{W}_{ideal}}{A_{wd}} = 2.9 \left[\frac{V(\text{mph})}{10}\right]^3 \text{ W/ft}^2 \tag{12.3.4}$$

The total energy obtained from the wind depends on the local wind characteristics. A typical daily average profile is shown in Fig. 12.3.1. The wind is generally low at night, and peaks in mid to late afternoon. There are also seasonal variations as shown in Fig. 12.3.2. The wind is higher in winter and lower in summer. The two curves are for a low and a high elevation, and show that since wind speed increases with altitude, tall machines are advantageous. These characteristics make wind use most attractive for daylight loads in winter.

These curves are for the average wind speed. As indicated by Eq. (12.3.4), the power depends on the average of the cube of the wind speed. Thus, the fluctuations that occur are important with higher speeds producing significantly more power. The fluctuations in wind speed about the average mean that the average power produced is about twice the amount produced at the average wind speed. In terms of averages, then, the average ideal power density is

$$\frac{\dot{W}_{ideal}}{A_{wd}} = 5.8 \left[\frac{V_{ave}(\text{mph})}{10}\right]^3 \text{ W/ft}^2 \tag{12.3.5}$$

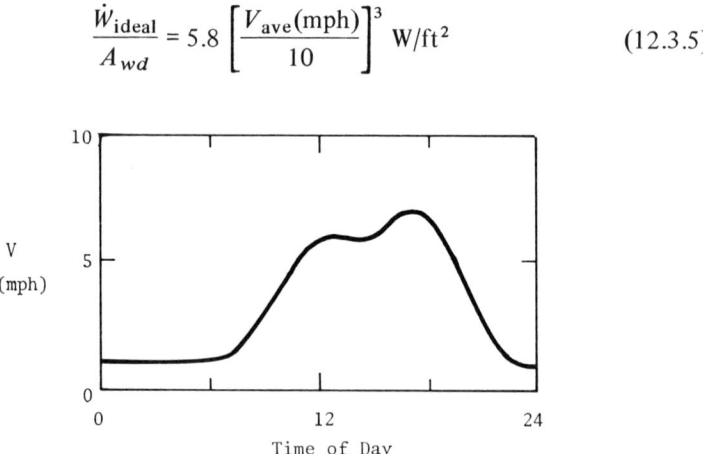

Fig. 12.3.1. Typical wind profile over the course of a day.

WIND   289

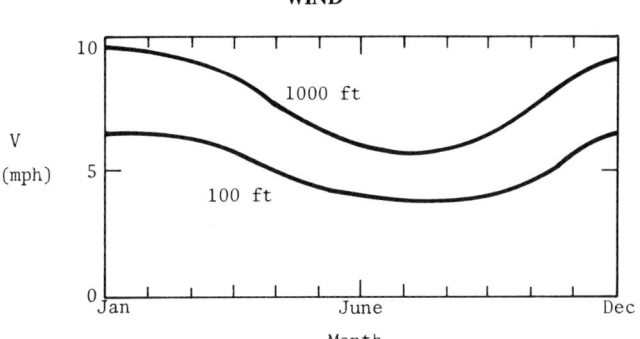

Fig. 12.3.2. Wind profiles over the course of a year.

A plot of power density as a function of wind speed is shown in Fig. 12.3.3. The world average wind speed at ground level is 5 mph, and many locations have close to this value. Average power densities are on the order of 2-5 W/ft². This means large intercept areas are required to generate significant amounts of power.

The efficiency of actual wind machines depends on the type of unit. These are given in Table 12.3.1 for the types commonly proposed. The high speed two- or three-bladed propeller has the highest efficiency, but is best at high wind speeds. Darrius machines are vertical axis machines that look like egg beaters. Savonius rotors are cylinders with curved blades and may be mounted with either a horizontal or vertical axis. The traditional windmill, which has been employed successfully for years for pumping water in remote areas, has the lowest efficiency. The performance of these machines is also shown in Fig. 12.3.3.

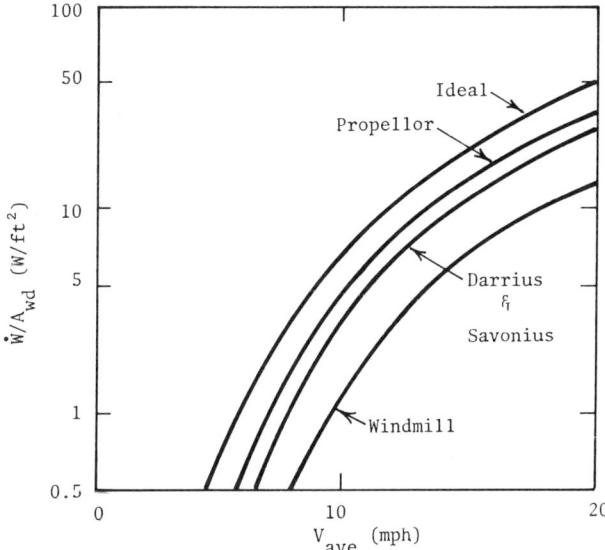

Fig. 12.3.3. Power outputs of wind machines as a function of wind speed.

TABLE 12.3.1
Efficiency of Wind Machines

Type	Efficiency
Ideal	0.59
High speed propeller	0.45
Darrius	0.35
Savonius	0.30
Windmill	0.15

As an example of the energy and economic savings of wind systems, a residential wind machine will be considered. The wind machine will be designed to supply 8000 kWh/yr and meet a peak load of 2 kW (20 A). It will be assumed that the system produces alternating current (ac) electricity and there is no storage provided to meet loads at low wind conditions. In practice, then, either a backup motor generator set or power from the utility would be required.

For a location with an average wind speed of 10 mph, the average power density for a Darrius or Savonius rotor of residential size is 3 W/ft^2. To produce 2 kW, an area of 670 ft^2 is needed. The vertical axis machines would be about 30 ft tall and 20 ft in diameter. The wind machine itself would cost about $5000, and the tower and controls would add another $2000-$5000. Wind machines are eligible for 40 percent tax credits up to a maximum of $10,000, and this helps reduce first costs.

The electricity displaced is assumed to cost $0.05/kWh. The values of P_1 and P_2 for this application are 22.17 and 0.6, respectively. For a 20-year life, the life cycle savings are $2800 for a first cost of $10,000. Wind machines can be cost effective for residences. However, they are probably limited to those areas where codes do not prohibit high towers, where liability insurance can be readily obtained, where the access to the wind is undisturbed by surrounding trees and buildings, and where there is sufficient land. This probably means that wind machines are most suitable for rural sites.

The role of electricity generated by wind for central power stations can also be assessed. Large systems could be built for an estimated $2000-$4000 per installed kilowatt, or about two to four times that of conventional coal or nuclear systems. Fuel is "free" and so the systems could be economic. However, the area requirements are large. For a power plant of average size, the peak power requirements are about 200,000 kW. If the wind system were located in an area with a power density of 5 W/ft^2, an intercept area of 1.4 miles^2 would be needed. This would be obtained by a wind tower system about 1000 ft tall and 8 miles long. Clearly, the land costs might be prohibitive, and aesthetic conditions might eliminate this system from consideration. Further, the utility would need a backup power plant equal to that of the wind system since wind velocities are not dependable.

In summary, wind systems are economically viable. However, wind is a low density energy source, and little power is produced by low wind speeds. Wind systems are probably best suited for rural applications where peak power requirements are in the range of 1-10 kW. They also have potential to add to existing utility power grids.

12.4. SOLAR ENERGY

12.4.1. Overview

Solar energy has been publicized as the most promising energy source of the future. In that the conventional, depletable supplies of coal, oil, and natural gas will be exhausted eventually, solar, in some form, will have to be the source in the future. The rate at which it displaces the conventional fuels depends on the current economics and the perceived supplies of conventional fuels.

The rate of solar energy striking the earth is 600×10^{15} Btu/hr. This is about 10 times the annual United States energy consumption per hour. The energy flux at the surface is on the order of 1 kW/m^2 or 1 hp/yd^2. Thus, there is a lot of solar energy impinging on the earth's surface, but it is a very diffuse energy source. Large areas are required to collect a significant amount of energy.

Solar energy varies over the course of the day and year in a regular manner. For a northern location such as Madison, Wisconsin, the daily average variation of energy incident on the ground over the year is as shown in Fig. 12.4.1. As expected, there is less solar energy in winter than in summer. The seasonal difference in daily flux is about 4 to 1. Average daily profiles are shown in Fig. 12.4.2 for July and December. The greater amount in July is due to a combination of longer day length and higher flux since the sun is more nearly overhead. The distribution on a daily and seasonal basis more nearly matches air conditioning than heating needs. However, it is more difficult to air condition using solar energy than to heat water or air.

In this section, two aspects of solar energy will be considered. The first is to produce electrical power using thermal or direct conversion devices. The second is for space and water heating. Many books have been written on these subjects and there is extensive literature available. The goal in this section will be to present the fundamental ideas and perform economic evaluations to evaluate the feasibility of different systems.

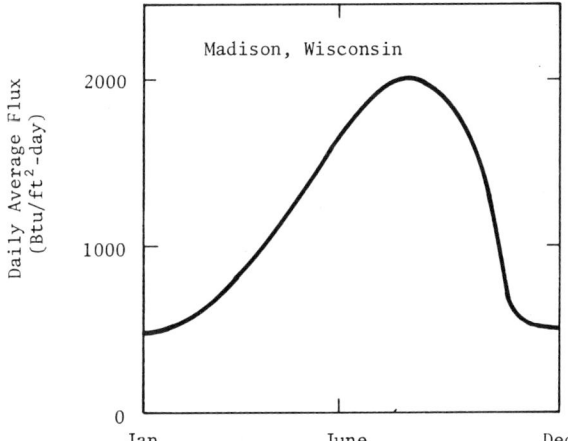

Fig. 12.4.1. Daily average solar flux over the course of a year.

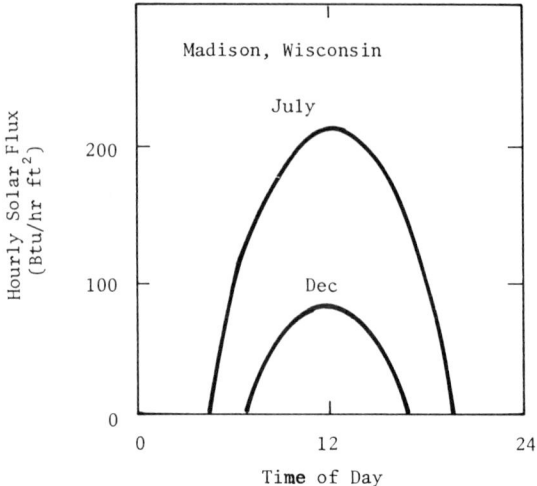

Fig. 12.4.2. Hourly flux over the course of the day for July and December.

12.4.2. Electric Power from Solar

There are two proposed methods for producing electric power from solar. In the direct conversion approach, solar energy is converted to electrical energy using photovoltaic cells. Most of the cells manufactured to date have been made from silicon, with gallium–arsenide cells holding the promise of higher efficiency. The cell is made as a *p–n* junction semiconductor. High energy photons from the sun are absorbed in the material and produce an electric charge across the cell which creates dc electricity.

The conversion efficiency of solar energy into electricity is 10–15 percent. The cell efficiency is strongly temperature dependent, and the cell must be cooled to remove the heat and maintain its temperature low. Flat plate arrays have been used as shown in Fig. 12.4.3a. The lines on the cells are the buss bars carrying current. Many systems reduce the number of cells and use concentrating collectors to focus the energy on the cell. A schematic of a concentrating system is shown in Fig. 12.4.3b. The concentration ratio of the collector is the ratio of energy flux at the focus to that at the intercept area. Silicon cells have been used in arrays with concentration ratios of 50:1, while gallium–arsenide cells may operate at concentration ratios of 500:1 at efficiencies of 20 percent.

It is technically feasible to produce electricity using photovoltaic cells. The technology to manufacture cells and collectors exists, and the transmission of dc electricity and its conversion to ac electricity are well known. The barrier to implementation is economic.

In order to assess the economic feasibility of photovoltaic systems, an example calculation will be performed. It will be assumed that a power plant with a capacity of about 200,000 kW is desired. In a southwestern location, the average daily flux would be about 2000 Btu/day ft^2 on the plane of the collector over a 10-hour day.

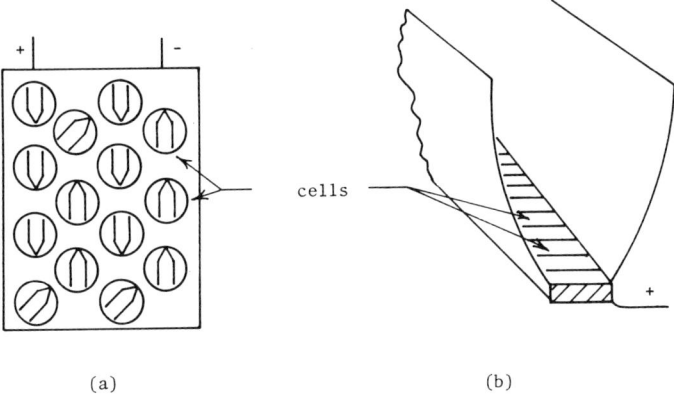

Fig. 12.4.3. Schematic of a photovoltaic power source. (a) Flat plate array. (b) Concentrating array.

For a preliminary assessment, a design value of 200 Btu/hr ft^2 would be reasonable for sizing the components.

The cell efficiency is 15 percent, and thus each square foot of intercept area produces 8.8 W. To produce 200,000 kW requires 23×10^6 ft^2 or 0.8 mile^2. The land area is large and the cost may be significant. The total daily output of the plant is 2×10^6 kWh for an annual output of 730×10^6 kWh.

The cost of solar cells mass produced is expected to be about \$20/ft^2 of cell area. For a system using a flat plate array, the total cost of photovoltaic cells would be \$460 $\times 10^6$. If a concentrating system were used with a concentration ratio of 50:1, the cell cost would be \$0.40/ft^2. However, collector costs will probably be \$10-\$20/ft^2, and so the system costs would be in the range of \$230-\$460 $\times 10^6$.

The cost per unit of installed capacity is \$1150-\$2300/kW. This is two to four times higher than that for conventional nuclear or coal plants and does not include costs of transmission lines and conversion equipment. If it is assumed that the plant will last 20 years, the energy charge to customers must be \$0.03-\$0.06/kWh to pay for the solar components. The cost of the other equipment will increase this charge. However, these prices are not out of line with expected electricity prices in general or current prices in some areas of the country.

Photovoltaic power plants can be cost effective if cell and collector prices are as low as expected. There also remain the technical problems of dissipating heat from the cells, of transmitting electricity from the cells to the distribution system efficiently, and of interfacing with the load management schemes for the central grid. The power produced is in phase with industrial and air conditioning demands which reduces the need for conventional peak power plants. It appears that photovoltaic systems will become a competitive electricity generating system.

The second method proposed for producing electric power uses focusing collectors to concentrate solar energy on the small absorbing area of a receiver. The heat is transferred by an intermediate fluid (a liquid metal) to a boiler. Steam is produced which is used in a conventional power plant. The boiler must be mounted on

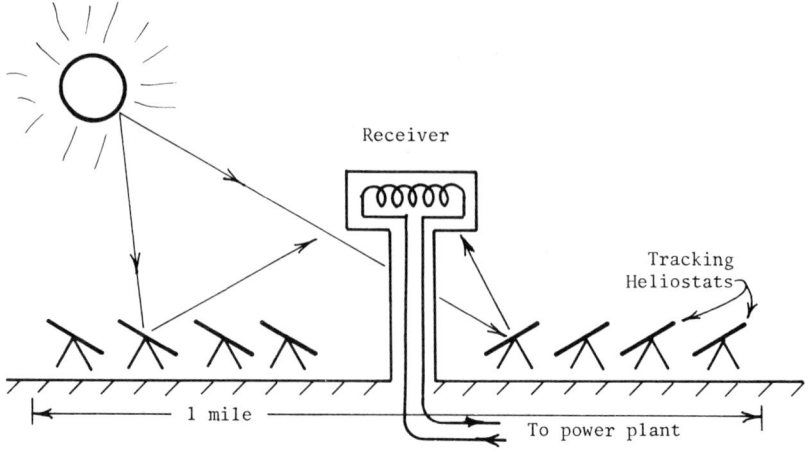

Fig. 12.4.4. Schematic of a power tower plant.

a tower 300-500 ft high, and thus the systems have been called "power towers." A schematic of this system is shown in Fig. 12.4.4. For a power plant the same size as considered earlier, about 60,000 tracking heliostats would be required.

The thermal efficiency of the power plant is about 30 percent based on the solar energy reaching the boiler. This is lower than a conventional system since another heat transfer loop is required to collect solar energy and transport it to the boiler. There also are significant losses in the mirror array, and tracking and focusing errors for the large array of relatively small area mirrors. These mirrors require a higher precision than do those for plotovoltaic cells, and the output is more sensitive to error. As a result, the mirror efficiency is about 50 percent, and thus the overall conversion efficiency of solar energy to electricity is about 15 percent, or comparable to that for the photovoltaic systems.

The area requirements, peak power, and energy output are comparable to that for the photovoltaic system. However, the tracking mirrors (heliostats) are expected to be more expensive by a factor of up to 2. Thus the installed cost of the mirror array alone is about $2000/kW. The tower, absorber, and heat exchanger loop may add another $1000/kW. These costs are in addition to that for the conventional power plant that is also needed. It does not appear as though power towers will be cost effective as central power stations.

12.4.3. Heating from Solar Energy

Solar energy is currently employed to heat homes, water, and swimming pools in many areas of the country. Solar heating techniques may be either passive or active. Passive heating techniques employ large south facing windows, attached sun spaces, night insulation, and internal storage capacity to store the sun's energy. The term passive usually implies that no fans or pumps are used to move fluids around. The methods of analysis are really those of conventional homes (Chapter 3) and have already been covered. Passive techniques are a cost competitive option at present, and some should be employed in all new buildings.

Active space, water, and swimming pool systems have separate devices for collecting solar energy, transporting it, storing it, and releasing it to the load. A general schematic of a solar energy system is shown in Fig. 12.4.5. A solar collector, usually flat and called a flat plate collector, absorbs the solar energy. This energy goes into heating up a fluid which is either air or an antifreeze solution. The warmed fluid is pumped into the store. The storage device is used to accommodate the phase difference between supply and demand. It stores excess energy when the demand is low and delivers energy to the load when supply is low. If air is used, the store is a bin of rocks, while a storage tank is used for liquid systems.

The load can be either the house, the hot water system, or both. In general, a system for a swimming pool would be separate. Since it is uneconomical to size the system to meet the greatest load under the worst conditions, an auxiliary energy supply is required. A solar energy system is thus more complicated than a conventional system and, in addition, requires a conventional auxiliary energy source.

The collector is the unique component in the solar system. It is a thermal device, and its thermal performance will govern the performance of the entire system. A representative modern flat plate collector is shown schematically in Fig. 12.4.6. The sun's energy is absorbed by an absorber plate which is coated with a material to absorb solar radiation. This coating may be either a flat black paint or a selective surface which has a low emittance to reduce radiation losses back to the ambient. The circulating fluid, either air, water, or an antifreeze solution, flows on the underside of the absorber to transfer the absorbed energy. The other side of the passage is well insulated to reduce back losses. The absorber plate is usually covered with one or two glass cover plates, although most swimming pool heaters have none. These act as insulating windows to reduce convection and radiation losses.

The thermal energy flows are shown in Fig. 12.4.7. The incident solar energy is the product of the instantaneous solar flux I_T and the collector area. This energy flow is partially reduced by transmission losses in the glass cover, and a fraction of this is absorbed at the absorber plate. The useful energy transfer at the absorber

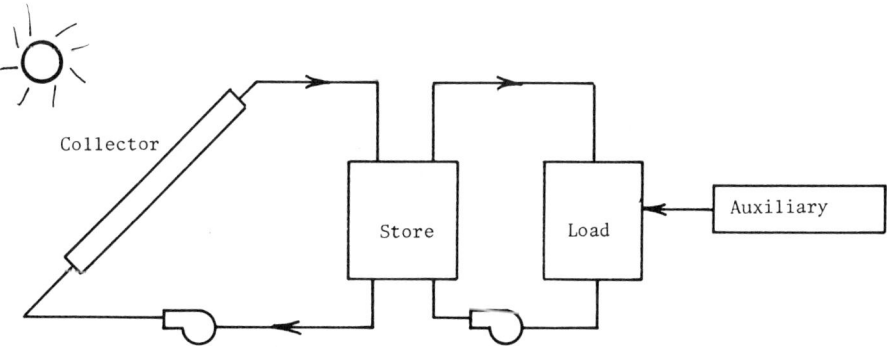

Fig. 12.4.5. Schematic of a solar system.

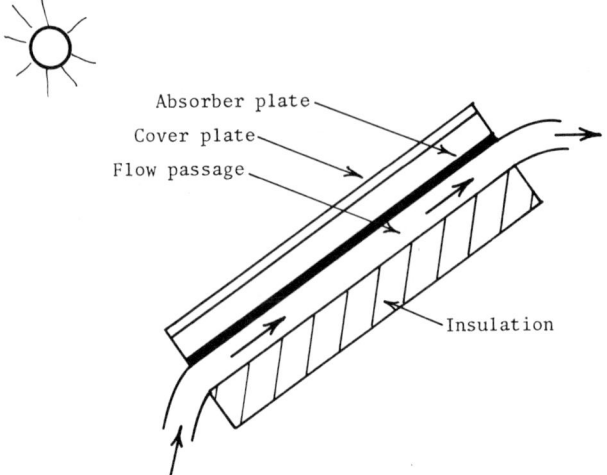

Fig. 12.4.6. Schematic of a flat plate collector.

plate goes into heating the circulating fluid. There are also conduction and convection losses back to the environment. The energy balance at the absorber surface is

$$q_{abs} - q_{loss} - q_{useful} = 0 \qquad (12.4.1)$$

The absorbed energy is given in terms of the incident flux, the transmittance of the glass cover τ, and the absorptance of the absorber plate α by

$$q_{abs} = I_T A_c \tau \alpha \qquad (12.4.2)$$

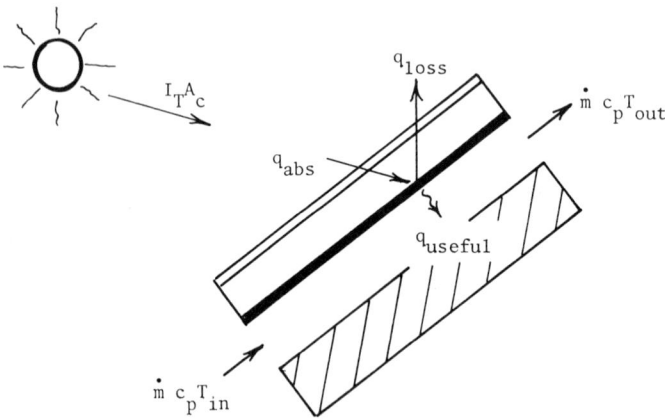

Fig. 12.4.7. Energy flows for a collector.

The loss is given in terms of an overall conductance and the average temperature difference between the absorber plate and the ambient

$$q_{loss} = A_c U_L (T_{ave} - T_a) \qquad (12.4.3)$$

This allows the useful energy gain to be calculated as

$$q_{useful} = A_c [I_T \tau\alpha - U_L(T_{ave} - T_a)] \qquad (12.4.4)$$

The useful energy flow is also given in terms of the energy transfer to the circulating fluid as

$$q_{useful} = \dot{m} C_p (T_{out} - T_{in}) \qquad (12.4.5)$$

Equation (12.4.4) shows that the useful energy transfer is the difference between a gain, due to incoming solar energy, and a loss, due to heat transfer to the ambient. For a given collector, the useful energy is largest when the loss is smallest, which occurs when the average temperature equals that of the ambient. In other words, the collector can collect the most energy when the absorber plate, and consequently the circulating fluid, are at ambient temperature. However, this energy will probably be at too low a temperature to really be useful. This equation also shows that as collection temperature, and consequently losses, increase, the useful energy decreases. At some temperature the losses equal the gain and although the temperature is high there is no useful energy flow. From Eq. (12.4.5), this would occur at zero mass flow rate. These fundamental considerations place a limit on solar energy system performance.

The values of $\tau\alpha$ and U_L govern the collector performance. Well-designed collectors with flat black or selective surfaces have values of $\tau\alpha$ between 0.8 and 0.9. The values of U_L depend on the abosrber surface and the number of covers. Typical values of collector parameters are given in Table 12.4.1. The values for flat black surfaces are essentially those for windows, while those for selective surfaces are lower due to the low value of long wave emittance of the absorber surface.

TABLE 12.4.1
Properties for Collectors[a]

No. of Covers	Black		Selective	
	$\tau\alpha$	U_L	$\tau\alpha$	U_L
0	0.95	6	0.9	5
1	0.9	1	0.85	0.5
2	0.85	0.6	0.8	0.3

[a] U_L in Btu/hr ft^2 °F.

The collector efficiency is defined as

$$\eta_c = \frac{q_{useful}}{A_c I_T} \tag{12.4.6}$$

Using Eq. (12.4.4), the efficiency can be written as

$$\eta_c = (\tau\alpha) - U_L \frac{(T_{ave} - T_a)}{I_T} \tag{12.4.7}$$

The collector efficiency as a function of the group $(T_{ave} - T_a)/I_T$ is plotted in Fig. 12.4.8 for three collectors using the values in Table 12.4.1. In order to show the range of operation, a typical winter day will be considered. For a collector tilted at an angle equal to the latitude, which will maximize year-round energy collection, the winter solar flux is on the order of 100–200 Btu/hr ft². For an ambient temperature of 20°F and an operating temperature of 120°F, the value of $(T_{ave} - T_a)/I_T$ is on the order of 0.5–1.0.

From Fig. 12.4.8, it is seen that a zero cover collector would be inadequate for wintertime heating. Such a collector could be used for swimming pool heating in summer. A single-cover black absorber collector would be marginal for this situation, and more appropriate in a milder climate. For climates with significant heating loads, either a single-cover selective surface absorber or a two-cover black absorber is needed. Figure 12.4.8 also demonstrates that collection efficiencies are on the order of 0.25–0.5 for wintertime operation.

The evaluation of a solar energy system to supply heating depends on many weather factors (level and daily distribution of insolation, ambient temperature),

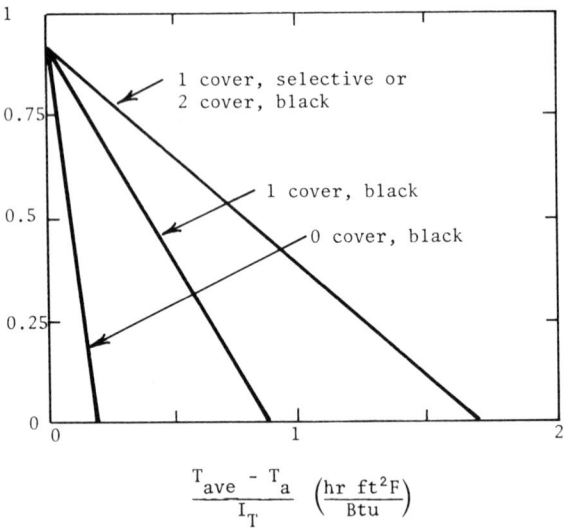

Fig. 12.4.8. Flat plate collector performance.

solar system parameters (collector type, storage capacity), and load characteristics (space heating, water heating, temperature level). One calculation procedure (*F*-CHART) allows a ready determination of the monthly energy contribution and the auxiliary energy requirements. The procedure is beyond the scope of this chapter, but the results of this calculation will be presented and the implications discussed.

For the example house of Chapter 3, the annual heating load was 46×10^6 Btu. The contribution of a solar energy system to meeting this load is given by the annual solar fraction F defined as

$$F = \frac{Q_{sol}}{Q_{loss}} \qquad (12.4.8)$$

This fraction is plotted as a function of the collector area in Fig. 12.4.9. At small areas the contribution increases rapidly with area, while at larger areas there is a reduced contribution. This reflects the fact that a large system is too big for months with medium loads (spring and fall), but is not large enough to meet the entire load in wintertime.

The economics of the solar energy system can also be evaluated. The cost of typical residential sized systems is the sum of a fixed cost for pipes, controls, and storage and an area dependent cost for the collectors, and can be represented as

$$C_{sol} = E + C_A A_c \qquad (12.4.9)$$

The life cycle savings for a solar system are the difference between the cost of fuel that the solar displaces and the equipment costs, and are given by

$$\text{LCS} = P_1 F \frac{Q_{loss}}{\eta_{fur}} - P_2(E + C_A A_c) \qquad (12.4.10)$$

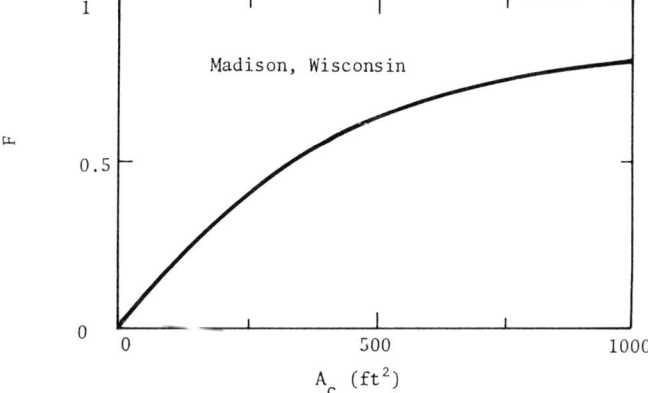

Fig. 12.4.9. Performance of a solar system as a function of collector area.

The current costs of solar systems are about $2000 for fixed costs and $20/ft² for collectors. Solar systems are eligible for 40 percent tax credits up to $10,000, and so P_2 is about 0.6. The savings will be calculated for a 20-year life ($P_1 = 22.17$) and with two fuels. Natural gas used in a furnace of 70 percent efficiency ($5/10⁶ Btu delivered) is one alternative, while either fuel oil at $1.10/gal in a 65 percent efficiency furnace or electricity at 4¢/kWh ($12/10⁶ Btu delivered) is the other. The life cycle savings as a function of collector area are given in Fig. 12.4.10. If natural gas at current prices is available, the solar system is not economical. If either fuel oil or electricity is the option, the solar system is competitive. The optimum size system is about 400 ft² and costs $10,000 which is the current limit on federal tax credits. The system will save about $1000 over its life. Clearly this is not a very large sum, and not too many people may invest. In practice, the savings will be increased in that the system would also heat the hot water. It could meet most of the load in spring, summer, and fall and save $100-$200 annually.

A national assessment of the role of various solar energy systems has been undertaken and the conclusions are similar to those for Madison, Wisconsin. Solar systems are cost competitive with oil or electric heating systems in all areas of the country, but not with natural gas. As is true for many conservation options, the price of natural gas is too low to encourage significant change. Solar energy systems do save resource energy however, and are better in this regard than even heat pump systems. In terms of national goals, solar energy could play a significant role in reducing conventional energy consumption.

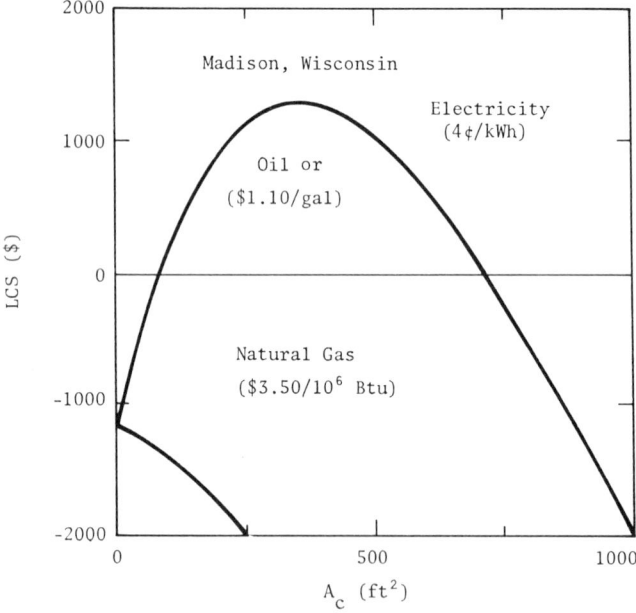

Fig. 12.4.10. Life cycle savings as a function of collector area.

12.5. SUMMARY

The technical and economic feasibility of renewable energy supplied has been explored in this chapter. It is possible to produce a significant fraction of the United States needs by bioconversion and solar methods, and wind energy can be a useful supplement in many situations. The economic costs are the major impediments to rapid growth of these technologies.

Bioconversion of sunlight into plants and then into fuels is potentially cost competitive. Synthetic natural gas, alcohol, fuel oil, and solid fuels have been produced in demonstration or pilot plants, and it is expected that these fuels will continue to develop. The use of wood for home heating has received considerable favorable attention recently, but it is not economically competitive with natural gas unless wood prices are low and a high efficiency furnace is used. There are serious pollution problems accompanying the use of wood.

Electricity produced by wind is most suitable to remote areas where conventional costs are high. Wind energy is diffuse, and large wind machines are needed to produce significant power. The present cost of wind equipment is two to four times that of conventional power plants.

Electricity produced by photovoltaic devices is promising. Costs are being reduced and the price is nearly competitive with current, conventionally generated electricity. Solar thermal power plants do not appear to be competitive.

Solar heating of homes, domestic water, and swimming pools is economically competitive with fuel oil and electricity. The price of solar components will probably not decrease substantially, and further expansion will depend on increased conventional fuel prices.

SUGGESTED READING

W. C. Dickinson and P. N. Cheremisinoff (eds.), *Solar Energy Technology Handbook*, Marcel Dekker, Inc., New York, 1980.

J. Shelton, *The Woodburners Encyclopedia*, Vermont Crossroads Press, Waitsfield, VT, 1976.

J. A. Duffie and W. A. Beckman, *Solar Engineering of Thermal Processes*, John Wiley & Sons, New York, 1980.

W. A. Beckman, S. A. Klein, and J. A. Duffie, *Solar Heating Design by the f-Chart Method*, John Wiley & Sons, New York, 1977.

■ APPENDIXES ■

A. WEATHER DATA FOR SELECTED CITIES

B. ENERGY CONTENT OF FUELS

APPENDIX A
WEATHER DATA FOR SELECTED CITIES

Madison, WI
Design Conditions: Winter $-11°F$
Summer $91°F$ DB, $74°F$ WB
$0.014\ lb_m/lb_m$
$\bar{T}_a = 45°F$

	J	F	M	A	M	J	J	A	S	O	N	D
Day length (hr)	9.2	10.3	11.8	13.1	14.5	15.1	14.8	13.2	12.2	10.7	9.5	8.9
Night length (hr)	14.8	13.7	12.2	10.9	9.5	8.9	9.2	10.8	11.8	13.3	14.5	15.1
T_{ave} (°F)	19.4	21.2	32.0	44.6	55.4	66.2	69.8	68.0	59.0	50.0	33.8	23.0

Incident Solar ($Btu/ft^2\ day$)

	J	F	M	A	M	J	J	A	S	O	N	D
S	906	1092	1077	902	848	823	857	969	1061	1101	770	682
E/W	372	551	730	847	1026	1136	1134	1030	819	611	348	277
N	159	231	322	431	589	700	657	505	360	258	168	134
Horizontal	514	803	1135	1397	1742	1946	1933	1707	1298	910	503	368

Chicago, IL
Design Conditions: Winter $-8°F$
Summer $91°F$ DB, $74°F$ WB
$0.014\ lb_m/lb_m$
$\bar{T}_a = 51.2°F$

	J	F	M	A	M	J	J	A	S	O	N	D
Day length (hr)	9.3	10.4	11.8	13.1	14.4	15.0	14.7	13.6	12.2	10.8	9.6	9.0
Night length (hr)	14.7	13.6	12.2	10.9	9.6	9.0	9.3	10.4	11.8	13.2	14.4	15.0
T_{ave} (°F)	26.6	28.4	37.4	50.0	60.8	69.8	75.2	73.4	66.2	55.4	41.0	30.2

Incident Solar ($Btu/ft^2\ day$)

	J	F	M	A	M	J	J	A	S	O	N	D
S	828	959	1010	921	846	820	839	950	1082	1150	863	662
E/W	355	505	689	879	1051	1161	1130	1030	847	645	390	278
N	164	233	324	440	595	712	656	508	368	268	180	140
Horizontal	507	759	1106	1458	1787	2005	1942	1718	1352	968	565	401

WEATHER DATA FOR SELECTED CITIES

Washington, DC
Design Conditions: Winter 10°F
Summer 92°F DB, 75°F WB
0.015 lb_m/lb_m
$\bar{T}_a = 56.5°F$

	J	F	M	A	M	J	J	A	S	O	N	D
Day length (hr)	9.6	10.6	11.8	13.0	14.1	14.7	14.4	13.4	12.2	10.9	9.9	9.3
Night time (hr)	14.4	13.4	12.2	11.0	9.9	9.3	9.6	10.6	11.8	13.1	14.1	14.7
T_{ave} (°F)	35.6	37.4	44.6	55.4	64.4	73.4	77.0	75.2	69.8	59.0	48.2	37.4
Incident Solar (Btu/ft^2 day)												
S	832	937	946	859	762	735	745	839	988	1078	904	726
E/W	382	523	690	859	983	1077	1035	944	813	643	433	322
N	186	253	338	446	580	684	630	499	378	287	203	165
Horizontal	571	814	1124	1457	1716	1899	1816	1616	1339	1003	650	480

Los Angeles, CA
Design Conditions: Winter 37°F
Summer 93°F DB, 70°F WB
0.011 lb_m/lb_m
$\bar{T}_a = 61°F$

	J	F	M	A	M	J	J	A	S	O	N	D
Day length (hr)	10.0	10.8	11.8	12.8	13.8	14.2	14.0	13.2	12.2	11.1	10.2	9.8
Night length (hr)	14.0	13.2	12.2	11.2	10.2	9.8	10.0	10.8	11.8	12.9	13.8	14.2
T_{ave} (°F)	53.6	55.4	55.4	59.0	60.8	64.4	68.0	68.0	68.0	64.4	59.0	55.4
Incident Solar (Btu/ft^2 day)												
S	1326	1348	1276	1002	764	688	754	916	1119	1311	1353	1308
E/W	612	769	983	1134	1159	1178	1294	1195	1003	823	657	569
N	241	306	386	493	636	731	722	553	413	329	257	223
Horizontal	925	1213	1617	1949	2058	2117	2306	2078	1680	1316	1003	847

APPENDIX B
ENERGY CONTENT OF FUELS

The energy content of a variety of fuels is given in this appendix. These values are reported on both the unit basis by which fuels are sold and on a per unit mass basis. The energy content is the lower heating value, which assumes that the water in the products of combustion is a vapor. (See Chapter 6.)

Fuels	Btu/ft^3	Btu/lb$_m$
Gaseous		
Methane	900	21500
Natural gas	1020	21500
Coal gas	600	13000
	Btu/gal	Btu/lb$_m$
Liquid		
Crude oil	138,000	19,500
Residual oil	150,000	19,000
Distillate oil	139,000	19,500
Automotive gasoline	125,000	19,300
Aviation gasoline	124,000	19,300
Diesel oil	139,000	19,500
Jet fuel (kerosene)	135,000	19,700
Liquefied petroleum gas	96,000	21,500
	Btu/ton	Btu/lb$_m$
Solid		
Anthracite or bituminous coal	26×10^6	13,000
Lignite coal	13×10^6	6,500
Wood	$12\text{-}16 \times 10^6$	6-8,000
Refuse derived fuels	$10\text{-}14 \times 10^6$	5-7,000

INDEX

Air changes:
  summer, 45
  winter, 87
Air conditioning:
  commercial buildings, 105
  residential, 79
Air conditioning systems:
  constant air volume, 109
  economizer, 111
  variable air volume, 112
Annual energy consumption:
  buildings, 32, 49, 56
  cooling, 91
  heating, 56
Appliances:
  energy use, 32
  heat generation, 48, 88
Automobile:
  energy use, 4, 260
  mileage, 262
Availability or available energy:
  of combustion, 186
  definition, 182
  of energy sources, 104, 209
  of petroleum fuels, 186
  transfers, 183, 190
Availability destruction:
  due to combustion, 187
  in power lants, 190

Bin method, 164
Bioconversion, 283
Break-even fuel cost, 233

Capacitance rate ratio, 230
Chimney, 136
Clothing levels, 122
Coefficient:
  of aerodynamics drag, 266
  of rolling friction, 264
  of Performance (COP), 159, 245
Coefficient of utilization, 116

Cogeneration:
  industrial, 204
  utility, 191, 197
Collectors:
  concentrating, 292
  flat plate, 295
Combustion:
  availability of, 186
  enthalpy of, 186
  furnace, 130
Comfort:
  clothing levels, 122
  physiological relations, 120
  standards, 120
Conductance-area product, 46
Conduction, 35
Constant air volume system, 109
Consumer price index, 6
Convection, 35

Daily average temperature:
  ambient, 46, 53
  standard deviation, 55
Degree days:
  annual, 74
  evaluation of, 54
  heating, 53
  variable base, 91
Depreciation, 21
Design cooling load, 89
Design heating load, 47
Dew point of products, 135
District heating, 195
Drag coefficient, 267

Economizer cycle, 112
Effectiveness of:
  boiler, 189
  cogeneration system, 193
  combustion, 189
  heat exchanger, 229, 250
Efficacy of light source, 116

INDEX

Efficiency of:
   boiler, 189
   furnace, 128, 144, 211
   solar collector, 298
   wind machines, 289
Electric ignition, 154
Electricity production, 3, 181, 292
End uses of energy, 5
Energy cost of materials, 223
Energy supplies, 9
Energy use:
   industrial, 3, 233
   major, 3
   projections, 7
   residential and commercial, 3, 31
   transportation, 3, 260
   utilities, 3
Engine-heatpump combination, 212
Engine efficiency, 214
Enthalpy of combustion, 132, 186
Exfiltration, 140

Fan power, 111, 250
Flue damper, 154
Forces on vehicles:
   aerodynamic, 266
   friction, 264
   gravitational, 270
   inertial, 271
Fuel:
   energy content, 7, 306
   heating value, 7, 133, 306
   mix for heating, 127
   prices, 6
   sources, 5
   wood, 285
Furnace efficiency, 128, 144, 211, 286

Gross national product (GNP), 2

Heat exchanger:
   effectiveness, 229, 250
   furnace, 134
   optimum, 235
   waste heat, 112, 228
Heat gains:
   appliances, 48, 88
   attic, 84
   basements, 85
   doors, 80
   infiltration, 87
   people, 47, 88, 122
   roof, 84
   walls, 80
   windows, 48, 85
Heating values of fuel, 7, 133, 306
Heat loss:
   basement, 42
   building, 34, 46
   crawl space, 42
   floor, 42
   infiltration, 44
   slab, 42
   wall, 35
   window, 41
Heat pump:
   balance point, 164
   capacity, 163
   coefficient of performance, 159, 245
   commercial application, 177
   degradation, 170
   engine combination, 212
   gas fired, 215
   ice maker, 176
   waste heat recovery, 242
   water source, 173
Heat recovery:
   buildings, 112
   heat pumps, 177, 242
   heat exchangers, 122, 228

Icemaker heat pump, 176
Ideal furnace, 209, 211
Illumination:
   concepts, 116
   levels, 106
Industrial energy use, 3, 223
Industrial refrigeration, 248
Infiltration, 44, 87
Inflation:
   general, 6
   fuel, 7, 16
Insulation:
   cost effectiveness, 71, 102
   economic optimum, 25, 225
   prices, 71
   thermal properties, 39
Interest rate, 14
Intermittent combustion furnace, 151

Lamp characteristics, 117
Latent load, 94
Life cycle cost, 16, 20
Life cycle savings, 18
Lighting:
   energy use, 4

INDEX

lamps, 117
levels, 106

Market discount rate, 14
Metabolic rates, 47, 88, 121
Moisture gains, 87
Mortgage rate, 20

Night setback, 47, 52, 72
Number of transfer units (Ntu), 229

Optimum heat exchanger, 235

P_1, 20
P_2, 20
Payback period, 21
Photovoltaic cells, 292
Pilot light, 148
Present worth factors, 14, 15, 17

Radiation, 35
Recycling, 253
Residential and commercial energy use, 3, 31
Return on capital, 21

Second Law of Thermodynamics:
   definition, 183
   effectiveness, 189
Sol-air temperature, 82
Solar:
   energy, 10, 291
   heating, 294
   gains, 64, 85
   radiation, 48
Space heating, 4, 31
Specific fuel consumption, 274
Stirling cycle, 216
Storage in:
   air conditioning systems, 113
   buildings, 81
   heat pump systems, 175

Tax credit, 20
Taxes, 20

Temperature:
   ambient, 46, 53
   balance, 51, 53, 92, 164
   combustion, 131
   daily average, 46, 53
   dew point of products, 135
   ground, 44
   room, 46
   sol-air, 82
   theoretical flame, 133
   ventilation, 93
Thermal:
   conductance, 37
   property values, 39
   resistance, 37
Transmittance — absorptance of:
   collectors, 297
   windows, 48, 86
Transportation energy use, 3, 259

Utilities energy use, 3

Variable air volume system, 111
Vehicle power requirements, 263
Ventilation, 44, 110

Waste heat:
   heat exchangers, 228
   heat pumps, 242
   utilization in cogeneration, 192
Water heating, 4, 31
Water source heat pump, 173
Weather data, 304
Wind energy, 287
Window:
   heat gain, 48, 85
   heat loss, 41
   properties, 49, 86
   orientation, 72
Wood, 285

Zones in buildings, 105